C*-ALGEBRAS

VOLUME 1: BANACH SPACES

North-Holland Mathematical Library

Board of Honorary Editors:

M. Artin, H. Bass, J. Eells, W. Feit, P.J. Freyd, F.W. Gehring, H. Halberstam, L.V. Hörmander, J.H.B. Kemperman, W.A.J.Luxemburg, F. Peterson, I.M. Singer and A.C. Zaanen

Board of Advisory Editors:

A. Björner, R.H. Dijkgraaf, A. Dimca, A.S. Dow, J.J. Duistermaat, E. Looijenga, J.P. May, I. Moerdijk, S.M. Mori, J.P. Palis, A. Schrijver, J. Sjöstrand, J.H.M. Steenbrink, F. Takens and J. van Mill

VOLUME 58

ELSEVIER
Amsterdam - London - New York - Oxford - Paris - Shannon - Tokyo

C*-Algebras

Volume 1: Banach Spaces

Corneliu Constantinescu
Departement Mathematik, ETH Zürich
CH-8092 Zürich
Switzerland

2001
ELSEVIER
Amsterdam - London - New York - Oxford - Paris - Shannon - Tokyo

ELSEVIER SCIENCE B.V.
Sara Burgerhartstraat 25
P.O. Box 211, 1000 AE Amsterdam, The Netherlands

First edition 2001

Library of Congress Cataloging in Publication Data
A catalog record from the Library of Congress has been applied for.

ISBN (this volume):	0 444 50749 3
ISBN (5 volume set):	0 444 50758 2
Series ISSN:	0924 6509

∞ The paper used in this publication meets the requirements of ANSI/NISO Z39.48-1992 (Permanence of Pape
Printed in The Netherlands.

Preface

Functional analysis plays an important role in the program of studies at the Swiss Federal Institute of Technology. At present, courses entitled Functional Analysis I and II are taken during the fifth and sixth semester, respectively. I have taught these courses several times and after a while typewritten lecture notes resulted that were distributed to the students. During the academic year 1987/88, I was fortunate enough to have an eager enthusiastic group of students that I had already encountered previously in other lecture courses. These students wanted to learn more in the area and asked me to design a continuation of the courses. Accordlingly, I proceeded during the academic year, following, with a series of special lectures, Functional Analysis III and IV, for which I again distributed typewritten lecture notes. At the end I found that there had accumulated a mass of textual material, and I asked myself if I should not publish it in the form of a book. Unfortunately, I realized that the two special lecture series (they had been given only once) had been badly organized and contained material that should have been included in the first two portions. And so I came to the conclusion that I should write everything anew – and if at all – then preferably in English. Little did I realize what I was letting myself in for! The number of pages grew almost imperceptibly and at the end it had more than doubled. Aslo, the English language turned out to be a stumbling block for me; I would like to take this opportunity to thank Prof. Imre Bokor and Prof. Edgar Reich for their help in this regard. Above all I must thank Mrs. Barbara Aquilino, who wrote, first a WordMARC™, and then a LaTeX™ version with great competence, angelic patience, and utter devotion, in spite of illness. My thanks also go to the Swiss Federal Institute of Technology that generously provided the infrastructure for this extensive enterprise and to my colleagues who showed their understanding for it.

Corneliu Constantinescu

Table of Contents of Volume 1

Contents of All Volumes

Table of Contents of Volume 1

Table of Contents of Volume 2

Table of Contents of Volume 3

Table of Contents of Volume 4

Table of Contents of Volume 5

Introduction

This book has evolved from the lecture course on Functional Analysis I had given several times at the ETH. The text has a strict logical order, in the style of "Definiton – Theorem – Proof – Example – Exercises". The proofs are rather thorough and there are many examples.

The first part of the book (the first three chapters, resp. the first two volumes) is devoted to the theory of Banach spaces in the most general sense of the term. The purpose of the first chapter (resp. first volume) is to introduce those results on Banach spaces which are used later or which are closely connected with the book. It therefore only contains a small part of the theory, and several results are stated (and proved) in a diluted form. The second chapter (which together with Chapter 3 makes the second volume) deals with Banach algebras (and involutive Banach algebras), which constitute the main topic of the first part of the book. The third chapter deals with compact operators on Banach spaces and linear (ordinary and partial) differential equations – applications of the theory of Banach algebras.

The second part of the book (the last four chapters, resp. the last three volumes) is devoted to the theory of Hilbert spaces, once again in the general sense of the term. It begins with a chapter (Chapter 4, resp. Volume 3) on the theory of C^*–algebras and W^*–algebras which are essentially the focus of the book. Chapter 5 (resp. Volume 4) treats Hilbert spaces for which we had no need earlier. It contains the representation theorems, i.e. the theorems on isometries between abstract C^*–algebras and the concrete C^*–algebras of operators on Hilbert spaces. Chapter 6 (which together with Chapter 7 makes Volume 5) presents the theory of $\mathcal{L}^p$–spaces of operators, its application to the self–adjoint linear (ordinary and partial) differential equations, and the von Neumann algebras. Finally, Chapter 7 presents examples of C^*–algebras defined with the aid of groups, in particular the Clifford algebras. Many important domains of C^*–algebras are ignored in the present book. It should be emphasized that the whole theory is constructed in parallel for the real and for the complex numbers, i.e. the C^*–algebras are real or complex.

In addition to the above (vertical) structure of the book, there is also a second (horizontal) division. It consists of a main strand, eight branches, and additional material. The results belonging to the main strand are marked with (0). Logically speaking, a reader could restrict himself/herself to these and ignore the rest. Results on the eight subsidiary branches are marked with $(1), (2), (3), (4), (5), (6), (7)$, and (8). The key is

1. Infinite Matrices
2. Banach Categories
3. Nuclear Maps
4. Locally Compact Groups
5. Differential Equations
6. Laurent Series
7. Clifford Algebras
8. Hilbert C^*–Modules

These are (logically) independent of each other, but all depend on the main strand. Finally, the results which belong to the additional material have no marking and – from a logical perspective – may be ignored. So the reader can shorten for himself/herself this very long book using the above marks. Also, since the proofs are given with almost all references, it is possible to get into the book at any level and not to read it linearly.

We assume that the reader is familiar with classical analysis and has rudimentary knowledge of set theory, linear algebra, point–set topology, and integration theory. The book addresses itself mainly to mathematicians, or to physicists interested in C^*–algebras.

I would like to apologize for any omissions in citations occasioned by the fact that my acquaintance with the history of functional analysis is, unfortunately, very restricted. For this history we recommand the following texts.

1. BIRKHOFF, G. and KREYSZIG, E., The Establishment of Functional Analysis, Historia Mathematica 11 (1984), 258–321.

2. BOURBAKI, N., Elements of the History of Mathematics, (21. Topological Vector Spaces), Springer–Verlag (1994).

3. DIEUDONNÉ, J., History of Functional Analysis, North–Holland (1981).

4. DIEUDONNÉ, J., A Panorama of Pure Mathematics (Chapter C III: Spectral Theory of Operators), Academic Press (1982).

5. HEUSER, H., Funktionalanalysis, 2. Auflage (Kapitel XIX: Ein Blick auf die werdende Functionalanalysis), Teubner (1986), 3. Auflage (1992).

6. KADISON, R.V., Operator Algebras, the First Forty Years, in: Proceedings of Symposia in Pure Mathematics 38 I (1982), 1–18.

7. MONNA, A.F., Functional Analysis in Historical Perspective, John Whiley & Sons (1973).

8. STEEN, L.A., Highlights in the History of Spectral Theory, Amer. Math. Monthly 80 (1973), 359–382.

There is no shortage of excellent books on C^*–algebras. Nevertheless, we hope that this book will be also of some utility to the mathematics commutity.

Some Notation and Terminology

We use in this book the notation and terminology which are usual in the current mathematical literature. In the following list we present some of those for which we felt that difficulties in interpretation may arise.

Any set theory (with the axiom of choise or equivalently with Zorn's Lemma) will do for the present book. $\exists$ and $\forall$ denote "there exists" and "for all", respectively; $\exists!$ means "there exists uniquely". We write iff for "if and only if". On special occasions (appearing very seldom) we choose the axiomatic setting of von Neumann: we call class a collection of sets (which need not be itself a set) and we define an ordinal number ξ as the set of ordinal numbers η strictly smaller than ξ, i.e.

$$\eta < \xi \Longleftrightarrow \eta \in \xi$$

$(0 := \emptyset)$. A cardinal number is the smallest ordinal number having a given cardinality; we denote for every set T by $\operatorname{Card} T$ its cardinal number, i.e. the cardinal number with the same cardinality as T.

If P is a proposition and x a variable (which may occur in P), then $\{x \mid P(x)\}$ denotes the class of x for which $P(x)$ holds. If in addition X is a set, then we put

$$\{x \in X \mid P(x)\} := \{x \mid x \in X \text{ and } P(x)\}.$$

If A and B are sets, then

$$A \backslash B := \{x \in A \mid x \notin B\},$$

$$A \triangle B := (A \backslash B) \cup (B \backslash A),$$

$$A \times B := \{(x, y) \mid x \in A \text{ and } y \in B\}.$$

A partition of a set X is a set of pairwise disjoint nonempty subsets of X the union of which is equal to X.

A function or a map is a triple $f := (X, Y, \Gamma)$ (denoted also by $f : X \to Y$), where X, Y are sets and Γ is a subset of $X \times Y$ such that

$$x \in X \Longrightarrow \exists!\, y_x \in Y,\ (x, y_x) \in \Gamma.$$

X, Y, and Γ are called the domain, the range of values (or codomain), and the graph of f, respectively. We set then

$$f(x) := y_x$$

for all $x \in X$ and

$$f(A) := \{y \in Y \mid \exists x \in A,\, f(x) = y\},$$

$$\overset{-1}{f}(B) := \{x \in X \mid f(x) \in B\},$$

$$\overset{-1}{f}(y) := \overset{-1}{f}(\{y\})$$

for all $A \subset X$, $B \subset Y$, and $y \in Y$. We call f a map of X into Y. If f, g are maps of X into Y, then we set

$$\{f = g\} := \{x \in X \mid f(x) = g(x)\},$$

$$\{f \neq g\} := \{x \in X \mid f(x) \neq g(x)\}.$$

If T is a term and x a variable (which may occur in T) and X, Y are sets such that

$$x \in X \Longrightarrow T(x) \in Y,$$

then we denote the map

$$f := (X, Y, \{(x, y) \in X \times Y \mid y = T(x)\})$$

by

$$f : X \longrightarrow Y, \quad x \longmapsto T(x).$$

If $f : X \to Y$ is a map and Z is a subset of X, then the restriction of f to Z (denoted $f|Z$) is the map

$$Z \longrightarrow Y, \quad x \longmapsto f(x).$$

The map $f : X \to Y$ is called injective (surjective), if

$$(x, y \in X,\, f(x) = f(y)) \Longrightarrow x = y$$

$$(f(X) = Y).$$

The expression "f is a map of X onto Y" means f is a surjective map of X into Y. f is called bijective if it is simultaneously injective and surjective, in which case we set

$$f^{-1} : Y \longrightarrow X, \quad f(x) \longmapsto x$$

and call f^{-1} the inverse of f.

If Y is a set and X is a subset of Y, then the inclusion map $X \to Y$ is the map

$$X \longrightarrow Y, \quad x \longmapsto x;$$

if $X = Y$ then we may call the inclusion map $X \to Y$ the identity map of Y. If X, Y, Z are sets and

$$f : X \longrightarrow Y, \quad g : Y \longrightarrow Z,$$

then we put

$$g \circ f : X \longrightarrow Z, \quad x \longmapsto g(f(x))$$

and call this map the composition of f and g. If X, Y, Z are sets and

$$f : X \times Y \longrightarrow Z,$$

then we put

$$f(a, \cdot) : Y \longrightarrow Z, \quad y \longmapsto f(a, y),$$

$$f(\cdot, b) : X \longrightarrow Z, \quad x \longmapsto f(x, b)$$

for all $a \in X$ and $b \in Y$.

A family $(x)_{\iota \in I}$ (indexed by I) is in fact the map $\iota \mapsto x_\iota$ defined on the set I for which the range of values (codomain) is not specified. Any set X defines the canonical family $(x)_{x \in X}$. If $(X_\iota)_{\iota \in I}$ is a family of sets, then we put

$$\prod_{\iota \in I} X_\iota := \{(x_\iota)_{\iota \in I} \mid \iota \in I \Longrightarrow x_\iota \in X_\iota\}$$

and call $\prod_{\iota \in I} X_\iota$ the product of the family $(X_\iota)_{\iota \in I}$.

Let X be a set. An equivalence relation on X is a binary relation $\sim$ on X such that we have for all $x, y, z \in X$:

$$x \sim x,$$

$$x \sim y \Longrightarrow y \sim x,$$

$$(x \sim y \text{ and } y \sim z) \Longrightarrow x \sim z.$$

An equivalence class of the equivalence relation $\sim$ is a nonmepty subset A of X such that

$$x, y \in A \Longrightarrow x \sim y\,,$$

$$(x \in A\,,\, y \in X\,,\, x \sim y) \Longrightarrow y \in A\,.$$

For every $x \in X$, the set $\{y \in X \mid x \sim y\}$ is an equivalence classe of $\sim$ called the equivalence class of x (with respect to $\sim$). The set of equivalence classes of $\sim$ is a partition of X which is denoted by $X/\sim$. The map $X \to X/\sim$ which sends every $x \in X$ into its equivalence class is called the quotient map.

A free ultrafilter on a set X is an ultrafilter on X possessing no one–point sets. If X is an infinite set then the filter on X,

$$\{X \backslash A \mid A \text{ finite set}\}$$

is called the filter of cofinite subsets of A. A totally ordered set is an ordered set X such that for all $x, y \in X$ either $x \leq y$ or $y \leq x$.

$\mathbb{N}, \mathbb{Z}, \mathbb{Q}, \mathbb{R}$, and $\mathbb{C}$ denote the sets of natural numbers, integers, rational numbers, real numbers, and complex numbers, respectively (we do not include 0 among the natural numbers). A sequence is a family indexed by $\mathbb{N}$. If $m, n \in \mathbb{Z}$ and $p \in \mathbb{N}$, then

$$m \equiv n \pmod p$$

means

$$\frac{m-n}{p} \in \mathbb{Z}$$

(in words: m equal n modulo p). If

$$\overline{\mathbb{R}} := \mathbb{R} \cup \{-\infty, \infty\}$$

and $\alpha, \beta \in \overline{\mathbb{R}}$, then

$$]\alpha, \beta[\; := \{\gamma \in \mathbb{R} \mid \alpha < \gamma < \beta\}\,,$$

$$[\alpha, \beta] := \{\gamma \in \overline{\mathbb{R}} \mid \alpha \leq \gamma \leq \beta\}\,,$$

$$[\alpha, \beta[\; := \{\gamma \in \overline{\mathbb{R}} \mid \alpha \leq \gamma < \beta\}\,,$$

$$]\alpha, \beta] := \{\gamma \in \overline{\mathbb{R}} \mid \alpha < \gamma \leq \beta\}\,.$$

If X is a set, $f : X \to \mathbb{R}$, and $\alpha \in \mathbb{R}$, then we set

$$\{f > \alpha\} := \{x \in X \mid f(x) > \alpha\}.$$

An additive group is a commutative group for which the composition law and the neutral element are denoted by $+$ and 0, respectively. If F is a commutative field and s, t are variables, then $F[t]$ and $F[s,t]$ denote the set of polynomials in the variables t and s and t, repsectively, with coefficients in F. The sets of polynomial in more variables (which will appear in the exercises only) will be denoted similarly.

If A is a square matrix over $\mathbb{C}$, then $\operatorname{Det} A$ denotes the determinant of A.

If T is a topological space and A is a subset of T, then $\overline{A}$ and $\overset{o}{A}$ denote the closure and the interior of A, respectively; their elements are called adherent points (or points of adherence) and interior points of A, respectively. If f is a function defined on T with values in a vector space then the support or the carrier of f (denoted by $\operatorname{Supp} f$) is the set $\overline{\{f \neq 0\}}$. The topological cardinality of T is the smallest cardinal number $\aleph$ such that there is a dense set of T of cardinality $\aleph$.

Let T be a Hausdorff topological space, $\mathfrak{K}$ the set of compact sets of T, and $\mathfrak{R}$ the set of relatively compact Borel sets of T. A Radon measure μ on T is a (real or complex) measure μ on $\mathfrak{R}$, such that for every $\varepsilon > 0$ there is a $K \in \mathfrak{K}$, $K \subset A$, such that

$$|\mu(L) - \mu(A)| < \varepsilon$$

for every $L \in \mathfrak{K}$ with $K \subset L \subset A$. A set A of $\mathfrak{R}$ is called a μ-null set if $\mu(B) = 0$ for all $B \in \mathfrak{R}$, $B \subset A$. A μ-null set is a subset A of T such that for every $K \in \mathfrak{K}$, there is a μ-null set $B \in \mathfrak{R}$ containing $A \cap K$. The carrier (or support) of μ (denoted $\operatorname{Supp} \mu$) is the smallest closed set F of T such that $T \backslash F$ is a μ-null set. The absolute value of μ (denoted $|\mu|$) is the smallest Radon measure ν on T such that

$$|\mu(A)| \leq \nu(A)$$

for all $A \in \mathfrak{R}$. A step function on T with respect to $\mathfrak{R}$ is a real or complex function f on T such that $f(T)$ is finite and $\overset{-1}{f}(\alpha) \in \mathfrak{R}$ for all $\alpha \neq 0$.

1. Banach Spaces

The theory of Banach spaces rears its head even when one wishes to devote attention exclusively to Hilbert spaces, for the set of operators on a Hilbert space only forms a Banach space and not a Hilbert space. Hence the first chapter of our book studies Banach spaces, focussing on those aspects of the theory which find application below.

1.1 Normed Spaces

The basic properties of normed spaces are presented, with emphasis on finite–dimensional spaces since the finite–dimensional subspaces play an important role in the theory of compact operators. The principal results are Minkowski's Theorem and Riesz's Theorem.

1.1.1 General Results

Convention 1.1.1.1 (0) *We use* $\mathbb{K}$ *throughout this book to denote either* $\mathbb{R}$, *the field of real numbers, or* $\mathbb{C}$, *the field of complex numbers. The elements of* $\mathbb{K}$ *are called* ***scalars***. *All* ***vector spaces*** *are vector spaces over* $\mathbb{K}$. *A* ***linear form*** *on the vector space* E *is a linear map of* E *into* $\mathbb{K}$. *The vector space of all linear forms on* E *is called* ***the algebraic dual of*** E. *Given a scalar,* $\alpha \in \mathbb{K}$, *we use* $\operatorname{re}\alpha$ *to denote its* ***real part***, $\operatorname{im}\alpha$ *to denote its* ***imaginary part***,

$$\overline{\alpha} := \operatorname{re}\alpha - i \operatorname{im}\alpha$$

to denote ***the conjugate of*** α, *and*

$$|\alpha| := \sqrt{(\operatorname{re}\alpha)^2 + (\operatorname{im}\alpha)^2}$$

to denote ***the absolute value of*** α.

If s, t *are variables, then* $\mathbb{K}[t]$ *and* $\mathbb{K}[s,t]$ *denote the set of polynomials in the variable* t *and* s *and* t, *respectively, with coefficients in* $\mathbb{K}$.

Definition 1.1.1.2 (0) *A* ***seminorm on the vector space*** E *is a map*

$$p : E \longrightarrow \mathbb{R}_+$$

such that for every $x, y \in E$ *and* $\alpha \in \mathbb{K}$

$$p(x + y) \leq p(x) + p(y)\,, \qquad \textit{\textbf{(Triangle Inequality)}}$$

$$p(\alpha x) = |\alpha| p(x)\,.$$

If, moreover,

$$p(x) = 0 \quad \textit{only if } x = 0\,,$$

then p *is said to be a* ***norm on*** E, *in which case it is customary to write*

$$\|x\| \quad \textit{for } p(x)$$

and to call the scalar $\|x\|$ ***the norm of*** x. *We define*

$$E^\# := \{x \in E \mid \|x\| \leq 1\}\,.$$

$E^\#$ *is called* ***the unit ball of*** E.

Two norms p, q *on a vector space are said to be* ***equivalent*** *if*

$$\frac{1}{\alpha} p \leq q \leq \alpha p$$

for some $\alpha > 0$.

A ***normed space*** *is a vector space endowed with a norm, which we usually denote by* $\|.\|$.

Given E *a normed space, the map*

$$E \times E \longrightarrow \mathbb{R}_+\,, \quad (x, y) \longmapsto \|x - y\|$$

is a metric on E, *called* ***the canonical metric of*** E. *Unless otherwise specified, we take every normed space to be endowed with its canonical metric. Its topology is called the* ***norm topology****. If this metric is complete, then we say that the norm is* ***complete*** *or that* E *is* ***complete****. In this case we call* E *a* ***Banach space****. A subset* A *of* E *is called* ***bounded*** *if*

$$\sup_{x \in A} \|x\| < \infty .$$

A sequence $(x_n)_{n\in\mathbb{N}}$ *in* E *is called* **bounded** *if the subset* $\{x_n \mid n \in \mathbb{N}\}$ *of* E *is bounded. Given a set* T, *a map* $x : T \to E$ *is called* **bounded** *if* $x(T)$ *is a bounded set of* E. *Let* F *be a vector subspace of* E. *The restriction of the norm of* E *to* F *is a norm on* F *(***the induced norm on*** F*);* F *endowed with this norm is called* **subspace of** E.

Normed spaces over $\mathbb{K}$ *are called* **real normed spaces** *if* $\mathbb{K} = \mathbb{R}$ *and* **complex normed spaces** *if* $\mathbb{K} = \mathbb{C}$. *If they are complete, we call them* **real** *(resp.* **complex***)* **Banach spaces**.

Given a metric space T, *we define for each* $t \in T$ *and* $\alpha > 0$,

$$U_\alpha(t) := U_\alpha^T(t) := \{s \in T \mid d(s,t) < \alpha\} ,$$

where d *denotes the metric of* T.

Remark. 1. In 1908 M. Fréchet and E. Schmidt introduced a norm on ℓ^2 (Example 1.1.2.5), without calling it a norm. F. Riesz did the same for $\mathcal{C}([a,b])$ in 1916 (Example 1.1.2.4), and E. Helly for spaces of sequences in 1921. The general notion of a norm on a vector space was introduced by H. Hahn and S. Banach in 1923. The notion of bounded set of a normed space was introduced by A. Kolmogoroff and J. von Neumann in 1935.

2. Let F be a subspace of the normed space E. If F is complete, then it is a closed set of E. The converse holds whenever E is complete.

3. The seminorm was introduced by H. Minkowski in 1896.

Proposition 1.1.1.3 (0) *Let* p *be a seminorm on the vector space* E. *Then*

$$p(0) = 0$$

and

$$|p(x) - p(y)| \leq p(x - y)$$

for every $x, y \in E$.

We have

$$p(0) = p(0.0) = |0| p(0) = 0 .$$

From

$$p(x) = p((x-y)+y) \leq p(x-y) + p(y)$$

we get

$$p(x) - p(y) \leq p(x-y)\,.$$

We deduce

$$p(y) - p(x) \leq p(y-x) = p(x-y)$$

and so

$$|p(x) - p(y)| \leq p(x-y)\,. \qquad \blacksquare$$

Corollary 1.1.1.4 (0) *Given a normed space E, the map*

$$E \longrightarrow \mathbb{R}_+\,, \quad x \longmapsto \|x\|$$

is uniformly continuous and the set $E^\#$ is closed. $\blacksquare$

Proposition 1.1.1.5 (5) *Let $(x_\iota)_{\iota\in I}$ be a finite family in the real (complex) normed space E. Put*

$$\alpha := \sup\left\{\left\|\sum_{\iota\in J} x_\iota\right\| \,\middle|\, J \subset I\right\}$$

and let $(\alpha_\iota)_{\iota\in I}$ be a family in $\mathbb{K}$ such that

$$\sup_{\iota\in I} |\alpha_\iota| \leq 1\,.$$

Then

$$\left\|\sum_{\iota\in I} \alpha_\iota x_\iota\right\| \leq 2\alpha \qquad \left(\left\|\sum_{\iota\in I} \alpha_\iota x_\iota\right\| \leq 4\alpha\right).$$

Step 1 $\{\alpha_\iota \mid \iota \in I\} \subset \mathbb{R}_+ \rightarrow \left\|\sum_{\iota\in I} \alpha_\iota x_\iota\right\| \leq \alpha$

We may assume that $I = \{k \in \mathbb{N} \mid k \leq n\}$ for some $n \in \mathbb{N}$ and that $(\alpha_k)_{k\in I}$ is decreasing. Then

$$\left\|\sum_{k=1}^{n} \alpha_k x_k\right\| = \left\|\sum_{k=1}^{n-1} (\alpha_k - \alpha_{k+1}) \sum_{i=1}^{k} x_i + \alpha_n \sum_{i=1}^{n} x_i\right\| \leq$$
$$\leq \sum_{k=1}^{n-1} (\alpha_k - \alpha_{k+1}) \left\|\sum_{i=1}^{k} x_i\right\| + \alpha_n \left\|\sum_{i=1}^{n} x_i\right\| \leq$$
$$\leq \sum_{k=1}^{n-1} (\alpha_k - \alpha_{k+1})\alpha + \alpha_n \alpha \leq \alpha\,.$$

Step 2 $\{\alpha_\iota \mid \iota \in I\} \subset \mathbb{R} \Longrightarrow \left\|\sum_{\iota\in I}\alpha_\iota x_\iota\right\| \leq 2\alpha$

We put

$$J := \{\iota \in I \mid \alpha_\iota \geq 0\}\,, \qquad K := \{\iota \in I \mid \alpha_\iota < 0\}\,.$$

By Step 1,

$$\left\|\sum_{\iota\in I}\alpha_\iota x_\iota\right\| \leq \left\|\sum_{\iota\in J}\alpha_\iota x_\iota\right\| + \left\|\sum_{\iota\in K}\alpha_\iota x_\iota\right\| \leq \alpha + \alpha = 2\alpha\,.$$

Step 3 $\{\alpha_\iota \mid \iota \in I\} \subset \mathbb{C} \Longrightarrow \left\|\sum_{\iota\in I}\alpha_\iota x_\iota\right\| \leq 4\alpha$

By Step 2,

$$\left\|\sum_{\iota\in I}\alpha_\iota x_\iota\right\| \leq \left\|\sum_{\iota\in I}(\operatorname{re}\alpha_\iota)x_\iota\right\| + \left\|\sum_{\iota\in I}(\operatorname{im}\alpha_\iota)x_\iota\right\| \leq 2\alpha + 2\alpha = 4\alpha\,. \qquad \blacksquare$$

1.1.2 Some Standard Examples

Definition 1.1.2.1 (0) *Given sets* E *and* T, *we denote by* E^T *the set of all maps of* T *into* E. *If* E *is a vector space (additive group), then* E^T *becomes a vector space (additive group) when the operations are defined pointwise.*

If E *is an additive group, we define*

$$E^{(T)} := \{x \in E^T \mid \{x \neq 0\} \quad \text{is finite}\}.$$

Given a subset A *of* T, *define*

$$e_A := e_A^T : T \longrightarrow \mathbb{K}, \quad t \longmapsto \begin{cases} 1 & \text{if } t \in A \\ 0 & \text{if } t \in T \setminus A \end{cases}$$

e_A^T *is called the* ***characteristic function of*** A ***on*** T. *If* A *is the singleton set* $\{t\}$, *then we write* e_t^T *(resp.* e_t*) instead of* $e_{\{t\}}^T$ *(resp.* $e_{\{t\}}$*).*

$\mathfrak{P}(T)$ *denotes the power set of* T *and* $\mathfrak{P}_f(T)$ *the set of finite subsets of* T. *Given* $x \in \mathbb{R}_+^T$, *define*

$$\sum_{t \in T} x(t) := \sup\left\{\sum_{t \in A} x(t) \,\middle|\, A \in \mathfrak{P}_f(T)\right\} \leq \infty,$$

adopting the convention that

$$\sum_{t \in \emptyset} x(t) := 0.$$

Example 1.1.2.2 (0) *Let* T *be a set and*

$$\ell^\infty(T) := \{x \in \mathbb{K}^T \mid x \text{ is bounded}\}.$$

$\ell^\infty(T)$ *is a vector subspace of* $\mathbb{K}^T$ *and*

$$\|\cdot\|_\infty : \ell^\infty(T) \longrightarrow \mathbb{R}_+, \quad x \longmapsto \sup_{t \in T} |x(t)|$$

is a norm, called ***the supremum norm.*** $\ell^\infty(T)$ *endowed with this norm is a Banach space. It is separable iff* T *is finite. We set* $\ell^\infty := \ell^\infty(\mathbb{N})$.

It is easy to see that $\ell^\infty(T)$ is a vector subspace of $\mathbb{K}^T$ and that $\|\cdot\|_\infty$ is a norm. Let $(x_n)_{n\in\mathbb{N}}$ be a Cauchy sequence in $\ell^\infty(T)$. For every $\varepsilon > 0$ there is an $n_\varepsilon \in \mathbb{N}$ such that

$$\|x_m - x_n\|_\infty < \varepsilon$$

for every $m, n \in \mathbb{N}$, with $m \geq n_\varepsilon$, $n \geq n_\varepsilon$. Hence, given $t \in T$, the sequence $(x_n(t))_{n\in\mathbb{N}}$ is a Cauchy sequence (and so a convergent sequence) in $\mathbb{K}$. Define

$$x : T \longrightarrow \mathbb{K}, \quad t \longmapsto \lim_{n\to\infty} x_n(t).$$

Then for all $n \in \mathbb{N}$, $n \geq n_\varepsilon$,

$$\sup_{t\in T} |x_n(t) - x(t)| \leq \varepsilon.$$

Thus $x_n - x \in \ell^\infty(T)$ and

$$\|x_n - x\|_\infty \leq \varepsilon$$

for every $\varepsilon > 0$ and $n \in \mathbb{N}$, $n \geq n_\varepsilon$. Hence $x \in \ell^\infty(T)$ and

$$\lim_{n\to\infty} x_n = x.$$

$\ell^\infty(T)$ is thus a Banach space.

It is obvious that $\ell^\infty(T)$ is separable whenever T is finite. Given two distinct subsets A, B of T, we have that

$$\|e_A - e_B\| = 1.$$

Hence the distance between any two distinct elements of the set $\{e_A \mid A \in \mathfrak{P}(T)\}$ is 1. If T is infinite, then this set is uncountable and $\ell^\infty(T)$ is not separable. ∎

Remark. The assertion that $\ell^\infty(T)$ is not separable whenever T is infinite will be generalized in Corollary 6.3.6.16 b).

Example 1.1.2.3 (0) *Let T be a set. If T is finite then put*

$$\ell^0(T) := c(T) := c_0(T) := \ell^\infty(T).$$

If T is infinite, let $\mathfrak{F}$ denote the filter on T consisting of the cofinite subsets of T, i.e.

$$\mathfrak{F} := \{A \subset T \mid T\backslash A \text{ is finite}\},$$

and define

$$c(T) := \{x \in \ell^\infty(T) \mid x(\mathfrak{F}) \text{ converges}\},$$

$$\ell^0(T) := c_0(T) := \{x \in \ell^\infty(T) \mid \lim x(\mathfrak{F}) = 0\}.$$

Then $c(T)$ *an* $c_0(T)$ *are closed vector subspaces of* $\ell^\infty(T)$ *and therefore Banach spaces with respect to the induced norm. Given* $x \in c_0(T)$, *if* $x \neq 0$, *then the set*

$$\{t \in T \mid |x(t)| = \|x\|\}$$

is finite and nonempty. $c(T)$ *and* $c_0(T)$ *are separable iff* T *is countable. The norm on* $c_0(T)$ *defined by the restriction of the norm on* $\ell^\infty(T)$ *to* $c_0(T)$ *is sometimes denotes by* $\|.\|_0$. *We set*

$$\ell^0 := c_0 := c_0(\mathbb{N}), \quad c := c(\mathbb{N}).$$

Only the last assertion needs proof. Assume that T is countable. The set of linear combinations of the vectors e_t $(t \in T)$ and e_T with coefficients in $\mathbb{Q}$ (resp. $\mathbb{Q} + i\mathbb{Q}$) is countable and dense in $c(T)$. Hence $c(T)$ and $c_0(T)$ are separable.

Given distinct $s, t \in T$,

$$\|e_s - e_t\| = 1.$$

Hence $c_0(T)$ and $c(T)$ are not separable if T is uncountable. ■

Example 1.1.2.4 (0) *Let* T *be a topological (measurable) space. Define*

$$\mathcal{C}(T) := \{x \in \ell^\infty(T) \mid x \text{ is continuous}\},$$

$$(\mathcal{B}(T) := \{x \in \ell^\infty(T) \mid x \text{ is measurable}\}).$$

Then $\mathcal{C}(T)$ $(\mathcal{B}(T))$ *is a closed vector subspace of* $\ell^\infty(T)$ *and consequently a Banach space with respect to the induced norm.* ■

Example 1.1.2.5 (0) *Let* T *be a set and* p *a real number,* $p \geq 1$. *Put*

$$\ell^p(T) := \left\{ x \in \mathbb{K}^T \,\middle|\, \sum_{t \in T} |x(t)|^p < \infty \right\}.$$

$\ell^p(T)$ *is a vector subspace of* $\mathbb{K}^T$ *and*

$$\|.\|_p : \ell^p(T) \longrightarrow \mathbb{R}_+, \quad x \longmapsto \left(\sum_{t \in T} |x(t)|^p \right)^{\frac{1}{p}}$$

defines a norm on $\ell^p(T)$, *called the* p–**norm**. $\ell^p(T)$ *with this norm is a Banach space. If* T *is infinite then* Card T *is the topological cardinality of* $\ell^p(T)$. *In particular,* $\ell^p(T)$ *is separable iff* T *is countable. We put*

$$\ell^p := \ell^p(\mathbb{N}).$$

By the Minkowski inequality,

$$\left(\sum_{t\in A}|x(t)+y(t)|^p\right)^{\frac{1}{p}} \le \left(\sum_{t\in A}|x(t)|^p\right)^{\frac{1}{p}} + \left(\sum_{t\in A}|y(t)|^p\right)^{\frac{1}{p}}$$

for every $x, y \in \mathbb{K}^T$ and $A \in \mathfrak{P}_f(T)$. Taking the supremum with respect to A on both sides, we may replace A by T in this inequality. It follows that $\ell^p(T)$ is a vector subspace of $\mathbb{K}^T$ and that $\|.\|_p$ is a norm.

Let $(x_n)_{n\in\mathbb{N}}$ be a Cauchy sequence in $\ell^p(T)$. Given $\varepsilon > 0$ there is an $n_\varepsilon \in \mathbb{N}$ with

$$\|x_m - x_n\|_p < \varepsilon$$

for $m, n \in \mathbb{N}$ with $m \ge n_\varepsilon$, $n \ge n_\varepsilon$. We deduce that for every $t \in T$, $(x_n(t))_{n\in\mathbb{N}}$ is a Cauchy sequence. Define

$$x : T \longrightarrow \mathbb{K}, \quad t \longmapsto \lim_{n\to\infty} x_n(t).$$

Then

$$\left|\sum_{t\subset T}|x_n(t) - x(t)|^p\right|^{\frac{1}{p}} \le \varepsilon$$

and so $x_n - x \in \ell^p(T)$ and

$$\|x_n - x\|_p \le \varepsilon$$

for every $n \in \mathbb{N}$, $n \ge n_\varepsilon$. Hence $x \in \ell^p(T)$ and

$$\lim_{n\to\infty} x_n = x.$$

$\ell^p(T)$ is thus a Banach space.

If T is infinite then the set

$$A := \{x \in \mathbb{K}^{(T)} \mid t \in T \Longrightarrow \operatorname{re} x(t), \operatorname{im} x(t) \in \mathbb{Q}\}$$

is a dense set of $\ell^p(T)$ and

$$\operatorname{Card} A = \operatorname{Card} T.$$

We have

$$\frac{1}{2}\|e_s - e_t\| = 2^{\frac{1}{p}-1} =: \alpha$$

for all distinct $s, t \in T$. Hence if B is a dense set of $\ell^p(T)$ then

$$B \cap U_\alpha(e_t) \neq \emptyset$$

for every $t \in T$, and so

$$\operatorname{Card} B \geq \operatorname{Card} T .$$ ■

Remark. The spaces $\ell^p(T)$ $(p \in [1, \infty])$ are special cases of the L^p–spaces of integration theory. They are precisely the L^p–spaces with respect to counting measure on T.

Proposition 1.1.2.6 (0) *Let T be a set and take $p, q \in [1, \infty[$, with $p < q$.*

a) $\mathbb{K}^{(T)}$ *is a vector subspace of* $\mathbb{K}^T$.

b) $\mathbb{K}^{(T)} \subset \ell^p(T) \subset \ell^q(T) \subset c_0(T) \subset c(T)$.

c) $\mathbb{K}^{(T)}$ *is dense in both* $\ell^p(T)$ *and* $c_0(T)$.

d) *The vector subspace of* $\ell^\infty(T)$ *generated by* $\{e_A \mid A \subset T\}$ *is dense in* $\ell^\infty(T)$.

e) $\|x\|_\infty \leq \|x\|_q \leq \|x\|_p$ *for every* $x \in \ell^p(T)$.

f) *If T is infinite, then the restrictions to* $\mathbb{K}^{(T)}$ *of the norms* $\|\cdot\|_p$, $\|\cdot\|_q$, $\|\cdot\|_\infty$ *are pairwise non–equivalent.*

a), b), c), and d) are easy to see.
e) The inequality

$$\|x\|_\infty \leq \|x\|_q$$

is trivial. For the second inequality, we may further assume that $\|x\|_q = 1$. Then

$$|x(t)|^q \leq |x(t)|^p$$

for every $t \in T$. Thus

$$1 = \sum_{t \in T} |x(t)|^q \leq \sum_{t \in T} |x(t)|^p ,$$

$$\|x\|_q = 1 \le \left|\sum_{t\in T} |x(t)|^p\right|^{\frac{1}{p}} = \|x\|_p\,.$$

f)

$$\|e_A\|_p = n^{\frac{1}{p}}\,,\quad \|e_A\|_\infty = 1$$

whenever $n \in \mathbb{N}$ and A is a subset of T containing precisely n elements. Hence the restrictions of the norms $\|.\|_p$, $\|.\|_q$, $\|.\|_\infty$ to $\mathbb{K}^{(T)}$ are not equivalent. ■

The next proposition generalizes the preceding examples.

Proposition 1.1.2.7 (0) *Let $(E_t)_{t\in T}$ be a family of normed spaces. Take $p \in [1,\infty] \cup \{0\}$. Let*

$$F_p := \left\{ x \in \prod_{t\in T} E_t \;\middle|\; (\|x(t)\|)_{t\in T} \in \ell^p(T) \right\},$$

and

$$q : F_p \longrightarrow \mathbb{R}_+\,,\quad x \longmapsto \|(\|x(t)\|)_{t\in T}\|_p\,.$$

a) *F_p is a vector subspace of $\prod_{t\in T} E_t$ and q is a norm on F_p.*

b) *If each E_t $(t \in T)$ is complete, then F_p is also complete.*

c) *If $p \neq \infty$, then $\{x \in F_p \mid \{t \in T \mid x(t) \neq 0\}$ is finite$\}$ is a dense vector subspace of F_p.*

d) *F_0 is a closed subspace of F_∞.*

a) follows from the Examples 1.1.2.2, 1.1.2.3, and 1.1.2.5.

b) Let $(x_n)_{n\in\mathbb{N}}$ be a Cauchy sequence in F_p. Given $\varepsilon > 0$ there is an $m_\varepsilon \in \mathbb{N}$ with

$$q(x_m - x_n) < \varepsilon$$

for $m, n \in \mathbb{N}$ with $m \ge m_\varepsilon$, $n \ge m_\varepsilon$. Then, given $t \in T$, $(x_n(t))_{n\in\mathbb{N}}$ is a Cauchy sequence in E_t . Let x denote the element of $\prod_{t\in T} E_t$ defined by

$$x(t) := \lim_{n\to\infty} x_n(t)$$

for every $t \in T$. Then $x_n - x \in F_p$ and

$$q(x_n - x) \le \varepsilon$$

for every $\varepsilon > 0$ and $n \ge m_\varepsilon$. Hence $x \in F_p$, $(x_n)_{n\in\mathbb{N}}$ converges to x, and F_p is complete.

c) follows from Proposition 1.1.2.6 c).

d) follows from Example 1.1.2.3. ■

Corollary 1.1.2.8 (0) *Let T be a topological space and E a normed space. Define*

$$\mathcal{C}(T,E) := \{x : T \longrightarrow E \mid x \text{ is continuous and bounded}\}$$

and

$$\|x\| := \sup_{t\in T} \|x(t)\|$$

for every $x \in \mathcal{C}(T,E)$. Then $\mathcal{C}(T,E)$ is a vector subspace of E^T and $\|.\|$ is a norm. If E is complete, then $\mathcal{C}(T,E)$ is complete. ■

Remark. Since

$$\mathcal{C}(T) = \mathcal{C}(T,\mathbb{K}),$$

the above example is a generalization of the topological part of Example 1.1.2.4.

Definition 1.1.2.9 (0) *The subset A of a metric space T is called* ***precompact****, if for every $\varepsilon > 0$ there is a finite subset B of T with*

$$A \subset \bigcup_{t\in B} U_\varepsilon(t).$$

A subset A of a topological space is called ***relatively compact*** *if it is contained in a compact set.*

Every compact set of a metric space is precompact and relatively compact. Subsets of precompact (relatively compact) sets are precompact (relatively compact). Every precompact set of a normed space is bounded.

Lemma 1.1.2.10 (0) *Let T be a metric space and A, B be subsets of T with $A \subset \overline{B}$. Then the following are equivalent:*

a) *A is precompact.*

b) *Every sequence in A contains a Cauchy subsequence.*

c) *Given any $\varepsilon > 0$, there is a finite subset C of B with*

$$A \subset \bigcup_{t \in C} U_\varepsilon(t) .$$

a $\Rightarrow$ b. Let $(t_n)_{n\in\mathbb{N}}$ be a sequence in A, and take $\varepsilon > 0$. Then there is a subsequence $(s_n)_{n\in\mathbb{N}}$ of $(t_n)_{n\in\mathbb{N}}$ with

$$d(s_m, s_n) < \varepsilon$$

for every $m, n \in \mathbb{N}$. Using the diagonal procedure, we can construct a Cauchy subsequence of $(t_n)_{n\in\mathbb{N}}$.

b $\Rightarrow$ c. Assume that if C is a finite subset of B, then

$$A \not\subset \bigcup_{t \in C} U_\varepsilon(t) .$$

Construct a sequence $(t_n)_{n\in\mathbb{N}}$ in A such that

$$d(t_m, t_n) > \frac{\varepsilon}{2}$$

whenever $m, n \in \mathbb{N}$ are distinct. Then $(t_n)_{n\in\mathbb{N}}$ has no Cauchy subsequence, which contradicts b).

c $\Rightarrow$ a is trivial. ■

Lemma 1.1.2.11 (0) *Let A be a subset of the metric space T. Then the following are equivalent:*

a) *$\overline{A}$ is compact.*

b) *A is relatively compact.*

c) *Every sequence in A has a subsequence, which converges in T.*

a $\Rightarrow$ b $\Rightarrow$ c is trival.

c $\Rightarrow$ a. Let $(t_n)_{n\in\mathbb{N}}$ be a sequence in $\overline{A}$. There is a sequence $(s_n)_{n\in\mathbb{N}}$ in A with

$$d(s_n, t_n) < \frac{1}{n}$$

for every $n \in \mathbb{N}$. By c), $(s_n)_{n\in\mathbb{N}}$ has a subsequence $(s_{k_n})_{n\in\mathbb{N}}$ which converges in T. Let

$$t := \lim_{n\to\infty} s_{k_n} .$$

Then $t \in \overline{A}$ and

$$\lim_{n\to\infty} t_{k_n} = t .$$

Hence $\overline{A}$ is compact. ■

Lemma 1.1.2.12 (0) *Every relatively compact set of a metric space is precompact. The converse implication holds whenever the metric space is complete.*

Take a relatively compact set A of the metric space T. Then $\overline{A}$ is compact (Lemma 1.1.2.11 b $\Rightarrow$ a). Hence A is precompact.

Conversely, let T be a complete metric space and A a precompact set of T. Take a sequence $(t_n)_{n\in\mathbb{N}}$ in A. By Lemma 1.1.2.10 a $\Rightarrow$ b, $(t_n)_{n\in\mathbb{N}}$ has a Cauchy subsequence $(s_n)_{n\in\mathbb{N}}$. Since T is complete, $(s_n)_{n\in\mathbb{N}}$ converges in T. Lemma 1.1.2.11 c $\Rightarrow$ b shows that A is relatively compact. ■

Proposition 1.1.2.13 (0) (Fréchet, 1907) *Take a set T. Take $p \in [1, \infty[$. Let A be a subset of $\ell^p(T)$ (resp. $c_0(T)$, resp. $c(T)$). Then the following are equivalent:*

a) *A is relatively compact.*

b) *Given $\varepsilon > 0$, there is a finite subset S of T with*

$$\left|\sum_{t\in T\setminus S} |x(t)|^p\right|^{\frac{1}{p}} < \varepsilon$$

$$(\textit{resp.} \sup_{t\in T\setminus S} |x(t)| < \varepsilon\,, \ \textit{resp.} \sup_{s,t\in T\setminus S} |x(s) - x(t)| < \varepsilon)$$

for every $x \in A$.

a $\Rightarrow$ b. A is precompact by Lemma 1.1.2.12, so that there is a finite subset B of $\ell^p(T)$ (resp. $c_0(T)$, resp. $c(T)$) with

$$A \subset \bigcup_{y\in B} U_{\frac{\varepsilon}{3}}(y) .$$

Take a finite subset S of T such that

$$\left(\sum_{t\in T\setminus S}|y(t)|^p\right)^{\frac{1}{p}}<\frac{\varepsilon}{2}$$

$$(\text{resp.}\quad \sup_{t\in T\setminus S}|y(t)|<\frac{\varepsilon}{2}\,,\qquad \text{resp.}\quad \sup_{s,t\in T\setminus S}|y(s)-y(t)|<\frac{\varepsilon}{3})$$

for every $y\in B$. Take $x\in A$. There is a $y\in B$ with $x\in U_{\frac{\varepsilon}{3}}(y)$. Then

$$\left(\sum_{t\in T\setminus S}|x(t)|^p\right)^{\frac{1}{p}}\le\left(\sum_{t\in T\setminus S}|x(t)-y(t)|^p\right)^{\frac{1}{p}}+\left(\sum_{t\in T\setminus S}|y(t)|^p\right)^{\frac{1}{p}}<$$

$$<\frac{\varepsilon}{3}+\frac{\varepsilon}{2}<\varepsilon\,.$$

$$(\text{resp.}\quad \sup_{t\in T\setminus S}|x(t)|\le\sup_{t\in T\setminus S}|x(t)-y(t)|+\sup_{t\in T\setminus S}|y(t)|<\frac{\varepsilon}{3}+\frac{\varepsilon}{2}<\varepsilon\,,$$

$$\text{resp.}\quad \sup_{s,t\in T\setminus S}|x(s)-x(t)|\le\sup_{s\in T\setminus S}|x(s)-y(s)|+$$

$$+\sup_{t\in T\setminus S}|x(t)-y(t)|+\sup_{s,t\in T\setminus S}|y(s)-y(t)|<2\frac{\varepsilon}{3}+\frac{\varepsilon}{3}=\varepsilon\,)\,.$$

b $\Rightarrow$ a. Take $\varepsilon>0$ and let S be a finite subset of T with

$$\left(\sum_{t\in T\setminus S}|x(t)|^p\right)^{\frac{1}{p}}<\frac{\varepsilon}{3}$$

$$(\text{resp.}\quad \sup_{t\in T\setminus S}|x(t)|<\frac{\varepsilon}{2}\,,\qquad \text{resp.}\quad \sup_{s,t\in T\setminus S}|x(s)-x(t)|<\frac{\varepsilon}{3})$$

for every $x\in A$. Take $t_0\in T\setminus S$. There is a finite subset B of A such that for each $x\in A$ there is a $y\in B$ with

$$\sum_{t\in S}|x(t)-y(t)|^p<\frac{\varepsilon^p}{3}$$

$$(\text{resp.}\quad \sup_{t\in S}|x(t)-y(t)|<\varepsilon\,,\qquad \text{resp.}\quad \sup_{t\in S\cup\{t_0\}}|x(t)-y(t)|<\frac{\varepsilon}{3})\,.$$

Take $x\in A$ and choose $y\in B$ fulfilling the above. Then

$$\begin{aligned}\|x-y\|_p^p &= \sum_{t\in T}|x(t)-y(t)|^p = \\ &= \sum_{t\in S}|x(t)-y(t)|^p + \sum_{t\in T\setminus S}|x(t)-y(t)|^p < \\ &< \frac{\varepsilon^p}{3} + \left(\left(\sum_{t\in T\setminus S}|x(t)|^p\right)^{\frac{1}{p}} + \left(\sum_{t\in T\setminus S}|y(t)|^p\right)^{\frac{1}{p}}\right)^p < \\ &< \frac{\varepsilon^p}{3} + \left(\frac{\varepsilon}{3}+\frac{\varepsilon}{3}\right)^p \le \varepsilon^p\end{aligned}$$

$$(\text{resp. } \|x-y\|_\infty = \sup_{t\in T}|x(t)-y(t)| < \varepsilon\,,$$

$$\text{resp. } \|x-y\|_\infty = \sup_{t\in T}|x(t)-y(t)| < \varepsilon\,).$$

Hence

$$A \subset \bigcup_{y\in B} U_\varepsilon(y)$$

and A is precompact. It now follows from Lemma 1.1.2.12 (and Examples 1.1.2.5 and 1.1.2.3) that A is relatively compact. ■

Definition 1.1.2.14 (0) (Ascoli, 1883) *Let $\mathcal{F}$ be a set of maps of the topological space T into the normed space E. $\mathcal{F}$ is called* ***equicontinuous at the point*** *t of T if for every $\varepsilon > 0$ there is a neighbourhood U of t such that*

$$\|x(s) - x(t)\| < \varepsilon$$

whenever $x \in \mathcal{F}$ and $s \in U$. $\mathcal{F}$ is called ***equicontinuous*** *if it is equicontinuous at every point of T.*

Proposition 1.1.2.15 (0) *Let T be a compact space, A a dense set of T, E a normed space, and $\mathcal{F}$ an equicontinuous set of maps of T into E. Then the topology on $\mathcal{F}$ of pointwise convergence in A coincides with the topology on $\mathcal{F}$ of uniform convergence on T.*

Take $x \in \mathcal{F}$ and $\varepsilon > 0$. For every $t \in T$ there is an open neighbourhood U_t of t with

$$\|y(s) - y(t)\| < \frac{\varepsilon}{5}$$

for every $y \in \mathcal{F}$ and $s \in U_t$. Since T is compact and $(U_t)_{t\in T}$ is an open covering of T, there is a finite subset B of T with

$$T = \bigcup_{t \in B} U_t .$$

Given $t \in B$, choose $s_t \in A \cap U_t$. Define

$$\mathcal{U} := \left\{ y \in \mathcal{F} \mid t \in B \Longrightarrow \|x(s_t) - y(s_t)\| < \frac{\varepsilon}{5} \right\} .$$

Then $\mathcal{U}$ is a neighbourhood of x in $\mathcal{F}$ in the topology of pointwise convergence in A. Take $y \in \mathcal{U}$ and $s \in T$. Then there is a $t \in B$ with $x \in U_t$. Thus

$$\|x(s) - y(s)\| \leq \|x(s) - x(t)\| + \|x(t) - x(s_t)\| +$$

$$+ \|x(s_t) - y(s_t)\| + \|y(s_t) - y(t)\| + \|y(t) - y(s)\| < 5\frac{\varepsilon}{5} = \varepsilon .$$

Hence

$$\mathcal{U} \subset \{ y \in \mathcal{F} \mid s \in T \Longrightarrow \|x(s) - y(s)\| < \varepsilon \} ,$$

so that the topology of $\mathcal{F}$ of pointwise convergence in A is finer than the topology on $\mathcal{F}$ of uniform convergence on T. Since the reverse relation is trivial, the two topologies on $\mathcal{F}$ coincide. ■

Theorem 1.1.2.16 (0) (Ascoli, 1883; Arzelà, 1889) *Let T be a compact space. A set of $\mathcal{C}(T)$ is relatively compact iff it is bounded and equicontinuous.*

First assume that $\mathcal{F}$ is a relatively compact set of $\mathcal{C}(T)$. Then $\mathcal{F}$ is precompact (Lemma 1.1.2.12) and it is therefore bounded. Choose $t \in T$ and $\varepsilon > 0$. Then there is a finite subset $\mathcal{G}$ of $\mathcal{C}(T)$ with

$$\mathcal{F} \subset \bigcup_{y \in \mathcal{G}} U_{\frac{\varepsilon}{3}}^{\mathcal{C}(T)}(y) .$$

Furthermore, there is a neighbourhood U of t with

$$|y(s) - y(t)| < \frac{\varepsilon}{3}$$

whenever $y \in \mathcal{G}$ and $s \in U$. Take $x \in \mathcal{F}$ and $s \in U$. Then there is a $y \in \mathcal{G}$ with

$$\|x - y\| < \frac{\varepsilon}{3} .$$

Thus

$$|x(s) - x(t)| \leq |x(s) - y(s)| + |y(s) - y(t)| + |y(t) - x(t)| < \frac{\varepsilon}{3} + \frac{\varepsilon}{3} + \frac{\varepsilon}{3} = \varepsilon ,$$

and so $\mathcal{F}$ is equicontinuous.

Now assume that $\mathcal{F}$ is bounded and equicontinuous. Take an ultrafilter $\mathfrak{F}$ on $\mathcal{F}$. Define

$$x : T \longrightarrow \mathbb{K}, \quad t \longmapsto \lim_{y,\mathfrak{F}} y(t).$$

Choose $t \in T$ and $\varepsilon > 0$. There is a neighbourhood U of t with

$$|y(s) - y(t)| < \varepsilon$$

whenever $y \in \mathcal{F}$ and $s \in U$. Then

$$|x(s) - x(t)| \leq \varepsilon$$

for every $s \in U$. Hence x is continuous and $\mathcal{F} \cup \{x\}$ is equicontinuous. $\mathfrak{F}$ converges to x in the topology of pointwise convergence. By Proposition 1.1.2.15, $\mathfrak{F}$ converges to x with respect to the norm topology. Hence $\mathcal{F}$ is relatively compact. ■

Lemma 1.1.2.17 (0) (Sierpiǹski, 1928) *Let T be an infinite set. Then there is a set, $\mathfrak{A}$, of infinite subsets of T of the cardinality of the continuum such that $A \cap B$ is finite whenever $A, B \in \mathfrak{A}$ are distinct.*

Since T is infinite and $\mathbb{Q}$ is countable, we may assume that $\mathbb{Q} \subset T$. Given $\alpha \in \mathbb{R} \backslash \mathbb{Q}$, let $(\alpha_n)_{n \in \mathbb{N}}$ be a sequence in $\mathbb{Q}$ converging to α. The set

$$\mathfrak{A} := \{\{\alpha_n \mid n \in \mathbb{N}\} \mid \alpha \in \mathbb{R} \backslash \mathbb{Q}\}$$

has the required properties. ■

Definition 1.1.2.18 (0) ***The (algebraic) dimension*** *of a vector space E is the cardinality of an algebraic basis for E. It does not depend on the particular choice of the basis and is denoted by* $\operatorname{Dim} E$. *A vector space is called **finite-dimensional (infinite-dimensional)** if its algebraic dimension is finite (infinite).*

Proposition 1.1.2.19 *Let E be a vector subspace of $\mathbb{K}^{\mathbb{N}}$ and take*

$$x \in E \backslash \mathbb{K}^{(\mathbb{N})}$$

with $xe_A \in E$ for every $A \subset \mathbb{N}$, where

$$xe_A : \mathbb{N} \to \mathbb{K}, \quad n \mapsto x(n)e_A(n).$$

Then the dimension of E and the cardinality of E are $2^{\aleph_0}$. In particular, for given $p \in [1, \infty] \cup \{0\}$ both the dimension of ℓ^p and the cardinality of ℓ^p are $2^{\aleph_0}$.

Let $\aleph$ be the dimension of E and let $\mathfrak{A}$ be a set of infinite subsets of $\overset{-1}{x}(\mathbb{K}\backslash\{0\})$ with

$$\operatorname{Card}\mathfrak{A} = 2^{\aleph_0}$$

such that $A \cap B$ is finite for distinct $A, B \in \mathfrak{A}$ (Lemma 1.1.2.17). Then $(xe_A)_{A\in\mathfrak{A}}$ is a linearly independent family in E so that

$$\operatorname{Card}\mathfrak{A} \leq \aleph .$$

Since

$$\aleph \leq \operatorname{Card} E \leq \operatorname{Card} \mathbb{K}^{\mathbb{N}} = (2^{\aleph_0})^{\aleph_0} = 2^{\aleph_0 \times \aleph_0} = 2^{\aleph_0}$$

it follows that

$$\aleph = \operatorname{Card} E = 2^{\aleph_0} . \qquad \blacksquare$$

Lemma 1.1.2.20 *Let E be a vector space of dimension $\aleph$. If $\aleph \neq 0$, then*

$$\operatorname{Card} E = \aleph 2^{\aleph_0} .$$

Let A be an algebraic basis for E. Given $B \in \mathfrak{P}_f(A)$, identify $\mathbb{K}^B$ canonically with the vector subspace of E generated by B. Then

$$E = \bigcup_{B\in\mathfrak{P}_f(A)} \mathbb{K}^B .$$

Since

$$\operatorname{Card} \mathbb{K}^B = 2^{\aleph_0}$$

for every $B \in \mathfrak{P}_f(A)$,

$$\operatorname{Card} E \leq \operatorname{Card} \mathfrak{P}_f(A) \times 2^{\aleph_0} = \aleph 2^{\aleph_0} .$$

Now

$$\aleph \leq \operatorname{Card} E , \quad 2^{\aleph_0} \leq \operatorname{Card} E ,$$

so that

$$\aleph 2^{\aleph_0} \leq \operatorname{Card} E ,$$

$$\operatorname{Card} E = \aleph 2^{\aleph_0} . \qquad \blacksquare$$

Proposition 1.1.2.21 *Let $p \in [1,\infty[\cup\{0\}$ and let $\aleph$ be an infinite cardinal. Then the dimension of $\ell^p(\aleph)$ and the cardinality of $\ell^p(\aleph)$ are $\aleph^{\aleph_0}$.*

Take $f \in \aleph^{\mathbb{N}}$ and $x \in \ell^p$. Given $t \in \aleph$, define

$$f_x(t) := \begin{cases} 0 & \text{if } \overset{-1}{f}(t) = \emptyset \\ x(n) & \text{if } \overset{-1}{f}(t) \neq \emptyset \end{cases}$$

where

$$n := \inf \overset{-1}{f}(t).$$

Then $f_x \in \ell^p(\aleph)$.

Since the map

$$\aleph^{\mathbb{N}} \times \ell^p \longrightarrow \ell^p(\aleph), \quad (f,x) \longmapsto f_x$$

is surjective,

$$\operatorname{Card} \ell^p(\aleph) \leq \operatorname{Card}(\aleph^{\mathbb{N}} \times \ell^p) = \aleph^{\aleph_0} \times 2^{\aleph_0} = \aleph^{\aleph_0}$$

(Proposition 1.1.2.19).

Given $f \in \aleph^{\mathbb{N}}$, define

$$\tilde{f} : \aleph \longrightarrow \mathbb{K}, \quad t \longmapsto \sum_{n \in \overset{-1}{f}(t)} \frac{1}{3^n}.$$

Since the map

$$\mathfrak{P}(\mathbb{N}) \longrightarrow \mathbb{K}, \quad A \longmapsto \sum_{n \in A} \frac{1}{3^n}$$

is injective, the map

$$\aleph^{\mathbb{N}} \longrightarrow \ell^p(\aleph), \quad f \longmapsto \tilde{f}$$

is also injective. Hence

$$\aleph^{\aleph_0} \leq \operatorname{Card} \ell^p(\aleph),$$

$$\operatorname{Card} \ell^p(\aleph) = \aleph^{\aleph_0}.$$

Let $\aleph'$ be the dimension of $\ell^p(\aleph)$. By Proposition 1.1.2.19

$$2^{\aleph_0} \leq \aleph'$$

so that, by Lemma 1.1.2.20,

$$\aleph' \leq \operatorname{Card} \ell^p(\aleph) = \aleph' 2^{\aleph_0} \leq \aleph'^2 = \aleph'$$

and

$$\aleph' = \operatorname{Card} \ell^p(\aleph) = \aleph^{\aleph_0} . \qquad \blacksquare$$

Corollary 1.1.2.22 *Let E be a vector space. Let $\aleph > 1$ be a cardinal number. If the dimension of E is $\aleph^{\aleph_0}$, then there is a set $\mathcal{P}$ of pairwise non-equivalent complete norms on E with* $\operatorname{Card} \mathcal{P} = 2^{\aleph_0}$.

Take $p \in [1, \infty]$. By Proposition 1.1.2.21, we may identify E algebraically with $\ell^p(\aleph)$. The assertion now follows from Example 1.1.2.5 and Proposition 1.1.2.6 f). ■

Remark. There is no Banach space whose dimension is $\aleph_0$ (Corollary 1.1.6.17).

Corollary 1.1.2.23 *Let $\xi \neq 0$ be an ordinal number such that*

$$\sup_{n \in \mathbb{N}} \xi_n \neq \xi$$

for every strictly increasing sequence $(\xi_n)_{n \in \mathbb{N}}$ of ordinal numbers and such that

$$\aleph^{\aleph_0} \leq \aleph_\xi$$

for every cardinal number $\aleph$ strictly smaller than $\aleph_\xi$. Then

$$\aleph_\xi^{\aleph_0} = \aleph_\xi$$

and the set of equivalence classes of norms on $\mathbb{K}^{(\aleph_\xi)}$ has cardinality at least that of the continuum.

Now

$$\ell^1(\aleph_\xi) = \bigcup_{\eta \in \xi} \ell^1(\aleph_\eta)$$

and

$$\operatorname{Card} \ell^1(\aleph_\xi) = \aleph_\xi^{\aleph_0} ,$$

$$\operatorname{Card} \ell^1(\aleph_\eta) = \aleph_\eta^{\aleph_0} \leq \aleph_\xi$$

for every $\eta \in \xi$ (Proposition 1.1.2.21). Hence

$$\aleph_\xi^{\aleph_0} \le \aleph_\xi \mathrm{Card}\,\xi = \aleph_\xi\,,$$

$$\aleph_\xi^{\aleph_0} = \aleph_\xi\,.$$

The last assertion follows from the above relation and Corollary 1.1.2.22, since the dimension of $\mathbb{K}^{(\aleph_\xi)}$ is $\aleph_\xi$. ∎

Proposition 1.1.2.24 *For every infinite cardinal number $\aleph$, the dimensions of $\ell^\infty(\aleph)$ and $\mathbb{K}^\aleph$, and the cardinalities of $\ell^\infty(\aleph)$ and $\mathbb{K}^\aleph$ are all $2^\aleph$.*

It is obvious that the cardinalities of $\ell^\infty(\aleph)$ and $\mathbb{K}^\aleph$ are $2^\aleph$. Let $\aleph'$ (resp. $\aleph''$) be the dimension of $\ell^\infty(\aleph)$ (resp. $\mathbb{K}^\aleph$). By Proposition 1.1.2.19,

$$2^{\aleph_0} \le \aleph'$$

and by Lemma 1.1.2.20,

$$2^\aleph = \mathrm{Card}\,\ell^\infty(\aleph) = \aleph' 2^{\aleph_0} \le \aleph'^2 = \aleph' \le \aleph'' \le \mathrm{Card}\,\mathbb{K}^\aleph = 2^\aleph\,.$$

Hence

$$\aleph' = \aleph'' = 2^\aleph\,.$$ ∎

Corollary 1.1.2.25 *For every infinite set T there is a set $\mathcal{P}$ of pairwise non–equivalent complete norms on $\mathbb{K}^T$ whose cardinality is $2^{\aleph_0}$.*

By Proposition 1.1.2.24 the dimension of $\mathbb{K}^T$ is $2^{\mathrm{Card}T}$ and the claim follows from Corollary 1.1.2.22 since

$$\left(2^{\mathrm{Card}T}\right)^{\aleph_0} = 2^{\mathrm{Card}T\times\aleph_0} = 2^{\mathrm{Card}T}\,.$$ ∎

Example 1.1.2.26 (0) *Let T be a Hausdorff space and $\mathcal{M}_b(T)$ the vector space of bounded Radon measures on T. Then*

$$\mathcal{M}_b(T) \longrightarrow \mathbb{R}_+\,, \quad \mu \longmapsto \|\mu\| := |\mu|(T)$$

is a norm which renders $\mathcal{M}_b(T)$ a Banach space.

Now

$$(\|\mu\| = 0) \Longrightarrow (|\mu|(T) = 0) \Longrightarrow (|\mu| = 0) \Longrightarrow (\mu = 0)\,,$$

$$\|\mu+\nu\| = |\mu+\nu|(T) \le (|\mu| + |\nu|)(T) = |\mu|(T) + |\nu|(T) = \|\mu\| + \|\nu\|\,,$$

$$\|\alpha\mu\| = |\alpha\mu|(T) = |\alpha|\,|\mu|(T) = |\alpha|\|\mu\|$$

for every $\mu, \nu \in \mathcal{M}_b(T)$ and $\alpha \in \mathbb{K}$, which proves that the above map is a norm.

Let $(\mu_n)_{n\in\mathbb{N}}$ be a Cauchy sequence in the normed space $\mathcal{M}_b(T)$. Given $\varepsilon > 0$, there is a $p_\varepsilon \in \mathbb{N}$ such that

$$\|\mu_m - \mu_n\| < \varepsilon$$

for every $m, n \in \mathbb{N}$ with $m \geq p_\varepsilon$, $n \geq p_\varepsilon$. Let $\mathfrak{R}$ be the σ–algebra of Borel sets of T and $\mathfrak{K}$ the set of compact sets of T. Then

$$|\mu_m(A) - \mu_n(A)| = |(\mu_m - \mu_n)(A)| \leq |\mu_m - \mu_n|(A) \leq$$

$$\leq |\mu_m - \mu_n|(T) = \|\mu_m - \mu_n\| < \varepsilon$$

for every $\varepsilon > 0$, $A \in \mathfrak{R}$, and $m, n \in \mathbb{N}$ with $m \geq p_\varepsilon$, $n \geq p_\varepsilon$. Thus $(\mu_n(A))_{n\in\mathbb{N}}$ is a Cauchy sequence whenever $A \in \mathfrak{R}$. Defining

$$\mu : \mathfrak{R} \longrightarrow \mathbb{K}, \quad A \longmapsto \lim_{n\to\infty} \mu_n(A),$$

it follows that

$$\mu(A \cup B) = \lim_{n\to\infty} \mu_n(A \cup B) = \lim_{n\to\infty} (\mu_n(A) + \mu_n(B)) = \mu(A) + \mu(B)$$

whenever $A, B \in \mathfrak{R}$ are disjoint and

$$|\mu(A) - \mu_n(A)| = \lim_{m\to\infty} |\mu_m(A) - \mu_n(A)| \leq \varepsilon$$

for every $\varepsilon > 0$, $A \in \mathfrak{R}$, and $n \in \mathbb{N}$ with $n \geq n_\varepsilon$. Let $(A_k)_{k\in\mathbb{N}}$ be a decreasing sequence in $\mathfrak{R}$ whose intersection is empty. Take $\varepsilon > 0$ and $n \in \mathbb{N}$ with $n \geq p_\varepsilon$. Then

$$|\mu(A_k)| \leq |\mu(A_k) - \mu_n(A_k)| + |\mu_n(A_k)| \leq \varepsilon + |\mu_n(A_k)|$$

for every $k \in \mathbb{N}$, so that

$$\limsup_{k\to\infty} |\mu(A_k)| \leq \varepsilon + \lim_{k\to\infty} |\mu_n(A_k)| = \varepsilon.$$

Since ε is arbitrary,

$$\lim_{k\to\infty} \mu(A_k) = 0.$$

Thus μ is a measure.

Take $A \in \mathfrak{R}$ and $\varepsilon > 0$. Take $n \in \mathbb{N}$ with $n \geq p_\varepsilon$. Since μ_n is a Radon measure on T, there is a $K \in \mathfrak{K}$, with $K \subset A$ and

$$|\mu_n(L) - \mu_n(A)| < \varepsilon$$

for every $L \in \mathfrak{K}$, $K \subset L \subset A$. Thus

$$|\mu(L) - \mu(A)| \leq |\mu(L) - \mu_n(L)| + |\mu_n(L) - \mu_n(A)| + |\mu_n(A) - \mu(A)| <$$
$$< \varepsilon + \varepsilon + \varepsilon = 3\varepsilon$$

for every $L \in \mathfrak{K}$, $K \subset L \subset A$. Hence μ is a Radon measure on T.

Take $\varepsilon > 0$ and $n \in \mathbb{N}$ with $n \geq p_\varepsilon$. Then

$$|\mu - \mu_n|(T) \leq 4\varepsilon\,,$$

so that $\mu - \mu_n \in \mathcal{M}_b(T)$ and

$$\|\mu - \mu_n\| \leq 4\varepsilon\,.$$

Hence $\mu \in \mathcal{M}_b(T)$ and

$$\lim_{n\to\infty} \mu_n = \mu\,.$$

Thus $\mathcal{M}_b(T)$ is a Banach space. ■

1.1.3 Minkowski's Theorem

Proposition 1.1.3.1 (0) *Let p and q be norms on the vector space E. Then the following are equivalent:*

a) *p and q are equivalent.*

b) *The Cauchy sequences with respect to p coincide with the Cauchy sequences with respect to q.*

c) *p and q generate the same topology on E.*

a $\Rightarrow$ b is trivial.

b $\Rightarrow$ c. Take $x \in E$ and let $(x_n)_{n \in \mathbb{N}}$ be a sequence in E converging to x with respect to p. For each $n \in \mathbb{N}$, put

$$y_{2n} := x_n , \quad y_{2n-1} := x .$$

Then $(y_n)_{n \in \mathbb{N}}$ is a Cauchy sequence with respect to p. By b), it is a Cauchy sequence with respect to q as well. Since x is point of adherence of $(y_n)_{n \in \mathbb{N}}$ with respect to q, $(y_n)_{n \in \mathbb{N}}$ converges to x with respect to q. It is now easy to see that p and q generate the same topology on E.

c $\Rightarrow$ a. Assume that for each $n \in \mathbb{N}$ there is an $x_n \in E$ with

$$p(x_n) > nq(x_n) .$$

Given $n \in \mathbb{N}$, put

$$y_n := \frac{1}{nq(x_n)} x_n .$$

Then, given $n \in \mathbb{N}$

$$p(y_n) = \frac{p(x_n)}{nq(x_n)} > 1 ,$$

$$q(y_n) = \frac{q(x_n)}{nq(x_n)} = \frac{1}{n} .$$

Hence

$$\lim_{n \to \infty} q(y_n) = 0$$

and so, by c),

$$\lim_{n \to \infty} p(y_n) = 0 ,$$

which is a contradiction. Hence there is some $\alpha > 0$ with $p \leq \alpha q$. Interchanging the roles of p and q, it follows that p and q are equivalent. ■

Remark. For a substantial part of the theory of normed spaces (which includes the first four chapters of this book) only the topology generated by the norm, and not the norm itself is significant. (Thus our interest is not really in normed spaces but, in "normable" spaces). In this "topological case" the norm may be replaced by any other equivalent norm. The theory of locally convex spaces is the best context for this theory. For more precise studies, the "geometrical aspect" is very important and we cannot exchange the norm without damaging the results. This is the case, e.g., with Hilbert spaces and C^*–algebras, which are treated in the last four chapters of this book.

Corollary 1.1.3.2 (0) *Let p, q be equivalent norms on the vector space E. If E is complete with respect to p, then it is also complete with respect to q.* ∎

Definition 1.1.3.3 (0) *Given $n \in \mathbb{N} \cup \{0, \infty\}$, we define*

$$\mathbb{N}_n := \{m \in \mathbb{N} \mid m \leq n\}.$$

Theorem 1.1.3.4 (0) ***(Minkowski's Theorem, 1896).*** *Norms of a finite–dimensional vector space are equivalent.*

Given $n \in \mathbb{N}$, let p be an arbitrary norm on $\mathbb{K}^n$ and $\|\cdot\|$ the Euclidean norm on $\mathbb{K}^n$. Furthermore, let $e_1, e_2, \ldots, e_n$ be the standard basis vectors of $\mathbb{K}^n$ and put

$$\alpha := \left(\sum_{k=1}^{n} p(e_k)^2\right)^{\frac{1}{2}} > 0.$$

Given $x := (x_k)_{k \in \mathbb{N}_n} \in \mathbb{K}^n$, we have, by Schwartz's Inequality, that

$$p(x) = p\left(\sum_{k=1}^{n} x_k e_k\right) \leq \sum_{k=1}^{n} |x_k| p(e_k) \leq \left(\sum_{k=1}^{n} |x_k|^2\right)^{\frac{1}{2}} \left(\sum_{k=1}^{n} p(e_k)^2\right)^{\frac{1}{2}} = \alpha \|x\|.$$

It follows that

$$|p(x) - p(y)| \leq p(x - y) \leq \alpha \|x - y\|$$

for all $x, y \in \mathbb{K}^n$ (Proposition 1.1.1.3). Hence p is continuous.

Define

$$K := \{x \in \mathbb{K}^n \mid \|x\| = 1\},$$

$$\beta := \inf\{p(x) \mid x \in K\}.$$

By the Bolzano–Weierstrass Theorem, K is compact. So, by Weierstrass' Theorem, there is an $x_0 \in K$ such that

$$p(x_0) = \beta .$$

But $x_0 \neq 0$. Hence $\beta > 0$. Take $x \in \mathbb{K}^n \backslash \{0\}$. Then

$$\|\frac{1}{\|x\|} x\| = 1 ,$$

so that

$$\frac{1}{\|x\|} x \in K .$$

Hence

$$\beta \leq p\left(\frac{1}{\|x\|} x\right) = \frac{1}{\|x\|} p(x) ,$$

i.e.

$$\beta \|x\| \leq p(x) .$$

Thus

$$\beta \|\cdot\| \leq p \leq \alpha \|\cdot\| ,$$

i.e. p and $\|\cdot\|$ are equivalent. It follows that all norms on $\mathbb{K}^n$ are equivalent. ∎

Remark. The crucial step in the above proof was applying the Bolzano–Weierstrass Theorem, which, like Minkowski's theorem, no longer holds when the field considered is not locally compact. For example, the "norms"

$$\mathbb{Q}^2 \longrightarrow \mathbb{R}_+ , \quad (\alpha, \beta) \longmapsto \sqrt{\alpha^2 + \beta^2} ,$$

$$\mathbb{Q}^2 \longrightarrow \mathbb{R}_+ , \quad (\alpha, \beta) \longmapsto |\alpha + \sqrt{2}\,\beta|$$

are not equivalent. In fact, if $(\alpha_n)_{n \in \mathbb{N}}$ is a sequence in $\mathbb{Q}$ converging to $-\sqrt{2}$ then

$$\lim_{n \to \infty} |\alpha_n + \sqrt{2} \cdot 1| = 0 ,$$

but

$$\lim_{n \to \infty} \sqrt{\alpha_n^2 + 1} = \sqrt{3} \neq 0 .$$

Corollary 1.1.3.5 (0) *Every finite-dimensional normed space is complete.*

Take $n \in \mathbb{N}$ and let p be a norm on $\mathbb{K}^n$. By Minkowski's Theorem, p and the Euclidean norm on $\mathbb{K}^n$ are equivalent. Since $\mathbb{K}^n$ is complete with respect to the Euclidean norm, it follows by Corollary 1.1.3.2 that $\mathbb{K}^n$ is also complete with respect to p. ■

Corollary 1.1.3.6 (0) *Let E be a normed space. Every finite-dimensional vector subspace of E is closed.*

Let F be a finite-dimensional vector subspace of E. By Corollary 1.1.3.5, F is complete with respect to the induced norm and so it is a closed subset of E. ■

1.1.4 Locally Compact Normed Spaces

Definition 1.1.4.1 (0) *Let A be a nonempty subset of the normed space E. Given $x \in E$, we set*

$$d_A(x) := \inf_{y \in A} \|x - y\|,$$

$d_A(x)$ is called ***the distance of x from A.***

Proposition 1.1.4.2 (0) *Let F be a proper closed vector subspace of the normed space E. Then there is an $x \in E$ with $\|x\| = 1$ and*

$$d_F(x) \geq \frac{1}{2}.$$

Take $y \in E \backslash F$. Since F is closed, $d_F(y) > 0$. There is a $z \in F$ with

$$\|y - z\| < 2d_F(y).$$

We set

$$x := \frac{1}{\|y - z\|}(y - z).$$

Then $\|x\| = 1$. Take $a \in F$. Then

$$x - a = \frac{1}{\|y - z\|}(y - z) - a = \frac{1}{\|y - z\|}(y - (z + \|y - z\|a)).$$

Since $z + \|y - z\|a$ belongs to F, we deduce that

$$\|x - a\| = \frac{1}{\|y - z\|}\|y - (z + \|y - z\|a)\| \geq \frac{1}{\|y - z\|}d_F(y) > \frac{1}{2},$$

$$d_F(x) = \inf_{a \in F} \|x - a\| \geq \frac{1}{2}.$$ ∎

Corollary 1.1.4.3 (0) *Every infinite–dimensional normed space E contains a sequence $(x_n)_{n \in \mathbb{N}}$ with*

$$\|x_n\| = 1, \quad \|x_m - x_n\| \geq \frac{1}{2}$$

for distinct elements $m, n \in \mathbb{N}$.

We construct the sequence recursively. Take $n \in \mathbb{N}$ and suppose that the sequence has been constructed up to $n - 1$. Let F be the vector subspace of E generated by $x_1, x_2, \ldots, x_{n-1}$. Then F is finite–dimensional and so $E \neq F$.

By Corollary 1.1.3.6, F is closed, so, by Proposition 1.1.4.2, there is an $x_n \in E$ with $\|x_n\| = 1$ and

$$d_F(x_n) \geq \frac{1}{2}.$$

Then

$$\|x_n - x_m\| \geq \frac{1}{2}$$

for every $m \in \mathbb{N}$, $m < n$, which completes the recursive construction. ■

Remark. The above corollary can be improved (see Exercise 1.3.5).

Theorem 1.1.4.4 (0) (F. Riesz, 1918) *Every locally compact normed space is finite–dimensional.*

Let E be a locally compact normed space. We assume that E is infinite–dimensional. By Corollary 1.1.4.3 there is a sequence $(x_n)_{n\in\mathbb{N}}$ in E such that

$$\|x_n\| = 1, \quad \|x_m - x_n\| \geq \frac{1}{2}$$

for distinct elements $m, n \in \mathbb{N}$. Since

$$\{x \in E \mid \|x\| = 1\}$$

is compact (Corollary 1.1.1.4), the sequence $(x_n)_{n\in\mathbb{N}}$ contains a convergent subsequence, which is obviously a contradiction. ■

Proposition 1.1.4.5 (3) *Let A be a nonempty subset of the normed space E. Then, given $x, y \in E$,*

$$|d_A(x) - d_A(y)| \leq \|x - y\|.$$

We have

$$d_A(x) \leq \|x - z\| \leq \|x - y\| + \|y - z\|$$

for any $z \in A$, so that

$$d_A(x) \leq \|x - y\| + d_A(y),$$

$$d_A(x) - d_A(y) \leq \|x - y\|,$$

$$|d_A(x) - d_A(y)| \leq \|x - y\|.$$

■

1.1.5 Products of Normed Spaces

Proposition 1.1.5.1 (0) *Let $(E_\iota)_{\iota\in I}$ be a finite family of normed spaces and p be a norm on $\mathbb{K}^I$ such that given $(\alpha_\iota)_{\iota\in I}, (\beta_\iota)_{\iota\in I} \in \mathbb{R}_+^I$ with $\alpha_\iota \leq \beta_\iota$ for all $\iota \in I$,*

$$p\left((\alpha_\iota)_{\iota\in I}\right) \leq p\left((\beta_\iota)_{\iota\in I}\right).$$

Then

$$\prod_{\iota\in I} E_\iota \longrightarrow \mathbb{R}_+, \quad (x_\iota)_{\iota\in I} \longmapsto p\left((\|x_\iota\|)_{\iota\in I}\right)$$

is a norm on $\prod_{\iota\in I} E_\iota$, which generates the product topology. If each E_ι $(\iota \in I)$ is separable (resp. complete) then $\prod_{\iota\in I} E_\iota$ is also separable (resp. complete).

It is easy to check that the above map is a norm. It is obviuous that a sequence in $\prod_{\iota\in I} E_\iota$ converges to a point $x \in \prod_{\iota\in I} E_\iota$ iff the projections of this sequence converge to the corresponding projections of x We deduce that the above norm generates the product topology of $\prod_{\iota\in I} E_\iota$. The last assertion is easy to prove. ∎

Definition 1.1.5.2 (0) *Let $(E_\iota)_{\iota\in I}$ be a finite family of normed spaces. The norm*

$$\prod_{\iota\in I} E_\iota \longrightarrow \mathbb{R}_+, \quad (x_\iota)_{\iota\in I} \longmapsto \sup_{\iota\in I} \|x_\iota\|$$

(Example 1.1.2.5, Proposition 1.1.5.1) is called **the supremum norm of the product** $\prod_{\iota\in I} E_\iota$ *and for every $p \in [1, \infty[$ the norm*

$$\prod_{\iota\in I} E_\iota \longrightarrow \mathbb{R}_+, \quad (x_\iota)_{\iota\in I} \longmapsto \left(\sum_{\iota\in I} \|x_\iota\|^p\right)^{\frac{1}{p}}$$

is called ***the p–norm of the product*** $\prod_{\iota\in I} E_\iota$ *(Example 1.1.2.5, Proposition 1.1.5.1). These norms are denoten sometimes by $\|\cdot\|_p$. The 2–norm is also called* ***the Euclidean norm of the product*** $\prod_{\iota\in I} E_\iota$*. Unless otherwise specified, we take the Euclidean norm on the product of normed space.*

All these norms generate the product topology (Proposition 1.1.5.1) and they are therefore equivalent (Proposition 1.1.3.1 c ⇒ a). When our interest is restricted to "topological aspects" of the theory (i.e. if we are concentrating on "normable" and not on normed spaces) any of these norms on products will do. But as soon as the "geometric aspects" are important, we have to choose a specific one of these norms. For example, the Euclidean norm will be the appropriate one in the case of Hilbert spaces while the supremum norm will be needed in the case of C^*–algebras.

Proposition 1.1.5.3 (0) *Let E be a normed space and take $\alpha, \beta \in \mathbb{K}$. Then the map*

$$E \times E \longrightarrow E\,, \quad (x,y) \longmapsto \alpha x + \beta y$$

is uniformly continuous.

We have

$$\begin{aligned}\|(\alpha x_1 + \beta y_1) - (\alpha x_2 + \beta y_2)\| &= \|\alpha(x_1 - x_2) + \beta(y_1 - y_2)\| \leq \\ &\leq |\alpha|\,\|x_1 - x_2\| + |\beta|\,\|y_1 - y_2\| \leq \\ &\leq (|\alpha|^2 + |\beta|^2)^{\frac{1}{2}}\,(\|x_1 - x_2\|^2 + \|y_1 - y_2\|^2)^{\frac{1}{2}} = \\ &= (|\alpha|^2 + |\beta|^2)^{\frac{1}{2}}\,\|(x_1, y_1) - (x_2, y_2)\|\end{aligned}$$

for all $x_1, x_2, y_1, y_2 \in E$, which proves the assertion. ■

Corollary 1.1.5.4 (0) *If F is a vector subspace of the normed space E then $\overline{F}$ is a vector subspace of E.*

Let $\alpha, \beta \in \mathbb{K}$ and let φ denote the map

$$E \times E \longrightarrow E\,, \quad (x,y) \longmapsto \alpha x + \beta y\,.$$

By Proposition 1.1.5.3, φ is continuous and so $\overset{-1}{\varphi}(\overline{F})$ is closed. Since

$$F \times F \subset \overset{-1}{\varphi}(F) \subset \overset{-1}{\varphi}(\overline{F})\,,$$

it follows that

$$\overline{F} \times \overline{F} = \overline{F \times F} \subset \overset{-1}{\varphi}(\overline{F})\,.$$

Hence

$$\alpha x + \beta y = \varphi(x,y) \in \overline{F}$$

for every $x, y \in \overline{F}$, i.e. $\overline{F}$ is a vector subspace of E. ■

Corollary 1.1.5.5 (0) *Let E be a normed space, A a subset of E, and F the vector subspace of E generated by A. Then $\overline{F}$ is the smallest closed vector subspace of E constaining A. It is called* ***the closed vector subspace of E generated by A****. If A is countable, then $\overline{F}$, endowed with the induced norm, is separable.*

By Corollary 1.1.5.4, $\overline{F}$ is a vector subspace of E and it is clear that it is the smallest closed vector subspace of E containing A.

Assume now that A is countable and let B be the set of linear combinations of elements of A with coefficients in $\mathbb{Q}$ (resp. $\mathbb{Q} + i\mathbb{Q}$). Then B is countable and

$$B \subset F \subset \overline{B}.$$

Thus

$$\overline{B} \subset \overline{F} \subset \overline{B},$$

so that $\overline{F} = \overline{B}$, i.e. $\overline{F}$ is separable. ∎

Example 1.1.5.6 *Let E be a normed (Banach) space and $\alpha, \beta, \gamma, \delta$ scalars such that*

$$\alpha\delta - \beta\gamma \neq 0.$$

Then

$$E \times E \longrightarrow \mathbb{R}_+, \quad (x, y) \longmapsto \sup\{\|\alpha x + \beta y\|, \|\gamma x + \delta y\|\}$$

is a (complete) norm.

The map

$$E \times E \longrightarrow E \times E, \quad (x, y) \longmapsto (\alpha x + \beta y, \gamma x + \delta y)$$

is linear and bijective and the map

$$E \times E \longrightarrow \mathbb{R}_+, \quad (x, y) \longmapsto \sup\{\|x\|, \|y\|\}$$

is a (complete) norm (Proposition 1.1.5.1). ∎

1.1.6 Summable Families

Definition 1.1.6.1 (0) *An ordered set T is called* ***upward (downward) directed*** *if for every $s,t \in T$, there is an $r \in T$ such that*

$$s \leq r, \quad t \leq r \quad (r \leq s, r \leq t).$$

If T is a nonempty set which is directed up (down) then the filter on T generated by the filter base

$$\{\{s \in T \mid s \geq t\} \mid t \in T\} \quad (\{\{s \in T \mid s \leq t\} \mid t \in T\})$$

is called ***the upper (lower) section filter of*** T.

Let I be a set and let $\mathfrak{P}_f(I)$ be ordered by the inclusion relation. We denote by $\mathfrak{F}_I$ the upper section filter of $\mathfrak{P}_f(I)$, i.e. the filter on $\mathfrak{P}_f(I)$ generated by the filter base

$$\{\{J \in \mathfrak{P}_f(I) \mid J \supset K\} \mid K \in \mathfrak{P}_f(I)\}.$$

Definition 1.1.6.2 (0) *Let $(x_\iota)_{\iota \in I}$ be a family in the normed space E. If the limit*

$$\lim_{J,\mathfrak{F}_I} \sum_{\iota \in J} x_\iota \qquad \left(\sum_{\iota \in \emptyset} x_\iota := 0\right)$$

exists, then the family $(x_\iota)_{\iota \in I}$ is called ***summable*** *and the above limit is called* ***the sum of the family*** $(x_\iota)_{\iota \in I}$*; we set*

$$\sum_{\iota \in I} x_\iota := \lim_{J,\mathfrak{F}_I} \sum_{\iota \in J} x_\iota \quad (in\ E).$$

We adopt the following convention: whenever we write $\sum_{\iota \in I} x_\iota$ we tacitly assume that $(x_\iota)_{\iota \in I}$ is summable.

If

$$I = \{n \in \mathbb{Z} \mid p \leq n \leq q\}$$

for $p, q \in \mathbb{Z} \cup \{-\infty, +\infty\}$ then we write

$$\sum_{n=p}^{q} x_n := \sum_{\iota \in I} x_\iota.$$

Remark. 1) Every finite family is summable and its sum in the above sense is its usual (algebraic) sum.

2) The assertion

$$x = \sum_{\iota \in I} x_\iota$$

is equivalent to the following one: for every $\varepsilon > 0$ there is a $K \in \mathfrak{P}_f(I)$ such that

$$\left\| \sum_{\iota \in J} x_\iota - x \right\| < \varepsilon$$

whenever $J \in \mathfrak{P}_f(I)$, and $K \subset J$.

Proposition 1.1.6.3 (0) *A family $(\alpha_\iota)_{\iota \in I}$ in $\mathrm{I\!R}_+$ is summable iff*

$$\sup_{J \in \mathfrak{P}_f(I)} \sum_{\iota \in J} \alpha_\iota < \infty$$

and in this case

$$\sum_{\iota \in I} x_\iota = \sup_{J \in \mathfrak{P}_f(I)} \sum_{\iota \in J} \alpha_\iota \,.$$

If $(\alpha_\iota)_{\iota \in I}$ is summable then

$$\infty > \sum_{\iota \in I} \alpha_\iota = \lim_{J \in \mathfrak{F}_I} \sum_{\iota \in J} \alpha_\iota = \sup_{J \in \mathfrak{P}_f(I)} \sum_{\iota \in J} \alpha_\iota \,.$$

Conversely, if

$$\sup_{J \in \mathfrak{P}_f(I)} \sum_{\iota \in J} \alpha_\iota < \infty$$

then

$$\sup_{J \in \mathfrak{P}_f(I)} \sum_{\iota \in J} \alpha_\iota = \lim_{J, \mathfrak{F}_I} \sum_{\iota \in J} \alpha_\iota \,,$$

i.e. $(\alpha_\iota)_{\iota \in I}$ is summable. ■

Remark. By the above result, given families of positive real numbers, the sums defined in Definition 1.1.2.1 and in Definition 1.1.6.2 coincide.

Proposition 1.1.6.4 (0) *Let $(E_\lambda)_{\lambda\in L}$ be a finite family of normed spaces and for each $\lambda_0 \in L$ let p_{λ_0} be the canonical projection $\prod_{\lambda\in L} E_\lambda \to E_{\lambda_0}$. A family $(x_\iota)_{\iota\in I}$ in $\prod_{\lambda\in L} E_\lambda$ is summable iff for each $\lambda \in L$ the family $(p_\lambda x_\iota)_{\iota\in I}$ is summable, and in this case*

$$p_\lambda \sum_{\iota\in I} x_\iota = \sum_{\iota\in I} p_\lambda x_\iota$$

for each $\lambda \in L$.

The assertion follows immediately from Proposition 1.1.5.1 and Definition 1.1.6.2. ■

Corollary 1.1.6.5 (0) *A family $(\alpha_\iota)_{\iota\in I}$ of complex numbers is summable iff the families $(\operatorname{re}\alpha_\iota)_{\iota\in I}$, $(\operatorname{im}\alpha_\iota)_{\iota\in I}$ are summable and in this case*

$$\sum_{\iota\in I} \alpha_\iota = \sum_{\iota\in I} \operatorname{re}\alpha_\iota + i\sum_{\iota\in I} \operatorname{im}\alpha_\iota\,. \qquad ■$$

Proposition 1.1.6.6 (0) *Let $(x_\iota)_{\iota\in I}$ be a family in the normed space E. If $(x_\iota)_{\iota\in I}$ is summable, then for every $\varepsilon > 0$ there is a $J \in \mathfrak{P}_f(I)$ such that*

$$\left\|\sum_{\iota\in K} x_\iota\right\| < \varepsilon$$

for every $K \in \mathfrak{P}_f(I\backslash J)$. Conversely, if this condition is fulfilled and if E is complete, then $(x_\iota)_{\iota\in I}$ is summable.

First assume that $(x_\iota)_{\iota\in I}$ is summable and let x be its sum. Then there is a $J \in \mathfrak{P}_f(I)$ with

$$\left\|\sum_{\iota\in K} x_\iota - x\right\| < \frac{\varepsilon}{2}$$

for every $K \in \mathfrak{P}_f(I), J \subset K$. Take $K \in \mathfrak{P}_f(I\backslash J)$. Then J and $J\cup K$ are finite subsets of I containing J, hence

$$\left\|\sum_{\iota\in K} x_\iota\right\| = \left\|\sum_{\iota\in J\cup K} x_\iota - \sum_{\iota\in J} x_\iota\right\| \le \left\|\sum_{\iota\in J\cup K} x_\iota - x\right\| + \left\|x - \sum_{\iota\in J} x_\iota\right\| < \frac{\varepsilon}{2} + \frac{\varepsilon}{2} = \varepsilon\,.$$

Now assume that E is complete and that the condition described above is fulfilled. Then there is an increasing sequence $(J_n)_{n\in\mathbb{N}}$ in $\mathfrak{P}_f(I)$ such that

$$\left\| \sum_{\iota \in K} x_\iota \right\| < \frac{1}{n}$$

for every $n \in \mathbb{N}$ and for every $K \in \mathfrak{P}_f(I \backslash J_n)$. For each $n \in \mathbb{N}$ we put

$$y_n := \sum_{\iota \in J_n} x_\iota .$$

Then $J_p \backslash J_n \in \mathfrak{P}_f(I \backslash J_n)$ and so

$$\|y_p - y_n\| = \left\| \sum_{\iota \in J_p} x_\iota - \sum_{\iota \in J_n} x_\iota \right\| = \left\| \sum_{\iota \in J_p \backslash J_n} x_\iota \right\| < \frac{1}{n}$$

for every $n, p \in \mathbb{N}$ with $n < p$. Hence $(y_n)_{n \in \mathbb{N}}$ is a Cauchy sequence. We put

$$x := \lim_{n \to \infty} y_n .$$

Take $\varepsilon > 0$. Take $n \in \mathbb{N}$ with

$$\frac{1}{n} < \frac{\varepsilon}{2}, \quad \|y_n - x\| < \frac{\varepsilon}{2} .$$

Then

$$\left\| \sum_{\iota \in K} x_\iota - x \right\| = \left\| \sum_{\iota \in K \backslash J_n} x_\iota + \sum_{\iota \in J_n} x_\iota - x \right\| \le \left\| \sum_{\iota \in K \backslash J_n} x_\iota \right\| + \|y_n - x\| < \frac{\varepsilon}{2} + \frac{\varepsilon}{2} = \varepsilon$$

for every $K \in \mathfrak{P}_f(I)$ with $J_n \subset K$. The family $(x_\iota)_{\iota \in I}$ is therefore summable and x is its sum. ■

Remark. Let φ be the map

$$\mathfrak{P}_f(I) \longrightarrow E, \quad J \longmapsto \sum_{\iota \in J} x_\iota .$$

The condition formulated in the proposition says that $\varphi(\mathfrak{F}_I)$ is a Cauchy filter, so if E is complete $\varphi(\mathfrak{F}_I)$ converges and $(x_\iota)_{\iota \in I}$ is summable.

Corollary 1.1.6.7 (0) *Let $(x_\iota)_{\iota \in I}$ be a summable family in the normed space E. Then for any $\varepsilon > 0$ the set*

$$\{\iota \in I \mid \|x_\iota\| \ge \varepsilon\}$$

is finite and the set

$$\{\iota \in I \mid x_\iota \ne 0\}$$

is countable.

There is a $J \in \mathfrak{P}_f(I)$ with

$$\left\| \sum_{\iota \in K} x_\iota \right\| < \varepsilon$$

for every $K \in \mathfrak{P}_f(I \backslash J)$ (Proposition 1.1.6.6). We get

$$\{\iota \in I \mid \|x_\iota\| \geq \varepsilon\} \subset J,$$

i.e.

$$\{\iota \in I \mid \|x_\iota\| \geq \varepsilon\}$$

is a finite set. From

$$\{\iota \in I \mid x_\iota \neq 0\} = \bigcup_{n \in \mathbb{N}} \left\{\iota \in I \,\middle|\, \|x_\iota\| \geq \frac{1}{n}\right\}$$

we deduce that

$$\{\iota \in I \mid x_\iota \neq 0\}$$

is countable. ∎

Corollary 1.1.6.8 *Let E be a Banach space. If $(x_\iota)_{\iota \in I}$ is a summable family in E, then $(\alpha_\iota x_\iota)_{\iota \in I}$ is summable in E for every $(\alpha_\iota)_{\iota \in I} \in \ell^\infty(I)$.*

Take $\varepsilon > 0$. Put

$$\delta := \sup_{\iota \in I} |\alpha_\iota|.$$

There is a $J \in \mathfrak{P}_f(I)$ such that

$$\left\| \sum_{\iota \in K} x_\iota \right\| < \frac{\varepsilon}{4(1+\delta)}$$

for every $K \in \mathfrak{P}_f(I \backslash J)$ (Proposition 1.1.6.6). By Proposition 1.1.1.5,

$$\left\| \sum_{\iota \in K} \alpha_\iota x_\iota \right\| < \varepsilon$$

for every $K \in \mathfrak{P}_f(I \backslash J)$, so that $(\alpha_\iota x_\iota)_{\iota \in I}$ is summable (Proposition 1.1.6.6). ∎

Definition 1.1.6.9 (0) *A family $(x_\iota)_{\iota\in I}$ in a normed space is called* ***absolutely summable*** *if*

$$\sum_{\iota\in I}\|x_\iota\| < \infty\,,$$

i.e. if $(\|x_\iota\|)_{\iota\in I}$ is summable in $\mathbb{R}$ (Proposition 1.1.6.3).

Corollary 1.1.6.10 (0) *Let E be a normed space. Then the following are equivalent:*

a) *E is complete.*

b) *every absolutely summable family in E is summable.*

c) *every absolutely summable countable family in E is summable.*

If these conditions are fulfilled and $(x_\iota)_{\iota\in I}$ is an absolutely summable family in E then

$$\left\|\sum_{\iota\in I} x_\iota\right\| \le \sum_{\iota\in I}\|x_\iota\|\,.$$

a ⇒ b & last assertion. Let $\varepsilon > 0$. There exists a $J \in \mathfrak{P}_f(I)$ such that

$$\sum_{\iota\in K}\|x_\iota\| < \varepsilon$$

for every $K \in \mathfrak{P}_f(I\backslash J)$. We get

$$\left\|\sum_{\iota\in K} x_\iota\right\| \le \sum_{\iota\in K}\|x_\iota\| < \varepsilon$$

for every $K \in \mathfrak{P}_f(I\backslash J)$. By Proposition 1.1.6.6 the family $(x_\iota)_{\iota\in I}$ is summable.

We have

$$\left\|\sum_{\iota\in J} x_\iota\right\| \le \sum_{\iota\in J}\|x_\iota\|$$

for every $J \in \mathfrak{P}_f(I)$ and therefore

$$\left\|\sum_{\iota\in I} x_\iota\right\| = \lim_{J,\mathfrak{F}_I}\left\|\sum_{\iota\in J} x_\iota\right\| \le \lim_{J,\mathfrak{F}_I}\sum_{\iota\in J}\|x_\iota\| = \sum_{\iota\in I}\|x_\iota\|\,.$$

b ⇒ c is trivial.

c ⇒ a. Let $(x_n)_{n\in\mathbb{N}}$ be a Cauchy sequence in E. There is a strictly increasing family $(k_p)_{p\in\mathbb{N}\cup\{0\}}$ in $\mathbb{N}$ such that

$$\|x_n - x_{k_p}\| < \frac{1}{2^p}$$

whenever $n \in \mathbb{N}$, $n \geq k_p$. Then

$$\sum_{p\in\mathbb{N}} \|x_{k_p} - x_{k_{p-1}}\| \leq \sum_{p\in\mathbb{N}} \frac{1}{2^p} < \infty .$$

By c), $(x_{k_p} - x_{k_{p-1}})_{p\in\mathbb{N}}$ is summable. Given $n \in \mathbb{N}$

$$x_{k_n} = x_{k_0} + \sum_{p=1}^{n} (x_{k_p} - x_{k_0})$$

so that $(x_{k_n})_{n\in\mathbb{N}}$ converges. It follows that $(x_n)_{n\in\mathbb{N}}$ converges. E is thus complete. ■

Proposition 1.1.6.11 (0) *Let E be a normed space, $(x_\iota)_{\iota\in I}, (y_\iota)_{\iota\in I}$ be summable families in E and $\alpha, \beta \in \mathbb{K}$. Then $(\alpha x_\iota + \beta y_\iota)_{\iota\in I}$ is summable and*

$$\sum_{\iota\in I} (\alpha x_\iota + \beta y_\iota) = \alpha \sum_{\iota\in I} x_\iota + \beta \sum_{\iota\in I} y_\iota .$$

We have (Proposition 1.1.5.3) that

$$\alpha \sum_{\iota\in I} x_\iota + \beta \sum_{\iota\in I} y_\iota = \alpha \lim_{J,\mathfrak{F}_I} \sum_{\iota\in J} x_\iota + \beta \lim_{J,\mathfrak{F}_I} \sum_{\iota\in J} y_\iota =$$

$$= \lim_{J,\mathfrak{F}_I} \left(\alpha \sum_{\iota\in J} x_\iota + \beta \sum_{\iota\in J} y_\iota \right) = \lim_{J,\mathfrak{F}_I} \sum_{\iota\in J} (\alpha x_\iota + \beta y_\iota) .$$ ■

Proposition 1.1.6.12 (0) *Let $(x_\iota)_{\iota\in I}$ be a family in the normed space E and*

$$I_0 := \{\iota \in I \mid x_\iota \neq 0\} .$$

Then $(x_\iota)_{\iota\in I}$ is summable iff $(x_\iota)_{\iota\in I_0}$ is summable and in this case

$$\sum_{\iota\in I} x_\iota = \sum_{\iota\in I_0} x_\iota .$$

We set

$$\varphi : \mathfrak{P}_f(I) \longrightarrow \mathfrak{P}_f(I_0), \quad J \longmapsto I_0 \cap J .$$

Then $\varphi(\mathfrak{F}_I) = \mathfrak{F}_{I_0}$ and

$$\sum_{\iota\in J} x_\iota = \sum_{\iota\in\varphi(J)} x_\iota$$

for every $J \in \mathfrak{P}_f(I)$. The proposition now follows. ■

Corollary 1.1.6.13 (0) *Let $(x_\iota)_{\iota\in I}$, $(y_\lambda)_{\lambda\in L}$ be summable families in the normed space E such that $I\cap L=\emptyset$. Define*

$$z_\mu := \begin{cases} x_\mu & \text{if } \mu\in I \\ y_\mu & \text{if } \mu\in L\,. \end{cases}$$

Then

$$\sum_{\mu\in I\cup L} z_\mu = \sum_{\iota\in I} x_\iota + \sum_{\lambda\in L} y_\lambda\,.$$

For each $\lambda\in L$ and each $\iota\in I$, define

$$x_\lambda := 0\,,\quad y_\iota := 0\,.$$

The corollary now follows from Proposition 1.1.6.12 and Proposition 1.1.6.11. ■

Proposition 1.1.6.14 (0) *The following assertions are equivalent for every family $(\alpha_\iota)_{\iota\in I}$ in $\mathbb{K}$:*

a) *$(\alpha_\iota)_{\iota\in I}$ is summable.*

b) *$(\alpha_\iota)_{\iota\in I}$ is absolutely summable.*

c) *For every injective map $\varphi:\mathbb{N}\to I$ the sequence $(\sum_{k=1}^{n}\alpha_{\varphi(k)})_{n\in\mathbb{N}}$ converges.*

d) $\sup_{J\in\mathfrak{P}_f(I)}\left|\sum_{\iota\in J}\alpha_\iota\right|<\infty$.

In particular, the summable and absolutely summable families coincide in finite-dimensional normed spaces.

a ⇒ b. Assume first $\mathbb{K}=\mathbb{R}$. We set

$$I_+ := \{\iota\in I \mid \alpha_\iota\geq 0\}\,,\quad I_- := \{\iota\in I \mid \alpha_\iota<0\}\,.$$

Then $(\alpha_\iota)_{\iota\in I_+}$, $(\alpha_\iota)_{\iota\in I_-}$, and $(-\alpha_\iota)_{\iota\in I}$ are summable. By Corollary 1.1.6.13, $(|\alpha_\iota|)_{\iota\in I}$ is summable, hence $(\alpha_\iota)_{\iota\in I}$ is absolutely summable.

Assume now $\mathbb{K}=\mathbb{C}$. Then $(\operatorname{re}\alpha_\iota)_{\iota\in I}$, $(\operatorname{im}\alpha_\iota)_{\iota\in I}$ are summable (Corollary 1.1.6.5), so by the above considerations they are absolutely summable. We deduce

$$\sup_{J\in\mathfrak{P}_f(I)}\sum_{\iota\in J}|\alpha_\iota| \leq \sup_{J\in\mathfrak{P}_f(I)}\sum_{\iota\in J}|\operatorname{re}\alpha_\iota| + \sup_{J\in\mathfrak{P}_f(I)}\sum_{\iota\in J}|\operatorname{im}\alpha_\iota|<\infty\,,$$

hence $(\alpha_\iota)_{\iota\in I}$ is absolutely summable (Proposition 1.1.6.3).

b $\Rightarrow$ c is trivial.

c $\Rightarrow$ d. Assume

$$\sup_{J\in\mathfrak{P}_f(I)}\left|\sum_{\iota\in J}\alpha_\iota\right| = \infty .$$

Then there is an increasing sequence $(J_n)_{n\in\mathbb{N}}$ in $\mathfrak{P}_f(I)$ such that

$$\lim_{n\to\infty}\left|\sum_{\iota\in J_n}\alpha_\iota\right| = \infty .$$

Let $\varphi : \mathbb{N}\to I$ be an injective map for which there is an increasing sequence $(p_n)_{n\in\mathbb{N}}$ in $\mathbb{N}$ with

$$\varphi(\mathbb{N}_{p_n}) = J_n$$

for every $n\in\mathbb{N}$. Then

$$\lim_{n\to\infty}\left|\sum_{k=1}^{p_n}\alpha_{\varphi(k)}\right| = \lim_{n\to\infty}\left|\sum_{\iota\in J_n}\alpha_\iota\right| = \infty ,$$

which contradicts c).

d $\Rightarrow$ a. Assume first $\mathbb{K}=\mathbb{R}$. We set

$$I_+ := \{\iota\in I \mid \alpha_\iota \geq 0\}, \quad I_- := \{\iota\in I \mid \alpha_\iota < 0\}.$$

Then

$$\sup_{J\in\mathfrak{P}_f(I_+)}\sum_{\iota\in J}\alpha_\iota < \infty , \qquad \sup_{J\in\mathfrak{P}_f(I_-)}\sum_{\iota\in J}(-\alpha_\iota) < \infty ,$$

so $(\alpha_\iota)_{\iota\in I_+}$ and $(-\alpha_\iota)_{\iota\in I_-}$ are summable (Proposition 1.1.6.3). Hence $(\alpha_\iota)_{\iota\in I}$ is summable (Proposition 1.1.6.11, Corollary 1.1.6.13).

Assume now $\mathbb{K}=\mathbb{C}$. Then

$$\sup_{J\in\mathfrak{P}_f(I)}\left|\sum_{\iota\in J}\operatorname{re}\alpha_\iota\right| = \sup_{J\in\mathfrak{P}_f(I)}\left|\operatorname{re}\sum_{\iota\in J}\alpha_\iota\right| \leq \sup_{J\in\mathfrak{P}_f(I)}\left|\sum_{\iota\in J}\alpha_\iota\right| < \infty ,$$

$$\sup_{J\in\mathfrak{P}_f(I)}\left|\sum_{\iota\in J}\operatorname{im}\alpha_\iota\right| = \sup_{J\in\mathfrak{P}_f(I)}\left|\operatorname{im}\sum_{\iota\in J}\alpha_\iota\right| \leq \sup_{J\in\mathfrak{P}_f(I)}\left|\sum_{\iota\in J}\alpha_\iota\right| < \infty .$$

By the above considerations $(\operatorname{re}\alpha_\iota)_{\iota\in I}$, $(\operatorname{im}\alpha_\iota)_{\iota\in I}$ are summable, hence $(\alpha_\iota)_{\iota\in I}$ is summable (Corollary 1.1.6.5). ■

Remark. a) Let $(\alpha_\iota)_{\iota\in I}$ be a family in $\mathbb{K}$,

$$f : I \longrightarrow \mathbb{K}, \quad \iota \longmapsto \alpha_\iota,$$

and μ be the counting measure on I, i.e.

$$\mu : \mathfrak{P}_f(I) \longrightarrow \mathbb{R}, \quad J \longmapsto \operatorname{Card} J.$$

Then $(\alpha_\iota)_{\iota \in I}$ is summable iff f is μ–integrable and in this case

$$\sum_{\iota \in I} \alpha_\iota = \int f \, d\mu.$$

b) A. Dworetzky and C.A. Rogers proved (1950) that if every summable family in E is absolutely summable, then E is finite–dimensional.

Example 1.1.6.15 (1) (3) *Let T be a set, $p \in [1, \infty[$, and $x \in \mathbb{K}^T$. Then $(x(t)e_t)_{t \in T}$ is summable in $\ell^p(T)$ (in $\ell^\infty(T)$) iff $x \in \ell^p(T)$ $(x \in c_0(T))$ and in this case the sum is x.*

Assume first $x \in \ell^p(T)$ $(x \in c_0(T))$. Then

$$\left\| \sum_{t \in A} x(t)e_\iota - x \right\|_p = \left(\sum_{t \in T \setminus A} |x(t)|^p \right)^{\frac{1}{p}}, \quad \left(\left\| \sum_{t \in A} x(t)e_t - x \right\|_\infty = \sup_{t \in T \setminus A} |x(t)| \right)$$

for every finite subset A of T. Hence $(x(t)e_t)_{t \in T}$ is summable in $\ell^p(T)$ (in $\ell^\infty(T)$) and its sum is x.

Assume now $(x(t)e_t)$ summable in $\ell^p(T)$ (in $\ell^\infty(T)$). Let $\varepsilon > 0$. Then there is a finite subset A of T such that

$$\left\| \sum_{t \in B} x(t)e_t \right\|_p < \varepsilon, \quad \left(\left\| \sum_{t \in B} x(t)e_t \right\|_\infty < \varepsilon \right)$$

for every $B \in \mathfrak{P}_f(T \setminus A)$ (Proposition 1.1.6.6). We get

$$\sum_{t \in T} |x(t)|^p = \sum_{t \in A} |x(t)|^p + \sup \left\{ \sum_{t \in B} |x(t)|^p \,|\, B \in \mathfrak{P}_f(T \setminus A) \right\} \le$$

$$\le \sum_{t \in A} |x(t)|^p + \varepsilon < \infty,$$

$$\left(\sup_{t \in T \setminus A} |x(t)| \le \varepsilon \right),$$

i.e. $x \in \ell^p(T)$ $(x \in c_0(T))$. ∎

Proposition 1.1.6.16 *Let E be a Banach space and $(E_n)_{n\in\mathbb{N}}$ be a strictly increasing sequence of closed vector subspaces of E. Then the dimension of $E/\bigcup_{n\in\mathbb{N}} E_n$ is at least $2^{\aleph_0}$.*

By Proposition 1.1.4.2, there is a sequence $(a_n)_{n\in\mathbb{N}}$ in E so that

$$a_n \in E_{n+1}\,, \quad \|a_n\| = 1\,, \quad d_{E_n}(a_n) \geq \frac{1}{2}$$

for every $n \in \mathbb{N}$. We set

$$u : \ell^\infty \longrightarrow E\,, \quad x \longmapsto \sum_{n\in\mathbb{N}} \frac{x(n)}{4^n} a_n$$

(Corollary 1.1.6.10 a $\Rightarrow$ c). u is linear (Proposition 1.1.6.11). Let $x \in \ell^\infty$ so that $ux \in \bigcup_{n\in\mathbb{N}} E_n$. Then there is a $p \in \mathbb{N}$ so that $ux \in E_p$. Let $q \in \mathbb{N}$, $q \geq p$, such that $x(q) \neq 0$. Then

$$\frac{4^q}{x(q)}\left(ux - \sum_{n=1}^{q-1} \frac{x(n)}{4^n} a_n\right) \in E_q\,,$$

so that

$$\frac{1}{2} \leq d_{E_q}(a_q) \leq \left\| \frac{4^q}{x(q)}\left(ux - \sum_{n=1}^{q-1} \frac{x(n)}{4^n} a_n\right) - a_q \right\| =$$

$$= \frac{4^q}{|x(q)|}\left\| ux - \sum_{n=1}^{q} \frac{x(n)}{4^n} a_n \right\| = \frac{4^q}{|x(q)|}\left\| \sum_{n=q+1}^{\infty} \frac{x(n)}{4^n} a_n \right\| \leq$$

$$\leq \frac{4^q}{|x(q)|} \sum_{n=q+1}^{\infty} \left\| \frac{x(n)}{4^n} a_n \right\| = \frac{4^q}{|x(q)|} \sum_{n=q+1}^{\infty} \frac{|x(n)|}{4^n}\,,$$

$$\frac{|x(q)|}{2} \leq \sum_{n=1}^{\infty} \frac{|x(q+n)|}{4^n}\,.$$

Let $\mathfrak{A}$ be a set of infinite subsets of $\mathbb{N}$ having the power of continuum such that $A \cap B$ is finite for every $A, B \in \mathfrak{A}$, $A \neq B$ (Lemma 1.1.2.17) and let F be the vector subspace of ℓ^∞ generated by $(e_A)_{A\in\mathfrak{A}}$. Since $(e_A)_{A\in\mathfrak{A}}$ is linearly independent the dimension of F is $2^{\aleph_0}$. Let $y \in F\backslash\{0\}$. There is a finite subset $\mathfrak{B}$ of $\mathfrak{A}$ and a family $(\alpha_B)_{B\in\mathfrak{B}}$ in $\mathbb{K}\backslash\{0\}$ such that

$$y = \sum_{B\in\mathfrak{B}} \alpha_B e_B\,.$$

Choose $A \in \mathfrak{B}$ with

$$|\alpha_A| = \sup_{B \in \mathfrak{B}} |\alpha_B|\,,$$

and $p \in \mathbb{N}$ with

$$\bigcup_{\substack{B,C \in \mathfrak{B} \\ B \neq C}} (B \cap C) \subset \mathbb{N}_p\,.$$

Given $q \in A$ with $q \geq p$, we have that

$$\sum_{n=1}^{\infty} \frac{|y(q+n)|}{4^n} \leq \sum_{n=1}^{\infty} \frac{|\alpha_A|}{4^n} = \frac{|\alpha_A|}{3} < \frac{|\alpha_A|}{2} = \frac{|y(q)|}{2}\,.$$

By the above considerations, $uy \notin \bigcup_{n \in \mathbb{N}} E_n$ and so the map

$$F \longrightarrow E/\bigcup_{n \in \mathbb{N}} E_n\,, \quad y \longmapsto vuy\,,$$

where

$$v : E \longrightarrow E/\bigcup_{n \in \mathbb{N}} E_n$$

is the quotient map, is injective. We conclude that the dimension of $E/\bigcup_{n \in \mathbb{N}} E_n$ is at least the dimension of F, i.e. at least $2^{\aleph_0}$. ■

Corollary 1.1.6.17 *The dimension of an infinite–dimensional Banach space is at least* $2^{\aleph_0}$.

By Corollary 1.1.3.6, every infinite–dimensional normed space has a strictly increasing sequence of closed vector subspaces and hence the corollary follows from Proposition 1.1.6.16. ■

Proposition 1.1.6.18 *Let F be a vector subspace of the normed space E. If $\aleph > 1$ is the dimension of F, then*

$$\operatorname{Card} \overline{F} \leq \aleph^{\aleph_0}\,.$$

By Lemma 1.1.2.20,

$$\operatorname{Card} F = \aleph 2^{\aleph_0}$$

so that the set of sequences in F has cardinality

$$(\aleph 2^{\aleph_0})^{\aleph_0} = \aleph^{\aleph_0} 2^{\aleph_0} = \aleph^{\aleph_0}\,.$$

Since every point of $\overline{F}$ is the limit of a convergent sequence in F, it follows that

$$\operatorname{Card} \overline{F} \leq \aleph^{\aleph_0}\,.$$

■

Proposition 1.1.6.19 *Let E be an infinite-dimensional normed space of dimension $\aleph$. If there is an increasing sequence $(\aleph_{(n)})_{n\in\mathbb{N}}$ of cardinal numbers with*

$$\aleph = \sup_{n\in\mathbb{N}} \aleph_{(n)}$$

such that

$$\aleph_{(n)}^{\aleph_0} < \aleph$$

for every $n \in \mathbb{N}$, then there is a strictly increasing sequence of closed vector subspaces of E whose union is E. In particular, E is not complete and

$$\aleph^{\aleph_0} \neq \aleph .$$

Let $(x_\xi)_{\xi\in\aleph}$ be an algebraic basis for E. For each $n \in \mathbb{N}$, let E_n denote the closed vector subspace of E generated by $(x_\xi)_{\xi\in\aleph_{(n)}}$. By Proposition 1.1.6.18, the dimension of E_n is at most $\aleph_{(n)}^{\aleph_0}$, so that $E_n \neq E$. $(E_n)_{n\in\mathbb{N}}$ contains a subsequence with the desired properties.

The last assertion follows from the first one, Proposition 1.1.6.16, and Corollary 1.2.2.22. ■

Corollary 1.1.6.20 *Let ξ be an ordinal number with*

$$\aleph_\xi^{\aleph_0} < \aleph_{\xi+\omega_0} .$$

Then

$$\aleph_{\xi+n}^{\aleph_0} < \aleph_{\xi+\omega_0}$$

for every $n \in \omega_0$,

$$\aleph_{\xi+\omega_0}^{\aleph_0} \neq \aleph_{\xi+\omega_0} ,$$

and there is no Banach space of dimension $\aleph_{\xi+\omega_0}$.

We have that

$$\ell^1(\aleph_{\xi+1}) = \bigcup_{\eta\in\aleph_{\xi+1}} \ell^1(\eta)$$

and for each $\eta \in \aleph_{\xi+1}$,

$$\operatorname{Card} \ell^1(\aleph_{\xi+1}) = \aleph_{\xi+1}^{\aleph_0} ,$$

$$\operatorname{Card} \ell^1(\eta) = (\operatorname{Card} \eta)^{\aleph_0} \le \aleph_\xi^{\aleph_0} < \aleph_{\xi+\omega_0}$$

(Proposition 1.1.2.21). Hence

$$\aleph_{\xi+1}^{\aleph_0} \le \aleph_\xi^{\aleph_0} \aleph_{\xi+1} < \aleph_{\xi+\omega_0} .$$

By complete induction

$$\aleph_{\xi+n}^{\aleph_0} < \aleph_{\xi+\omega_0}$$

for each $n \in \omega_0$.

The last assertions follow from the above relation and Proposition 1.1.6.19.

■

Corollary 1.1.6.21 *Let E be an infinite-dimensional vector space. If*

$$\aleph_\xi^{\aleph_0} < \aleph_{\xi+\omega_0}$$

for every ordinal number ξ (this condition follows from the generalized continuum hypothesis), then either there is no complete norm on E or there is a set $\mathcal{P}$ of pairwise non-equivalent complete norms on E which is of the power of continuum.

Let ξ be an ordinal number such that the dimension of E is $\aleph_\xi$. If $\xi = 0$ then E admits no complete norm (Corollary 1.1.6.17). If there is an ordinal number η such that

$$\xi = \eta + \omega_0$$

then, by Proposition 1.1.6.19, E once again admits no complete norm. If ξ does not fulfill either of the above hypotheses, then, by Corollary 1.1.2.23, there is a set $\mathcal{P}$ of pairwise non-equivalent complete norms on E which is of the power of continuum.

The generalized continuum hypothesis implies that

$$\aleph_\xi^{\aleph_0} \le 2^{\aleph_\xi} = \aleph_{\xi+1} < \aleph_{\xi+\omega_0}$$

for every ordinal number ξ. ■

Remark. The assumption in the corollary does not follow from the usual axioms of the set theory, since

$$2^{\aleph_0} \le \aleph_\xi^{\aleph_0} .$$

Definition 1.1.6.22 (0) *Let E be a Banach space. A* **power series in** E *is an expression of the form* $\sum\limits_{n=0}^{\infty} t^n x_n$, *where t is a variable and* $(x_n)_{n\in\mathbb{N}\cup\{0\}}$ *is a family in E . If the family* $(\alpha^n x_n)_{n\in\mathbb{N}\cup\{0\}}$ *is absolutely summable for an* $\alpha \in \mathbb{K}$ $(0^0 := 1)$ *then we set*

$$\sum_{n=0}^{\infty} \alpha^n x_n := \sum_{n\in\mathbb{N}\cup\{0\}} \alpha^n x_n$$

(Definition 1.1.6.2). The number

$$\frac{1}{\limsup\limits_{n\to\infty} \|x_n\|^{\frac{1}{n}}} \qquad \left(\frac{1}{0} := \infty\,, \quad \frac{1}{\infty} := 0\right)$$

is called **the radius of convergence of the power series** $\sum\limits_{n=0}^{\infty} t^n x_n$.

Theorem 1.1.6.23 (0) *Let* $\sum\limits_{n=0}^{\infty} t^n x_n$ *be a power series in the Banach space E , and let r be its radius of convergence. Then, given* $\alpha \in U_r^{\mathbb{K}}(0)$, *the family* $(\alpha^n x_n)_{n\in\mathbb{N}\cup\{0\}}$ *is absolutely summable and the partial sums converge uniformly on the compact sets of* $U_r^{\mathbb{K}}(0)$ *so that*

$$U_r^{\mathbb{K}}(0) \longrightarrow E\,, \quad \alpha \longmapsto \sum_{n=0}^{\infty} \alpha^n x_n$$

is continuous. The family $(\alpha^n x_n)_{n\in\mathbb{N}\cup\{0\}}$ *is not summable is* $\alpha \in \mathbb{K}\setminus\overline{U_r^{\mathbb{K}}(0)}$.

Take $\rho \in [0, r[$ and $\alpha \in U_\rho^{\mathbb{K}}(0)$. Then

$$\limsup_{n\to\infty} \|\alpha^n x_n\|^{\frac{1}{n}} = \limsup_{n\to\infty} |\alpha|\, \|x_n\|^{\frac{1}{n}} \leq \frac{\rho}{r} < 1\,.$$

By the Cauchy Root Test, the family $(\alpha^n x_n)_{n\in\mathbb{N}\cup\{0\}}$ is absolutely summable for every $\alpha \in U_r^{\mathbb{K}}(0)$ and the partial sums converge uniformly on the compact sets of $U_r^{\mathbb{K}}(0)$ to the sum

$$U_r^{\mathbb{K}}(0) \longrightarrow E\,, \quad \alpha \longmapsto \sum_{n=0}^{\infty} \alpha^n x_n\,.$$

Now take $\alpha \in \mathbb{K}\setminus\overline{U_r^{\mathbb{K}}(0)}$. Then

$$\limsup_{n\to\infty} \|\alpha^n x_n\|^{\frac{1}{n}} = |\alpha| \limsup_{n\to\infty} \|x_n\|^{\frac{1}{n}} = \frac{|\alpha|}{r} > 1\,.$$

By Cauchy Root Test, the family $(\alpha^n x_n)_{n\in\mathbb{N}\cup\{0\}}$ is not summable. ■

Definition 1.1.6.24 *Let E be a Banach space, U a subset of $\mathbb{K}$, α a non-isolated point of U, and $f : U \to E$. We say that f is* ***differentiable at*** *α if*

hier

$$\lim_{\substack{\beta \to \alpha \\ \beta \in U \setminus \{\alpha\}}} \frac{1}{\beta - \alpha}\big(f(\beta) - f(\alpha)\big)$$

exists and this limit is called ***the derivative of*** *f* ***at*** *α. It is denoted by $f'(\alpha)$. f is called* ***differentiable on*** *U if it is differentiable at every nonisolated point of U.*

If f is differentiable at α then it is continuous at α.

Proposition 1.1.6.25 *Let $\sum_{n=0}^{\infty} t^n x_n$ be a power series in the Banach space E, r its radius of convergence, and*

$$f : U_r^{\mathbb{K}}(0) \longrightarrow E, \quad \alpha \longmapsto \sum_{n=0}^{\infty} \alpha^n x_n .$$

Then the power series $\sum_{n=0}^{\infty} n t^{n-1} x_n$ has radius of convergence r, f is differentiable, and

$$f'(\alpha) = \sum_{n=0}^{\infty} n \alpha^{n-1} x_n$$

for every $\alpha \in U_r^{\mathbb{K}}(0)$.

We have that

$$\limsup_{n\to\infty} \Big(n\|x_n\|\Big)^{\frac{1}{n}} = \Big(\lim_{n\to\infty} n^{\frac{1}{n}}\Big) \limsup_{n\to\infty} \|x_n\|^{\frac{1}{n}} = \frac{1}{r},$$

so that the radius of convergence of the power series $\sum_{n=0}^{\infty} n t^{n-1} x_n$ is r.

Take $\alpha \in U_r^{\mathbb{K}}(0)$ and $\beta \in U_r^{\mathbb{K}}(0) \setminus \{\alpha\}$. Then

$$\frac{1}{\beta - \alpha}\Big(f(\beta) - f(\alpha)\Big) = \frac{1}{\beta - \alpha}\left(\sum_{n=0}^{\infty} \beta^n x_n - \sum_{n=0}^{\infty} \alpha^n x_n\right) =$$

$$= \sum_{n=0}^{\infty} \frac{\beta^n - \alpha^n}{\beta - \alpha} x_n = \sum_{n=0}^{\infty} \left(\sum_{m=0}^{n-1} \beta^m \alpha^{n-m-1}\right) x_n$$

(Proposition 1.1.6.11). Let $\varepsilon > 0$ and $r_0 \in]|\alpha|, r[$. There is a $p \in \mathbb{N}$ with

$$\sum_{n=p+1}^{\infty} n r_0^{n-1} \|x_n\| < \frac{\varepsilon}{3}.$$

Suppose that $\beta \in U_{r_0-|\alpha|}^{\mathbb{K}}(\alpha)$. Then $|\beta| < r_0$, and so

$$\left|\sum_{m=0}^{n-1} \beta^m \alpha^{n-m-1}\right| \le \sum_{m=0}^{n-1} |\beta|^m |\alpha|^{n-m-1} \le \sum_{m=0}^{n-1} r_0^m r_0^{n-m-1} = n r_0^{n-1},$$

$$\left\|\frac{1}{\beta-\alpha}\Big(f(\beta) - f(\alpha)\Big) - \sum_{n=0}^{p}\left(\sum_{m=0}^{n-1} \beta^m \alpha^{n-m-1}\right) x_n\right\| =$$

$$\left\|\sum_{n=p+1}^{\infty}\left(\sum_{m=0}^{n-1} \beta^m \alpha^{n-m-1}\right) x_n\right\| \le \sum_{n=p+1}^{\infty}\left|\sum_{m=0}^{n-1} \beta^m \alpha^{n-m-1}\right| \|x_n\| \le$$

$$\le \sum_{n=p+1}^{\infty} n r_0^{n-1} \|x_n\| < \tfrac{\varepsilon}{3}$$

(Corollary 1.1.6.13, Corollary 1.1.6.10). There is a $\delta' > 0$ with

$$\left\|\sum_{n=0}^{p}\left(\sum_{m=0}^{n-1} \beta^m \alpha^{n-m-1}\right) x_n - \sum_{n=0}^{p} r_0 \alpha^{n-1} x_n\right\| < \frac{\varepsilon}{3}$$

for every $\beta \in U_{\delta'}^{\mathbb{K}}(\alpha)$. Put

$$\delta := \inf\{\delta', r_0 - |\alpha|\}.$$

Then, given $\beta \in U_{\delta}^{\mathbb{K}}(\alpha)$,

$$\left\|\frac{1}{\beta-\alpha}(f(\beta) - f(\alpha)) - \sum_{n=0}^{\infty} n \alpha^{n-1} x_n\right\| \le$$

$$\le \left\|\frac{1}{\beta-\alpha}(f(\beta) - f(\alpha)) - \sum_{n=0}^{p}\left(\sum_{m=0}^{n-1} \beta^m \alpha^{n-m-1}\right) x_n\right\| +$$

$$+ \left\|\sum_{n=0}^{p}\left(\sum_{m=0}^{n-1} \beta^m \alpha^{n-m-1}\right) x_n - \sum_{n=0}^{p} n \alpha^{n-1} x_n\right\| +$$

$$+ \left\|\sum_{n=p+1}^{\infty} n \alpha^{n-1} x_n\right\| < \frac{\varepsilon}{3} + \frac{\varepsilon}{3} + \frac{\varepsilon}{3} = \varepsilon.$$

Since ε is arbitrary

$$f'(\alpha) = \sum_{n=0}^{\infty} n \alpha^{n-1} x_n. \qquad \blacksquare$$

Proposition 1.1.6.26 (5) *Let I be an interval in $\mathbb{R}$ and $(x_\iota)_{\iota\in I}$ a family of continuously differentiable scalar functions on I. If $(x_\iota)_{\iota\in I}$ and $(x'_\iota)_{\iota\in I}$ are summable families in $\mathcal{C}(I)$ then $\sum_{\iota\in I} x_\iota$ is differentiable and*

$$\Big(\sum_{\iota\in I} x_\iota\Big)' = \sum_{\iota\in I} x'_\iota .$$

Take $a, b \in I$. Then

$$\sum_{\iota\in I} x_\iota(b) - \sum_{\iota\in I} x_\iota(a) = \sum_{\iota\in I}(x_\iota(b) - x_\iota(a)) = \sum_{\iota\in I} \int_a^b x'_\iota(t)dt = \int_a^b \Big(\sum_{\iota\in I} x'_\iota(t)\Big)dt .$$

Hence $\sum_{\iota\in I} x_\iota$ is differentiable and

$$\Big(\sum_{\iota\in I} x_\iota\Big)' = \sum_{\iota\in I} x'_\iota . \qquad \blacksquare$$

Exercises

E 1.1.1 Prove that all norms on $\mathbb{K}$ are proportional.

E 1.1.2 Show that the map

$$\mathbb{C} \longrightarrow \mathbb{R}_+\,, \quad \alpha \longmapsto |\mathrm{re}\,\alpha| + |\mathrm{im}\,\alpha|$$

is a norm when $\mathbb{C}$ is endowed with its canonical structure as real vector space, but it is not a norm if $\mathbb{C}$ is taken to be complex vector space.

E 1.1.3 Let $n \in \mathbb{N}$. Show that there is a norm p on $\mathbb{K}^2$ such that

$$p\Big((1,0)\Big) < \frac{1}{n}\|(1,0)\|\,,$$
$$p\Big((0,1)\Big) > n\|(0,1)\|\,.$$

E 1.1.4 Let E be the vector space of continuously differentiable maps $]-1,1[\rightarrow \mathbb{K}$ with bounded derivatives. Show that

$$E \longrightarrow \mathbb{R}_+\,, \quad x \longmapsto |x(0)| + \sup_{t\in]-1,1[} \left|\frac{dx}{dt}(t)\right|,$$
$$E \longrightarrow \mathbb{R}_+\,, \quad x \longmapsto \sup_{t\in]-1,1[} \left(|x(t)| + \left|\frac{dx}{dt}(t)\right|\right)$$

are equivalent norms.

E 1.1.5 Show that a normed space E is separable iff it possesses a dense vector subspace with a countable algebraic basis.

E 1.1.6 Show that $\mathbb{K}^{\mathbb{N}}$ admits no norm which generates the product topology.

(Hint: Given a sequence $(\alpha_n)_{n\in\mathbb{N}}$ in $\mathbb{K}$, the sequence $(\alpha_n e_n^{\mathbb{N}})_{n\in\mathbb{N}}$ converges to 0 with respect to the product topology.)

E 1.1.7 Let T be an infinite set and given $f \in]0,\infty[^T$, define

$$q_f : \mathbb{K}^{(T)} \longrightarrow \mathbb{R}_+\,, \quad x \longmapsto \sum_{t\in T} f(t)|x(t)|\,.$$

Prove the following:

a) For each $f \in]0,\infty[^T$, q_f defines a norm. We write $\mathbb{K}_f^{(T)}$ for the vector space $\mathbb{K}^{(T)}$ endowed with this norm.

b) If $f, g \in]0,\infty[^T$ then:

b$_1$) $\mathbb{K}_f^{(T)}$ and $\mathbb{K}_g^{(T)}$ are isometric iff $f = g$.

b$_2$) q_f and q_g are equivalent iff

$$\inf_{t\in T} \frac{f(t)}{g(t)} \neq 0\,, \quad \inf_{t\in T} \frac{g(t)}{f(t)} \neq 0\,.$$

c) $]0,\infty[^T$ contains a family $(f_\alpha)_{\alpha\in\mathbb{R}}$ such that the norms $(q_{f_\alpha})_{\alpha\in\mathbb{R}}$ are pairwise inequivalent (Hint: use Lemma 1.1.2.17).

d) Given $f \in]0,\infty[^T$ and $p \in]1,\infty]$, the norm q_f and the norm induced on $\mathbb{K}^{(T)}$ by $\ell^p(T)$ are not equivalent (Hint: use Proposition 1.1.2.6 f)).

E 1.1.8 Take $n \in \mathbb{N}$ and $p \in]0,1[$. Show that the map

$$\mathbb{K}^n \longrightarrow \mathbb{R}_+\,, \quad (x_k)_{k\in\mathbb{N}_n} \longmapsto \left(\sum_{k=1}^{n} |x_k|^p\right)^{\frac{1}{p}}$$

is a norm iff $n = 1$.

E 1.1.9 Take $p \in [1,\infty[$ and let $\mathcal{C}([0,1])$ be endowed with the norm

$$\mathcal{C}([0,1]) \longrightarrow \mathbb{R}_+\,, \quad x \longmapsto \left(\int_0^1 |x(t)|^p dt\right)^{\frac{1}{p}}\,.$$

Show that $\mathcal{C}([0,1])$ is separable but not complete.

E 1.1.10 Let T be a subset of $\mathbb{R}$, endowed with the induced topology. Show that $\mathcal{C}(T)$ is separable iff T is compact.

E 1.1.11 Take $x \in \ell^\infty([0,1])$ and consider

$$p : \mathcal{C}([0,1]) \longrightarrow \mathbb{R}_+\,, \quad y \longmapsto \sup_{t\in[0,1]} |x(t)y(t)|\,.$$

Show that:

a) p is a norm iff $\{x \neq 0\}$ is dense.

b) If p is a norm then $\mathcal{C}([0,1])$ is separable with respect to p.

c) If p is a norm then $\mathcal{C}([0,1])$ is complete with respect to p iff there exists an $\alpha > 0$ such that $\{|x| \geq \alpha\}$ is dense.

E 1.1.12 Let $\aleph$ be a cardinal number such that

$$\operatorname{Card}\{A \subset \aleph \mid \operatorname{Card} A < \aleph\} = \aleph\,.$$

Show:

a) There exists a set $\mathfrak{A}$ of subsets of $\aleph$ such that

$$\operatorname{Card} \mathfrak{A} = 2^\aleph\,, \quad \operatorname{Card} A = \aleph\,, \quad \operatorname{Card}(A \cap B) < \aleph$$

for all $A, B \in \mathfrak{A}$, $A \neq B$ (generalization of Lemma 1.1.2.17).

b) The dimension of $\ell^\infty(\aleph)$ and the cardinal number of $\ell^\infty(\aleph)$ is $2^\aleph$.

E 1.1.13 Let F be a closed proper vector subspace of the normed space E and $\varepsilon > 0$. Show that there is an $x \in E$ with $\|x\| = 1$ and

$$d_F(x) > 1 - \varepsilon .$$

E 1.1.14 Let $(x_\iota)_{\iota \in I}$ be a summable family in the normed space E and $\mathfrak{F}$ the filter of cofinite subsets of I. Show that

$$\lim_{\iota, \mathfrak{F}} x_\iota = 0 .$$

E 1.1.15 Let $(x_{\iota\lambda})_{(\iota,\lambda) \in I \times L}$ be a summable family in the Banach space E. Prove that

$$\sum_{\iota \in I} \Big(\sum_{\lambda \in L} x_{\iota\lambda} \Big) = \sum_{(\iota,\lambda) \in I \times L} x_{\iota\lambda} .$$

E 1.1.16 Let $(x_\iota)_{\iota \in I}$ be a summable family in the normed space E and $f : L \to I$ a bijective map. Prove that

$$\sum_{\lambda \in L} x_{f(\lambda)} = \sum_{\iota \in I} x_\iota .$$

E 1.1.17 Let $(x_n)_{n \in \mathbb{N}}$ be a sequence in the Banach space E. Show that the following are equivalent:

a) $(x_n)_{n \in \mathbb{N}}$ is summable.

b) The sequence $\left(\sum_{k=1}^{n} x_{f(k)} \right)_{n \in \mathbb{N}}$ is convergent for every bijective map $f : \mathbb{N} \to \mathbb{N}$.

Show further that if these conditions are fulfilled, then for every bijective map $f : \mathbb{N} \to \mathbb{N}$,

$$\sum_{n \in \mathbb{N}} x_n = \lim_{n \to \infty} \sum_{k=1}^{n} x_{f(k)} .$$

1.2 Operators

In the language of category theory, normed spaces are the objects and operators the morphisms of the category of normed spaces, where an operator is a continuous linear map between normed spaces. It is the operators which are the subject of this section. An important and useful feature of the set of operators between two normed spaces is that it too admits a natural normed space structure.

1.2.1 General Results

Proposition 1.2.1.1 (0) (F. Riesz, 1911) *Let $u : E \to F$ be a linear map from the normed space E to the normed space F. Then the following are equivalent:*

a) *u is continuous.*

b) *u is continuous at 0.*

c) *u is continuous at a point of E.*

d) *There is an $\alpha \in \mathbb{R}_+$ with*

$$\|ux\| \leq \alpha\|x\|$$

for every $x \in E$.

e) *$u(E^{\#})$ is bounded.*

f) *u is uniformly continuous.*

f $\Rightarrow$ a $\Rightarrow$ b $\Rightarrow$ c and d $\Leftrightarrow$ e are trivial.

c $\Rightarrow$ d. Suppose u is continuous at $x_0 \in E$. Then there is a $\delta > 0$ such that

$$\|ux - ux_0\| \leq 1$$

for every $x \in E$ with $\|x - x_0\| \leq \delta$. We put

$$\alpha := \frac{1}{\delta}.$$

Take $x \in E \backslash \{0\}$ and let

$$y := x_0 + \frac{\delta}{\|x\|}x.$$

Then

$$\|y - x_0\| = \delta$$

and so

$$\frac{\delta}{\|x\|}\|ux\| = \left\|u\left(\frac{\delta}{\|x\|}x\right)\right\| = \|uy - ux_0\| \le 1\,,$$

$$\|ux\| \le \frac{1}{\delta}\|x\| = \alpha\|x\|\,.$$

d $\Rightarrow$ f. Since

$$\|ux - uy\| = \|u(x-y)\| \le \alpha\|x-y\|$$

for every $x, y \in E$, u is uniformly continuous. ■

Corollary 1.2.1.2 *Every infinite–dimensional normed space admits a discontinuous linear form.*

Let A be an algebraic base of the infinite–dimensional normed space E. Let $f : A \to \mathbb{K}$ be unbounded and let x' be the linear form on E defined by

$$x'(x) := f(x)\|x\|$$

for $x \in A$. Since $\frac{1}{\|x\|}x \in E^{\#}$ whenever $x \in A$,

$$f(A) \subset x'(E^{\#})\,.$$

Hence $x'(E^{\#})$ is not bounded and x' is not continuous (Proposition 1.2.1.1 a $\Rightarrow$ e). ■

Definition 1.2.1.3 (0) *Let E and F be normed spaces. We define*

$$\mathcal{L}(E,F) := \{u : E \longrightarrow F \mid u \textit{ is linear and continuous}\}\,,$$

$$\mathcal{L}_f(E,F) := \{u : \mathcal{L}(E,F) \mid u(E) \textit{ is finite–dimensional}\}\,,$$

and

$$\|u\| := \inf\{\alpha \in \mathbb{R}_+ \mid x \in E \Longrightarrow \|ux\| \le \alpha\|x\|\} \quad (\textbf{\textit{the norm of}}\ u)$$

*for $u \in \mathcal{L}(E,F)$ (Proposition 1.2.1.1 a $\Rightarrow$ d). The elements of $\mathcal{L}(E,F)$ are called **operators from** E **to** F. We further define*

$$E' := \mathcal{L}(E,\mathbb{K})\,, \quad \mathcal{L}(E) := \mathcal{L}(E,E)\,, \quad \mathcal{L}_f(E) = \mathcal{L}_f(E,E)\,,$$

$$1 := 1_E : E \longrightarrow E\,, \quad x \longmapsto x \quad (\textbf{\textit{the identity operator on}}\ E)$$

and

$$\langle x, x' \rangle := \langle x', x \rangle := x'(x)$$

for $(x, x') \in E \times E'$. E' *is called* ***the dual of*** E (H. Hahn, 1927). *The elements of* $\mathcal{L}(E)$ *are called* ***operators on*** E.

Remark. a) Because of the equivalence a $\Leftrightarrow$ e in Proposition 1.2.1.1, the operators from E to F are often referred to in the literature as **bounded operators.**

b) It will be proved in Theorem 1.2.1.9 a) that the map

$$\mathcal{L}(E, F) \longrightarrow \mathbb{R}_+\,, \quad u \longmapsto \|u\|$$

is a norm.

Proposition 1.2.1.4 (0) *Let* E *and* F *be normed spaces and take* $u \in \mathcal{L}(E, F)$.

a) $x \in E \Rightarrow \|ux\| \le \|u\|\,\|x\|$.

b) $\|u\| = \sup\limits_{x \in U_1(0)} \|ux\| = \sup\limits_{x \in E^\#} \|ux\| = \sup\limits_{\substack{x \in E \\ \|x\|=1}}$,

where in the last equality it is assumed that $E \ne \{0\}$.

c) *If* G *is a subspace of* E, *then*

$$\|u \mid G\| \le \|u\|\,.$$

a) We put

$$A := \{\alpha \in \mathbb{R}_+ \mid x \in E \Longrightarrow \|ux\| \le \alpha \|x\|\}\,.$$

Then

$$\|ux\| \le \inf_{\alpha \in A} \alpha \|x\| = \|u\|\,\|x\|$$

for every $x \in E$.

b) We put

$$\alpha := \sup_{x \in U_1(0)} \|ux\|\,.$$

Take $x \in E$ and $\beta > \|x\|$. Then

$$\left\| \frac{1}{\beta} x \right\| = \frac{1}{\beta}\|x\| < 1\,,$$

so that we deduce successively

$$\frac{1}{\beta}\|ux\| = \left\|u(\frac{1}{\beta}x)\right\| \leq \alpha\,, \quad \|ux\| \leq \alpha\beta\,, \quad \|ux\| \leq \alpha\|x\|\,, \quad \|u\| \leq \alpha\,,$$

since β and x are arbitrary.

Now take $x \in E^{\#}$. By a),

$$\|ux\| \leq \|u\|\,\|x\| \leq \|u\|\,.$$

Thus

$$\|u\| \leq \alpha \leq \sup_{x \in E^{\#}} \|ux\| \leq \|u\|$$

and

$$\|u\| = \sup_{x \in U_1(0)} \|ux\| = \sup_{x \in E^{\#}} \|ux\|\,.$$

Now assume that $E \neq \{0\}$. Take $x \in E^{\#}\backslash\{0\}$. Then

$$\|ux\| \leq \frac{1}{\|x\|}\|ux\| = \left\|u\left(\frac{1}{\|x\|}x\right)\right\| \leq \sup_{\substack{y \in E \\ \|y\|=1}} \|uy\|\,.$$

so that

$$\sup_{x \in E^{\#}} \|ux\| \leq \sup_{\substack{x \in E \\ \|ux\|}}\,.$$

The reverse inequality is trivial.

c) We have, by b), that

$$\|u \mid G\| = \sup_{x \in G^{\#}} \|ux\| \leq \sup_{x \in E^{\#}} \|ux\| = \|u\|\,.$$ ■

Corollary 1.2.1.5 (0) *Let E, F, and G be normed spaces and take $u \in \mathcal{L}(E,F)$, $v \in \mathcal{L}(F,G)$. Then $v \circ u \in \mathcal{L}(E,G)$ and*

$$\|v \circ u\| \leq \|v\|\,\|u\|\,.$$

$v \circ u$ is linear and

$$\|v \circ u(x)\| = \|v(ux)\| \leq \|v\|\,\|ux\| \leq \|v\|\,\|u\|\,\|x\|$$

for every $x \in E$ (Proposition 1.2.1.4 a)). The corollary now follows from Proposition 1.2.1.1 d $\Rightarrow$ a. ■

Corollary 1.2.1.6 (0) *Let E be a normed space and take $x' \in E'$. Then $\mathrm{re} \circ x'$ is a continuous $\mathbb{R}$–linear form on E and*

$$\|\mathrm{re} \circ x'\| = \|x'\| .$$

$\mathrm{re} \circ x$ is obviously continuous. We have that

$$\mathrm{re} \circ x'(\alpha x + \beta y) = \mathrm{re}(x'(\alpha x + \beta y)) = \mathrm{re}(\alpha x'(x) + \beta x'(y)) =$$

$$= \alpha\, \mathrm{re}(x'(x)) + \beta\, \mathrm{re}(x'(y)) = \alpha\, \mathrm{re} \circ x'(x) + \beta\, \mathrm{re} \circ x'(y)$$

for $x, y \in E$ and $\alpha, \beta \in \mathbb{R}$, i.e. $\mathrm{re} \circ x'$ is $\mathbb{R}$–linear.

It follows from Proposition 1.2.1.4 b) that

$$\|\mathrm{re} \circ x'\| = \sup_{x \in E^{\#}} |\mathrm{re} \circ x'(x)| \le \sup_{x \in E^{\#}} |x'(x)| = \|x'\| .$$

By Proposition 1.2.1.4 a),

$$|x'(x)|^2 = \overline{x'(x)} x'(x) = x'(\overline{x'(x)} x) = \mathrm{re}(x'(\overline{x'(x)} x)) =$$

$$= \mathrm{re} \circ x'(\overline{x'(x)} x) \le \|\mathrm{re} \circ x'\| \, \|\overline{x'(x)} x\| = \|\mathrm{re} \circ x'\| \, |x'(x)| \, \|x\| ,$$

so that

$$|x'(x)| \le \|\mathrm{re} \circ x'\| \, \|x\|$$

for $x \in E$. It follows that

$$\|x'\| \le \|\mathrm{re} \circ x'\|$$

and

$$\|\mathrm{re} \circ x'\| = \|x'\| .$$ ■

Proposition 1.2.1.7 (0) *Let E and F be normed spaces. Let T be a set, $\mathfrak{F}$ a filter on T, $(u_t)_{t \in T}$ a family in $\mathcal{L}(E, F)$ such that*

$$\lim_{t, \mathfrak{F}} u_t x$$

exists for every $x \in E$, and

$$u : E \longrightarrow F , \quad x \longmapsto \lim_{t, \mathfrak{F}} u_t x .$$

u is linear and if

$$\liminf_{t, \mathfrak{F}} \|u_t\| < \infty ,$$

then u is continuous and

$$\|u\| \le \liminf_{t, \mathfrak{F}} \|u_t\| .$$

We have that

$$u(\alpha x + \beta y) = \lim_{t,\mathfrak{F}} u_t(\alpha x + \beta y) = \lim_{t,\mathfrak{F}} (\alpha u_t x + \beta u_t y)) = \alpha ux + \beta uy$$

for $x, y \in E$ and $\alpha, \beta \in \mathbb{K}$ (Proposition 1.1.5.3), i.e. u is linear. Moreover

$$\|ux\| = \lim_{t,\mathfrak{F}} \|u_t x\| \leq \liminf_{t,\mathfrak{F}} \|u_t\| \, \|x\| = \|x\| \liminf_{t \in \mathfrak{F}} \|u_t\|$$

for every $x \in E$ (Corollary 1.1.1.4, Proposition 1.2.1.4 a)), which proves the last assertion. ■

Lemma 1.2.1.8 (0) *Let E and F be vector spaces. For every $x \in E$, the map*

$$F^E \longrightarrow F, \quad u \longmapsto ux$$

is linear (it is called ***the evaluation of*** *x) and*

$$\{u \in F^E \mid u \text{ is linear}\}$$

is a vector subspace of F^E.

The first assertion is a consequence of the definition of the vector space structure on F^E.

Let u, v be linear maps of E into F and put

$$w := u + v\,.$$

Then

$$\begin{aligned} w(\alpha x + \beta y) &= u(\alpha x + \beta y) + v(\alpha x + \beta y) = \alpha ux + \beta uy + \alpha vx + \beta vy = \\ &= \alpha(ux + vx) + \beta(uy + vy) = \alpha wx + \beta wy \end{aligned}$$

for $x, y \in E$ and $\alpha, \beta \in \mathbb{K}$, i.e. w is linear.

Let $u : E \to F$ be linear. Take $\gamma \in \mathbb{K}$, and put $v := \gamma u$. Then

$$v(\alpha x + \beta y) = \gamma u(\alpha x + \beta y) = \gamma(\alpha ux + \beta uy) = \alpha vx + \beta vy$$

for $x, y \in E$ and $\alpha, \beta \in \mathbb{K}$, i.e. v is linear.

It follows from the above considerations that

$$\{u \in F^E \mid u \text{ is linear}\}$$

is a vector subspace of F^E. ■

Theorem 1.2.1.9 (0) *Let E and F be normed spaces.*

a) *$\mathcal{L}(E,F)$ is a vector subspace of F^E and the map*

$$\mathcal{L}(E,F) \longrightarrow \mathbb{R}_+ , \quad u \longmapsto \|u\|$$

is a norm; it is called the ***canonical norm of*** *$\mathcal{L}(E,F)$.*

b) *If F is complete, then $\mathcal{L}(E,F)$ is also complete.*

a) Take $u, v \in \mathcal{L}(E,F)$. Then

$$\|(u+v)x\| = \|ux+vx\| \leq \|ux\| + \|vx\| \leq \|u\|\,\|x\| + \|v\|\,\|x\| =$$

$$= (\|u\| + \|v\|)\|x\|$$

for every $x \in E$ (Proposition 1.2.1.4 a)), so that $u+v \in \mathcal{L}(E,F)$ (Lemma 1.2.1.8, Proposition 1.2.1.1 d $\Rightarrow$ a) and

$$\|u+v\| \leq \|u\| + \|v\| .$$

Take $\alpha \in \mathbb{K}$. Then

$$\sup_{x \in E^\#} |(\alpha u)x| = |\alpha| \sup_{x \in E^\#} \|ux\| = |\alpha|\,\|u\|$$

(Proposition 1.2.1.4 b)), so that $\alpha u \in \mathcal{L}(E,F)$ (Proposition 1.2.1.1 d $\Rightarrow$ a, Lemma 1.2.1.8) and

$$\|\alpha u\| = |\alpha|\,\|u\|$$

(Proposition 1.2.1.4 b)). Finally, it is easy to see that $u = 0$ whenever $\|u\| = 0$.

b) Let $(u_n)_{n\in\mathbb{N}}$ be a Cauchy sequence in $\mathcal{L}(E,F)$. For every $\varepsilon > 0$ there is an m_ε such that

$$\|u_n - u_p\| < \varepsilon$$

for every $n, p \in \mathbb{N}$, with $n, p \geq m_\varepsilon$. Take $x \in E$. Then

$$\|u_n x - u_p x\| = \|(u_n - u_p)x\| \leq \|u_n - u_p\|\,\|x\| \leq \varepsilon \|x\|$$

for every $\varepsilon > 0$ and each $n, p \in \mathbb{N}$ with $n, p \geq m_\varepsilon$ (Proposition 1.2.1.4 a)). Hence $(u_n x)_{n\in\mathbb{N}}$ is a Cauchy sequence and therefore a convergent sequence in F. Define

$$u : E \longrightarrow F , \quad x \longmapsto \lim_{n\to\infty} u_n x .$$

Then

$$(u_n - u)x = u_n x - ux = u_n x - \lim_{p\to\infty} u_p x =$$

$$= \lim_{p\to\infty} (u_n x - u_p x) = \lim_{p\to\infty} (u_n - u_p)x$$

whenever $n \in \mathbb{N}$ and $x \in E$. By Proposition 1.2.1.7, $u_n - u \in \mathcal{L}(E,F)$ and

$$\|u_n - u\| \leq \varepsilon$$

for $\varepsilon > 0$ and $n \in \mathbb{N}$, $n \geq m_\varepsilon$. It follows that $u \in \mathcal{L}(E,F)$ and

$$\lim_{n\to\infty} u_n = u.$$

$\mathcal{L}(E,F)$ is thus complete. ∎

Corollary 1.2.1.10 (0) *The dual space of a normed space is a Banach space.* ∎

Corollary 1.2.1.11 (0) *If E is a Banach space, then so is $\mathcal{L}(E)$.* ∎

Definition 1.2.1.12 (0) *Let E, F be normed spaces and $u : E \to F$ a bijective linear map. u is called an* ***isometry of normed spaces*** *if*

$$\|ux\| = \|x\|$$

for every $x \in E$. u is called an ***isomorphism of normed spaces*** *if u and u^{-1} are continuous. E and F are called* ***isometric (isomorphic)*** *if there exists an isometry (isomorphism) $E \to F$.*

If $u : E \to F$ is an isometry (isomorphism), then so is u^{-1}. Every isometry is an isomorphism and every isomorphism is a homeomorphism. Isometry is equivalence in the category of normed spaces (geometrical aspect), whereas isomorphism is the equivalence relation in the category of normable spaces (topological aspect).

We may replace "bijective" by "surjective" in the definition of isometry, since the main property of isometry implies that the injectivity follows.

Let T be a finite set with at least two elements and take $p, q \in [1, \infty]$ with $p \neq q$. Then $\ell^p(T)$ and $\ell^q(T)$ are isomorphic (Minkowski Theorem), but they are not isometric. This can be seen by considering their unit balls. If T is infinite then $\ell^p(T)$ and $\ell^q(T)$ are not even isomorphic.

There exists an infinite–dimensional Banach space E not isomorphic to any of its proper subspaces (W.T. Gowers, Bull. London Math. Soc. 26 (1994), 523–530). Then E is not isomorphic to $E \times \mathbb{K}$.

Proposition 1.2.1.13 (0) *Let F be a dense subspace of the normed space E and G a Banach space. Then the map*

$$\mathcal{L}(E,G) \longrightarrow \mathcal{L}(F,G), \quad u \longmapsto u|F$$

is an isometry.

It is obvious that the above map is linear. By Proposition 1.2.1.4 b),

$$\|u|F\| = \sup_{x \in F^{\#}} \|ux\| = \sup_{x \in E^{\#}} \|ux\| = \|u\|$$

for every $u \in \mathcal{L}(E,G)$, since $F^{\#}$ is dense in $E^{\#}$. Take $v \in \mathcal{L}(F,G)$. By Proposition 1.2.1.1 a $\Rightarrow$ f, v is uniformly continuous. Since G is complete, there exists a continuous extension $u : E \to G$ of v. u is then linear, i.e. $u \in \mathcal{L}(E,G)$ and $u|F = v$. Hence the map

$$\mathcal{L}(E,G) \longrightarrow \mathcal{L}(F,G), \quad u \longmapsto u|F$$

is surjective and so it is an isometry. ■

Proposition 1.2.1.14 *Let E, F be isomorphic normed spaces. If E is complete then so is F.*

Let $u : E \to F$ be an isomorphism and let $(y_n)_{n\in\mathbb{N}}$ be a Cauchy sequence in F. Then $\left(u^{-1}y_n\right)_{n\in\mathbb{N}}$ is a Cauchy sequence (Proposition 1.2.1.1 a $\Rightarrow$ f) and so a convergent sequence in E. Hence

$$u\Big(\lim_{n\to\infty} u^{-1}y_n\Big) = \lim_{n\to\infty} u(u^{-1}y_n) = \lim_{n\to\infty} y_n .$$

Thus $(y_n)_{n\in\mathbb{N}}$ converges and F is complete. ■

Proposition 1.2.1.15 *Let p and q be norms on the vector space E. Let E_p (resp. E_q) be the vector space E endowed with the norm p (resp. q), and define*

$$u : E_p \longrightarrow E_q, \quad x \longmapsto x .$$

Then the following are equivalent:

a) *u is an isomorphism.*

b) *p and q are equivalent.*

a $\Rightarrow$ b. Given $x \in E$,

$$q(x) = q(ux) \le \|u\| p(x),$$

$$p(x) = p(u^{-1}x) \le \|u^{-1}\| q(x)$$

(Proposition 1.2.1.4 a)), i.e. p and q are equivalent.

b $\Rightarrow$ a. For some $\alpha > 0$

$$\frac{1}{\alpha} p \le q \le \alpha p.$$

Thus, given $x \in E$,

$$q(ux) = q(x) \le \alpha p(x),$$

$$p(u^{-1}x) = p(x) \le \alpha q(x).$$

Therefore both u and u^{-1} are continuous (Proposition 1.2.1.1 d $\Rightarrow$ a). ■

Proposition 1.2.1.16 (0) *Let E and F be normed spaces and $(x_\iota)_{\iota \in I}$ a summable family in E. Take $u \in \mathcal{L}(E,F)$. Then*

$$\sum_{\iota \in I} u x_\iota = u\left(\sum_{\iota \in I} x_\iota\right).$$

We have

$$u\left(\sum_{\iota \in I} x_\iota\right) = u\left(\lim_{J,\mathfrak{F}_I} \sum_{\iota \in J} x_\iota\right) = \lim_{J,\mathfrak{F}_I} u\left(\sum_{\iota \in J} x_\iota\right) = \lim_{J,\mathfrak{F}_I} \sum_{\iota \in J} u x_\iota.$$ ■

Corollary 1.2.1.17 (0) *Let E and F be Banach spaces. Take*

$$u \in \mathcal{L}(E,F).$$

Let $\sum_{n=0}^{\infty} t^n x_n$ be a power series in E, and r, r' be the radii of convergence of $\sum_{n=0}^{\infty} t^n x_n$ and $\sum_{n=0}^{\infty} t^n u x_n$, respectively. Then $r \le r'$ and

$$u\left(\sum_{n=0}^{\infty} \alpha^n x_n\right) = \sum_{n=0}^{\infty} \alpha^n u x_n$$

for every $\alpha \in U_r^{\mathbb{K}}(0)$.

Given $n \in \mathbb{N}$,

$$\|ux_n\| \leq \|u\| \, \|x_n\|$$

(Proposition 1.2.1.4 a)), so that

$$\frac{1}{r'} = \limsup_{n\to\infty} \|ux_n\|^{\frac{1}{n}} \leq \limsup_{n\to\infty} \|u\|^{\frac{1}{n}} \|x_n\|^{\frac{1}{n}} \leq \limsup_{n\to\infty} \|x_n\|^{\frac{1}{n}} = \frac{1}{r}, \quad r \leq r'.$$

The equality now follows from Proposition 1.2.1.16. ∎

Proposition 1.2.1.18 (0) *Let E and F be normed spaces, $u : E \to F$ a linear map, and α a strictly positive real number such that*

$$\|ux\| \geq \alpha \|x\|$$

for every $x \in E$. (This property of u is sometimes called ***lower bounded****.)*

a) *u is injective.*

b) *If u is surjective, then u^{-1} is continuous and $\|u^{-1}\| \leq \frac{1}{\alpha}$.*

c) *If E is complete and u is continuous, then $u(E)$ is closed.*

d) *If E is complete, u is continuous, and $u(E)$ is dense then u is an isomorphism of normed spaces.*

a) is trivial.

b) Take $y \in F$ and let $x := u^{-1}y$. Then

$$\|u^{-1}y\| = \|x\| \leq \frac{1}{\alpha}\|ux\| = \frac{1}{\alpha}\|y\|$$

and the assertion follows (Proposition 1.2.1.1 d $\Rightarrow$ a).

c) Take $y \in \overline{u(E)}$. There is a sequence $(x_n)_{n\in\mathbb{N}}$ in E with

$$\lim_{n\to\infty} ux_n = y.$$

Take $\varepsilon > 0$. There is an $m \in \mathbb{N}$ such that

$$\|ux_m - ux_n\| < \alpha\varepsilon$$

for every $n \in \mathbb{N}$ with $n \geq m$. Then

$$\|x_m - x_n\| \leq \frac{1}{\alpha}\|u(x_m - x_n)\| = \frac{1}{\alpha}\|ux_m - ux_n\| \leq \frac{\alpha\varepsilon}{\alpha} = \varepsilon$$

for every $n \in \mathbb{N}$ with $n \geq m$. Hence $(x_n)_{n\in\mathbb{N}}$ is a Cauchy sequence and so a convergent sequence in E. It follows that

$$y = \lim_{n\to\infty} ux_n = u\big(\lim_{n\to\infty} x_n\big) \in u(E).$$

Thus $u(E)$ is closed.

d) follows from b) and c). ∎

Proposition 1.2.1.19 (0) *Let $(E_\iota)_{\iota\in I}$ be a finite family of normed spaces. For each $x' = (x'_\iota)_{\iota\in I} \in \prod_{\iota\in I} E'_\iota$, define*

$$\widetilde{x'} : \prod_{\iota\in I} E_\iota \longrightarrow \mathbb{K}, \quad (x_\iota)_{\iota\in I} \longmapsto \sum_{\iota\in I} x'_\iota(x_\iota).$$

a) $\widetilde{x'} \in (\prod_{\iota\in I} E_\iota)'$ *and* $\|\widetilde{x'}\| = \|x'\| := \left(\sum_{\iota\in I} \|x'_\iota\|^2\right)^{\frac{1}{2}}$ *for every* $x' \in \prod_{\iota\in I} E'_\iota$.

b) *The map*

$$\prod_{\iota\in I} E'_\iota \longrightarrow \left(\prod_{\iota\in I} E_\iota\right)', \quad x' \longmapsto \widetilde{x'}$$

is an isometry.

a) Given $x = (x_\iota)_{\iota\in I} \in \prod_{\iota\in I} E_\iota$,

$$|\widetilde{x'}(x)| = \left|\sum_{\iota\in I} x'_\iota(x_\iota)\right| \leq \sum_{\iota\in I} |x'_\iota(x_\iota)| \leq \sum_{\iota\in I} \|x'_\iota\|\,\|x_\iota\| \leq$$

$$\leq \left(\sum_{\iota\in I} \|x'_\iota\|^2\right)^{\frac{1}{2}} \left(\sum_{\iota\in I} \|x_\iota\|^2\right)^{\frac{1}{2}} = \|x'\|\,\|x\|$$

(Proposition 1.2.1.4 a)), so that $\widetilde{x'} \in \left(\prod_{\iota\in I} E_\iota\right)'$ and $\|\widetilde{x'}\| \leq \|x'\|$ (Proposition 1.2.1.1 d $\Rightarrow$ a).

Take $\varepsilon > 0$. For each $\iota \in I$ there is an $x_\iota \in E_\iota$ such that

$$\|x_\iota\| \leq \|x'_\iota\|, \quad x'_\iota(x_\iota) \geq \|x'_\iota\|^2 - \varepsilon$$

(Proposition 1.2.1.4 b)). Put $x := (x_\iota)_{\iota\in I}$. Then

$$\|\widetilde{x'}\|^2 \geq \|\widetilde{x'}\|\,\|x\| \geq |x'(x)| =$$

$$= \sum_{\iota\in I} x'_\iota(x_\iota) \geq \sum_{\iota\in I} \|x'_\iota\|^2 - \varepsilon \operatorname{Card} I = \|x'\|^2 - \varepsilon \operatorname{Card} I$$

(Proposition 1.2.1.4 a)). Since ε is arbitrary,

$$\|\widetilde{x'}\| \geq \|x'\|, \quad \|\widetilde{x'}\| = \|x'\|.$$

b) Take $y' \in \left(\prod_{\iota\in I} E_\iota\right)'$. For each $\iota \in I$ let $u_\iota : E_\iota \to \prod_{\lambda\in L} E_\lambda$ denote the canonical imbedding and put $x'_\iota := y' \circ u_\iota$, $x' := (x'_\iota)_{\iota\in I}$. Then $x' \in \prod_{\iota\in I} E'_\iota$ and

$$\widetilde{x'}(x) = \sum_{\iota \in I} x'_\iota(x_\iota) = \sum_{\iota \in I} y' u_\iota x_\iota = y'(x)$$

for every $x := (x_\iota)_{\iota \in I} \in \prod_{\iota \in I} E_\iota$. Hence $\widetilde{x'} = y'$ and the map

$$\prod_{\iota \in I} E'_\iota \longrightarrow \Big(\prod_{\iota \in I} E_\iota\Big)', \quad x' \longmapsto \widetilde{x'}$$

is surjective. By a) it is an isometry. ■

1.2.2 Standard Examples

Definition 1.2.2.1 (0) *Let $p \in [1,\infty] \cup \{0\}$. The* ***conjugate exponent of*** *p is the number $q \in [1,\infty]$ defined as follows:*

$$q := \infty \quad \text{if } p = 1,$$

$$q := \tfrac{p}{p-1} \quad \text{if } p \in]1,\infty[,$$

$$q := 1 \quad \text{if } p \in \{0,\infty\}.$$

Two numbers p, q are called ***conjugate exponents (weakly conjugate exponents)*** *if $p,q \in [1,\infty]$, $(p,q \in [1,\infty] \cup \{0\})$ and p is the conjugate exponent of q or q ist the conjugate exponent of p.*

If p, q are conjugate exponents, then p is the conjugate exponent of q and q is the conjugate exponent of p.

Proposition 1.2.2.2 (0) *Let T be a set. Let p, q be conjugate exponents. Take $x \in \mathbb{K}^T$ and $\alpha \in \mathbb{R}_+$. If*

$$\left| \sum_{t \in T} x(t) y(t) \right| \leq \alpha \|y\|_p$$

for every $y \in \mathbb{K}^{(T)}$, then $x \in \ell^q(T)$ and

$$\|x\|_q \leq \alpha .$$

First assume that $p \neq 1$. Put

$$y : T \to \mathbb{K}, \quad t \longrightarrow \begin{cases} \overline{x(t)} |x(t)|^{q-2} & \text{if } x(t) \neq 0 \\ 0 & \text{if } x(t) = 0 . \end{cases}$$

If $p = \infty$, then $q = 1$ and for every $A \in \mathfrak{P}_f(T)$,

$$\sum_{t \in A} |x(t)| = \sum_{t \in T} x(t) y(t) e_A(t) \leq \alpha \|y e_A\|_\infty \leq \alpha$$

so that $x \in \ell^1(T)$ and

$$\|x\|_1 \leq \alpha .$$

If $p \neq \infty$, then for every $A \in \mathfrak{P}_f(T)$,

$$\sum_{t\in A}|x(t)|^q=\sum_{t\in T}x(t)y(t)e_A(t)\le\alpha\|ye_A\|_p=$$
$$=\alpha\Big(\sum_{t\in A}|y(t)|^p\Big)^{\frac{1}{p}}=\alpha\Big(\sum_{t\in A}|x(t)|^{p(q-1)}\Big)^{\frac{1}{p}}=$$
$$=\alpha\Big(\sum_{t\in A}|x(t)|^q\Big)^{\frac{1}{p}}$$

and so

$$\Big(\sum_{t\in A}|x(t)|^q\Big)^{\frac{1}{q}}=\Big(\sum_{t\in A}|x(t)|^q\Big)^{1-\frac{1}{p}}\le\alpha\,.$$

Thus $x\in\ell^q(T)$ and

$$\|x\|_q\le\alpha\,.$$

Now take $p=1$. Then

$$|x(s)|=\Big|\sum_{t\in T}x(t)e_s^T(t)\Big|\le\alpha\|e_s^T\|_1=\alpha$$

for every $s\in T$, so that $x\in\ell^\infty(T)$ and

$$\|x\|_\infty\le\alpha\,.$$

■

Example 1.2.2.3 (0) *Let T be a set and p,q conjugate exponents.*

a) $xy\in\ell^1(T)$ *and*

$$\|xy\|_1\le\|x\|_q\|y\|_p$$

for $x\in\ell^q(T)$ and $y\in\ell^p(T)$, where

$$xy:T\longrightarrow\mathbb{K}\,,\quad t\longmapsto x(t)y(t)\,;$$

we define

$$\widetilde{x}:\ell^p(T)\longrightarrow\mathbb{K}\,,\quad y\longmapsto\sum_{t\in T}x(t)y(t)$$

for every $x\in\ell^q(T)$.

b) $\widetilde{x}\in\ell^p(T)'$ *and $\|\widetilde{x}\|=\|x\|_q$ for every $x\in\ell^q(T)$.*

c) $\|\widetilde{x}\mid c_0(T)\|=\|\widetilde{x}\|=\|x\|_1$ *for every $x\in\ell^1(T)$.*

d) *If* $p \neq \infty$ *then the map*

$$\ell^q(T) \longrightarrow \ell^p(T)', \quad x \longmapsto \widetilde{x}$$

is an isometry.

e) *The map*

$$\ell^1(T) \longrightarrow c_0(T)', \quad x \longmapsto \widetilde{x} \mid c_0(T)$$

is an isometry.

f) *If* $\mathfrak{F}$ *is a free ultrafilter on* T *(i.e. an ultrafilter on* T *possessing no one–point set), then the map*

$$x' : \ell^\infty(T) \quad (resp.\ c(T)) \longrightarrow \mathbb{K}, \quad y \longmapsto \lim y(\mathfrak{F})$$

belongs to $\ell^\infty(T)'$, *(resp.* $c(T)'$*) and* $x' \neq \widetilde{x}$ *(resp.* $x' \neq \widetilde{x} \mid c(T)$*) for every* $x \in \ell^1(T)$.

g) *The map*

$$\ell^1(T) \longrightarrow \ell^\infty(T)', \quad x \longmapsto \widetilde{x},$$

$$(resp.\ \ell^1(T) \longrightarrow c(T)', \quad x \longmapsto \widetilde{x} \mid c(T))$$

is an isometry iff T *is finite.*

a) By the Hölder inequality,

$$\sum_{t \in A} |x(t)y(t)| \leq \|x\|_q \|y\|_p$$

for every $A \in \mathfrak{P}_f(T)$. Thus

$$\sum_{t \in T} |x(t)y(t)| \leq \|x\|_q \|y\|_p .$$

Hence $xy \in \ell^1(T)$ and

$$\|xy\|_1 \leq \|x\|_q \|y\|_p .$$

b) By a), Proposition 1.1.6.11, and Proposition 1.2.1.1 d $\Rightarrow$ a, $\widetilde{x} \in \ell^p(T)'$ and

$$\|\widetilde{x}\| \leq \|x\|_q .$$

We have that

$$\left|\sum_{t\in T} x(t)y(t)\right| = |\widetilde{x}(y)| \leq \|\widetilde{x}\| \, \|y\|_p$$

for every $y \in \mathbb{K}^{(T)}$ (Proposition 1.2.1.4 a)). By Proposition 1.2.2.2,

$$\|x\|_q \leq \|\widetilde{x}\| .$$

Hence

$$\|\widetilde{x}\| = \|x\|_q .$$

c) We have that

$$\left|\sum_{t\in T} x(t)y(t)\right| = |\widetilde{x}(y)| \leq \|\widetilde{x} \mid c_0(T)\| \, \|y\|_\infty$$

for every $y \in \mathbb{K}^{(T)}$ (Proposition 1.2.1.4 a)). By Proposition 1.2.2.2

$$\|x\|_1 \leq \|\widetilde{x} \mid c_0(T)\| .$$

Hence by b) and Proposition 1.2.1.4 c)

$$\|\widetilde{x}\| = \|x\|_1 \leq \|\widetilde{x} \mid c_0(T)\| \leq \|\widetilde{x}\| , \quad \|\widetilde{x} \mid c_0(T)\| = \|\widetilde{x}\| = \|x\|_1 .$$

d) (resp. e)). Let $x' \in \ell^p(T)'$, (resp. $x' \in c_0(T)'$). Define

$$x : T \longrightarrow \mathbb{K} , \quad t \longmapsto x'(e_t^T) .$$

Then

$$\left|\sum_{t\in T} x(t)y(t)\right| = \left|\sum_{t\in T} x'(e_t^T)y(t)\right| = |x'(y)| \leq \|x'\| \, \|y\|_p$$

$$(\text{resp. } \leq \|x'\| \, \|y\|_\infty)$$

for every $y \in \mathbb{K}^{(T)}$ (Proposition 1.2.1.4 a)). By Proposition 1.2.2.2, $x \in \ell^q(T)$ (resp. $x \in \ell^1(T)$). We have that

$$\widetilde{x}(e_t^T) = x(t) = x'(e_t^T)$$

for every $t \in T$, so that $\widetilde{x} = x'$ on $\mathbb{K}^{(T)}$. Since $\mathbb{K}^{(T)}$ is dense in $\ell^p(T)$ and $c_0(T)$ (Proposition 1.1.2.6 c)), $\widetilde{x} = x'$. Hence the map

$$\ell^q(T) \longrightarrow \ell^p(T)', \quad x \longmapsto \widetilde{x}$$

$$(\text{resp. } \ell^1(T) \longrightarrow c_0(T)', \quad x \longmapsto \widetilde{x} \mid c_0(T))$$

is surjective. By b) (resp. c)) it is an isometry.

f) It is obvious that $x' \in \ell^\infty(T)'$, $x'(e_T) = 1$, and $x' = 0$ on $\mathbb{K}^{(T)}$. Hence $x' = 0$ on $c_0(T)$ (Proposition 1.1.2.6 c)). Assume that there is an $x \in \ell^1(T)$ with $\widetilde{x} = x'$. Then $\widetilde{x} = 0$ on $c_0(T)$ and so, by c),

$$\|x\|_1 = \|\widetilde{x} \mid c_0(T)\| = 0, \quad x = 0.$$

This leads to the contradiction

$$1 = x'(e_T) = \widetilde{x}(e_T) = 0.$$

g) follows from e) and f). ■

Remark. A representation of $\ell^\infty(T)'$ is given in Corollary 1.2.2.14.

Example 1.2.2.4 *Let T be an infinite set and $\mathfrak{F}$ the filter of cofinite subsets of T (i.e.*

$$\mathfrak{F} := \{A \subset T \mid T \setminus A \text{ is finite}\}),$$

$$x' : c(T) \longrightarrow \mathbb{K}, \quad x \longmapsto \lim x(\mathfrak{F}),$$

$t_0 \in T$. Let $\varphi : T\setminus\{t_0\} \to T$ be a bijection. We set

$$\widetilde{x} : T \longrightarrow \mathbb{K}, \quad t \longmapsto \begin{cases} x'(x) & \text{if } t = t_0 \\ x(\varphi(t)) - x'(x) & \text{if } t \neq t_0 \end{cases}$$

for every $x \in c(T)$.

a) *$x' \in c(T)'$, $\|x'\| = 1$.*

b) *$\widetilde{x} \in c_0(T)$ for every $x \in c(T)$; define*

$$u : c(T) \longrightarrow c_0(T), \quad x \longmapsto \widetilde{x}.$$

c) *u is an isomorphism and*

$$\|u\| = \|u^{-1}\| = 2.$$

d) *$c_0(T)$ and $c(T)$ are not isometric.*

a) and b) are obvious.

c) It is clear that u is linear and injective. We have that

$$|\widetilde{x}(t_0)| = |x'(x)| \leq \|x\|_\infty ,$$

$$\sup_{t\in T\setminus\{t_0\}} |\widetilde{x}(t)| = \sup_{t\in T\setminus\{t_0\}} |x(\varphi(t)) - x'(x)| \leq |x'(x)| + \sup_{t\in T\setminus\{t_0\}} |x(\varphi(t))| \leq 2\|x\|_\infty ,$$

so that

$$\|\widetilde{x}\|_\infty \leq 2\|x\|_\infty$$

for every $x \in c(T)$. Hence u is continuous (Proposition 1.2.1.1 d $\Rightarrow$ a) and $\|u\| \leq 2$. From $e_T - 2e_{t_0} \in c(T)$,

$$\|e_T - 2e_{t_0}\|_\infty = 1 , \quad \|u(e_T - 2e_{t_0})\|_\infty = 2 ,$$

we see that $\|u\| = 2$.

Let $x \in c_0(T)$. Define

$$y := x(t_0)e_T + x \circ \varphi^{-1} \in c(T) .$$

Then

$$x'(y) = x(t_0) ,$$

so that

$$\widetilde{y} = x .$$

Hence u is surjective and

$$u^{-1}(x) = x(t_0)e_T + x \circ \varphi^{-1} .$$

We deduce that

$$\|u^{-1}x\|_\infty \leq 2\|x\|_\infty$$

for every $x \in c_0(T)$. Thus u^{-1} is continuous (Proposition 1.2.1.1 d $\Rightarrow$ a) and $\|u^{-1}\| \leq 2$. From $e_{t_0} + e_{\varphi^{-1}(t_0)} \in c_0(T)$,

$$\|e_{t_0} + e_{\varphi^{-1}(t_0)}\|_\infty = 1 , \quad \|u^{-1}(e_{t_0} + e_{\varphi^{-1}(t_0)})\|_\infty = 2$$

it follows that $\|u^{-1}\| = 2$.

d) Assume that

$$v : c(T) \longrightarrow c_0(T)$$

is an isometry. Put

$$A := \{|ve_T| = 1\}$$

and

$$A_t := \{|v(e_T + e_t)| = 2\}$$

for every $t \in T$. Since v is an isometry, A and A_t are finite, nonempty, and $A_t \subset A$ for every $t \in T$ (Example 1.1.2.3). Hence there are distinct $t', t'' \in T$ with $A_{t'} \cap A_{t''} \neq \emptyset$. Take $t \in A_{t'} \cap A_{t''}$. Then

$$(ve_{t'})(t) = (ve_T)(t) = (ve_{t''})(t)\,,$$

so that

$$2 = |(ve_{t'} + ve_{t''})(t)| \leq \|v(e_{t'} + e_{t''})\|_\infty = \|e_{t'} + e_{t''}\|_\infty = 1\,,$$

which is a contradiction. Hence $c_0(T)$ and $c(T)$ are not isometric. ■

Remark. d) will be generalized in Corollary 1.2.7.18.

Example 1.2.2.5 (0) (F. Riesz, 1910) *Let μ be a positive measure and p, q conjugate exponents.*

a) *$xy \in L^1(\mu)$ and*

$$\|xy\|_1 \leq \|x\|_q \|y\|_p \quad \textit{(Hölder inequality)}$$

for every $x \in L^q(\mu)$ and $y \in L^p(\mu)$. For each $x \in L^q(\mu)$, put

$$\widetilde{x} : L^p(\mu) \longrightarrow \mathbb{K}\,, \quad y \longmapsto \int xy\,d\mu\,.$$

b) *$\widetilde{x} \in L^p(\mu)'$ and $\|\widetilde{x}\| = \|x\|_q$ for every $x \in L^q(\mu)$.*

c) *If $p \in]1, \infty[$ then the map*

$$L^q(\mu) \longrightarrow L^p(\mu)'\,, \quad x \longmapsto \widetilde{x}$$

is an isometry.

d) *The map*

$$L^\infty(\mu) \longrightarrow L^1(\mu)', \quad x \longmapsto \widetilde{x}$$

is an isometry iff $L^\infty(\mu)$ is order complete.

The proofs can be found in most books on integration theory. ■

Remark. If T is a set and μ counting measure on T, then Example 1.2.2.3 a),b),d) becomes a special case of the above example.

Definition 1.2.2.6 (0) *We set*

$$\delta(s,t) := \delta_{st} := \begin{cases} 1 & \textit{if } s = t \\ 0 & \textit{if } s \neq t \end{cases} \qquad \textbf{\textit{(the Kronecker's symbol)}}$$

for all s,t.

Example 1.2.2.7 (0) *Let $[\alpha_{ij}]_{i\in\mathbb{N}_m, j\in\mathbb{N}_n}$ be a matrix over $\mathbb{K}$. Let $u : \mathbb{K}^n \to \mathbb{K}^m$ be the associated linear map and A the set of the roots of the equation*

$$\operatorname{Det}\left[\sum_{i=1}^{m} \alpha_{ij}\overline{\alpha}_{ik} - t\delta_{jk}\right]_{j,k\in\mathbb{N}_n} = 0 .$$

If we endow $\mathbb{K}^m$ and $\mathbb{K}^n$ with the Euclidean norm, then:

a) $A \subset \mathbb{R}_+$.

b) *Given $\lambda \in A$ and a solution $x := (x_j)_{j\in\mathbb{N}_n}$ of the system of linear equations*

$$(*) \qquad \sum_{j=1}^{n}\left(\sum_{i=1}^{m} \alpha_{ij}\overline{\alpha}_{ik}\right) x_j = \lambda x_k \qquad (k \in \mathbb{N}_n),$$

we have that

$$\|ux\|^2 = \lambda \|x\|^2 .$$

c) $\|u\|^2 = \sup_{\lambda\in A} \lambda \leq \sum_{i=1}^{m}\sum_{j=1}^{n} |\alpha_{ij}|^2 .$

d) *If* $n = 2$, *then*

$$\|u\|^2 = \frac{1}{2}\left(\sum_{i=1}^{m} |\alpha_{i1}|^2 + \sum_{i=1}^{m} |\alpha_{i2}|^2 + \right.$$

$$\left. + \sqrt{\begin{array}{c} \left(\sum_{i=1}^{m} |\alpha_{i1}|^2 + \sum_{i=1}^{m} |\alpha_{i2}|^2\right)^2 - \\ -4\left(\left(\sum_{i=1}^{m} |\alpha_{i1}|^2\right)\left(\sum_{i=1}^{m} |\alpha_{i2}|^2\right) - \left|\sum_{i=1}^{m} \alpha_{i1}\overline{\alpha}_{i2}\right|^2\right)\end{array}}\right) =$$

$$= \frac{1}{2}\left(\begin{array}{c} \sum_{i=1}^{m}\left(|\alpha_{i1}|^2 + |\alpha_{i2}|^2\right) + \\ + \sqrt{\left(\sum_{i=1}^{m} (|\alpha_{i1}|^2 + |\alpha_{i2}|^2)\right)^2 - 2\sum_{i,j=1}^{m} |\alpha_{i1}\alpha_{j2} - \alpha_{i2}\alpha_{j1}|^2}\end{array}\right).$$

a & b. We have that

$$\lambda\|x\|^2 = \sum_{k=1}^{n} \lambda x_k \overline{x}_k = \sum_{j,k=1}^{n}\sum_{i=1}^{m} \alpha_{ij}\overline{\alpha}_{ik} x_j \overline{x}_k =$$

$$= \sum_{i=1}^{m}\left(\sum_{j=1}^{n} \alpha_{ij} x_j\right)\left(\overline{\sum_{j=1}^{n} \alpha_{ik} x_k}\right) = \|ux\|^2 .$$

c) Take $x = (x_k)_{k \in \mathbb{N}_n}$ such that $\|ux\|^2 = \sup\{\|uy\|^2 \mid y \in \mathbb{K}^n, \|y\| = 1\}$. By Lagrange's Theorem, there is a $\lambda \in \mathbb{R}$ such that

$$\frac{\partial\|uy\|^2}{\partial \overline{y}_k} = \lambda \frac{\partial \|y\|^2}{\partial \overline{y}_k}$$

for every $k \in \mathbb{N}_n$ when $y = x$. Thus

$$\sum_{j=1}^{n}\left(\sum_{i=1}^{m} \alpha_{ij}\overline{\alpha}_{ik}\right) x_j = \lambda x_k$$

for every $k \in \mathbb{N}_n$. By a), b), and Proposition 1.2.1.4 b),

$$\|u\|^2 = \sup_{\lambda \in A} \lambda .$$

Take $\lambda \in A$ with $\lambda = \|u\|^2$ and $x \in \mathbb{K}^n$ such that $(*)$ is fulfilled. Then for every $k \in \mathbb{N}_n$,

$$\|u\|^2|x_k| \le \sum_{j=1}^{n} \left| \sum_{i=1}^{m} a_{ij}\overline{\alpha}_{ik} \right| |x_j| \le$$

$$\le \sum_{j=1}^{n} \left(\sum_{i=1}^{m} |\alpha_{ij}|^2 \right)^{\frac{1}{2}} \left(\sum_{i=1}^{m} |\alpha_{ik}|^2 \right)^{\frac{1}{2}} |x_j| \le$$

$$\le \left(\sum_{i=1}^{m} |\alpha_{ik}|^2 \right)^{\frac{1}{2}} \left(\sum_{j=1}^{n} \sum_{i=1}^{m} |\alpha_{ij}|^2 \right)^{\frac{1}{2}} \|x\| .$$

It follows

$$\|u\|^4\|x\|^2 = \sum_{k=1}^{n} \|u\|^4|x_k|^2 \le \left(\sum_{k=1}^{n} \sum_{i=1}^{m} |\alpha_{ik}|^2 \right) \left(\sum_{j=1}^{n} \sum_{i=1}^{m} |\alpha_{ij}|^2 \right) \|x\|^2 ,$$

$$\|u\|^2 \le \sum_{i=1}^{m} \sum_{j=1}^{n} |\alpha_{ij}|^2 .$$

d) We have that

$$\begin{vmatrix} \sum\limits_{k=1}^{m} |\alpha_{k1}|^2 - \lambda & \sum\limits_{k=1}^{m} \overline{\alpha}_{k1}\alpha_{k2} \\ \sum\limits_{k=1}^{m} \overline{\alpha}_{k2}\alpha_{k1} & \sum\limits_{k=1}^{m} |\alpha_{k2}|^2 - \lambda \end{vmatrix} =$$

$$= \lambda^2 - \left(\sum_{k=1}^{m} |\alpha_{k1}|^2 + \sum_{k=1}^{m} |\alpha_{k2}|^2 \right) \lambda + \left(\sum_{k=1}^{m} |\alpha_{k1}|^2 \right) \left(\sum_{k=1}^{m} |x_{k2}|^2 \right) - \left| \sum_{k=1}^{m} \overline{\alpha}_{k1}\alpha_{k2} \right|^2 ,$$

$$\sum_{i,j=1}^{m} |\alpha_{i1}\alpha_{j2} - \alpha_{i2}\alpha_{j1}|^2 = \sum_{i,j=1}^{m} \Big(\alpha_{i1}\alpha_{j2} - \alpha_{i2}\alpha_{j1}\Big)\Big(\overline{\alpha}_{i1}\overline{\alpha}_{j2} - \overline{\alpha}_{i2}\overline{\alpha}_{j1}\Big) =$$

$$= \sum_{i,j=1}^{m} \Big(|\alpha_{i1}|^2|\alpha_{j2}|^2 + |\alpha_{i2}|^2|\alpha_{j1}|^2\Big) - \sum_{i,j=1}^{m} \Big(\overline{\alpha}_{i1}\alpha_{i2}\alpha_{j1}\overline{\alpha}_{j2} + \alpha_{i1}\overline{\alpha}_{i2}\overline{\alpha}_{j1}\alpha_{j2}\Big) =$$

$$= 2 \left(\sum_{k=1}^{m} |\alpha_{k1}|^2 \right) \left(\sum_{k=1}^{m} |\alpha_{k2}|^2 \right) - 2 \left| \sum_{k=1}^{m} \overline{\alpha}_{k1}\alpha_{k2} \right|^2 .$$

The assertion now follows from c). ■

Example 1.2.2.8 (0) *Take* $\mathbb{K}^2$ *endowed with the Euclidean norm. Take* $u \in \mathcal{L}(\mathbb{K}^2)$ *and let*

$$\begin{bmatrix} \alpha & \beta \\ \gamma & \delta \end{bmatrix}$$

be the matrix associated to u.

a)

$$\|u\|^2 =$$

$$= \frac{1}{2}\left(|\alpha|^2 + |\beta|^2 + |\gamma|^2 + |\delta|^2 + \sqrt{\left(|\alpha|^2 + |\beta|^2 + |\gamma|^2 + |\delta|^2\right)^2 - 4|\alpha\delta - \beta\gamma|^2}\right).$$

b)

$$\alpha, \beta \in \mathbb{C} \Longrightarrow \left\|\begin{bmatrix} \alpha & \beta \\ \beta & \alpha \end{bmatrix}\right\| = \sup\{|\alpha + \beta|\,, \,|\alpha - \beta|\}\,.$$

c)

$$\alpha, \beta \in \mathbb{R} \Longrightarrow \left\|\begin{bmatrix} \alpha & \beta \\ \beta & \alpha \end{bmatrix}\right\| = |\alpha| + |\beta|\,.$$

a) follows from Example 1.2.2.7 d).
b) Take $\varphi, \psi \in \mathbb{R}$ such that

$$\alpha = |\alpha|e^{i\varphi}\,, \quad \beta = |\beta|e^{i\psi}$$

and put

$$\theta := 2(\psi - \varphi)\,.$$

Then

$$|\alpha \pm \beta|^2 = \Big|\,|\alpha|e^{i\varphi} \pm |\beta|e^{i\psi}\|^2 = \|\,|\alpha| \pm |\beta|e^{i\frac{\theta}{2}}\Big|^2 = |\alpha|^2 + |\beta|^2 \pm 2|\alpha|\,|\beta| \cos\frac{\theta}{2}$$

so that

$$\sup\{|\alpha + \beta|^2, |\alpha - \beta|^2\} = |\alpha|^2 + |\beta|^2 + 2|\alpha|\,|\beta|\,|\cos\frac{\theta}{2}|\,.$$

On the other side,

$$|\alpha^2 - \beta^2|^2 = \big|\,|\alpha|^2 e^{2i\varphi} - |\beta|^2 e^{2i\psi}\big|^2 = \big|\,|\alpha|^2 - |\beta|^2 e^{i\theta}\big|^2 =$$

$$= |\alpha|^4 + |\beta|^4 - 2|\alpha|^2|\beta|^2 \cos\theta\,,$$

$$\big|2|\alpha|^2 + 2|\beta|^2\big|^2 - 4\left|\alpha^2 - \beta^2\right|^2 =$$

$$= 4\left(|\alpha|^4 + 2|\alpha|^2|\beta|^2 + |\beta|^4 - |\alpha|^4 - |\beta|^4 + 2|\alpha|^2|\beta|^2\cos\theta\right) =$$

$$= 8|\alpha|^2|\beta|^2(1+\cos\theta) = 16|\alpha|^2|\beta|^2\cos^2\frac{\theta}{2}\,.$$

By a),

$$\left\|\begin{bmatrix}\alpha & \beta\\ \beta & \alpha\end{bmatrix}\right\|^2 = \frac{1}{2}\left(2|\alpha|^2 + 2|\beta|^2 + 4|\alpha|\,|\beta|\,|\cos\frac{\theta}{2}|\right) =$$

$$= |\alpha|^2 + |\beta|^2 + 2|\alpha|\,|\beta|\,|\cos\frac{\theta}{2}| = \sup\{|\alpha+\beta|^2, |\alpha-\beta|^2\}\,.$$

c) follows from b). ■

Remark. b) will be generalized in Example 7.1.4.10 h).

Example 1.2.2.9 *Take* $p \in [1,\infty] \cup \{0\}$. *Given* $x \in \ell^p$, *define*

$$x_r : \mathbb{N} \longrightarrow \mathbb{K}\,, \quad n \longmapsto \begin{cases} 0 & \text{if } n = 1\\ x(n-1) & \text{if } n \neq 1\,,\end{cases}$$

$$x_\ell : \mathbb{N} \longrightarrow \mathbb{K}\,, \quad n \longmapsto x(n+1)\,.$$

a) $x_r\,,\ x_\ell \in \ell^p$ *and*

$$\|x_r\|_p = \|x\|_p\,, \quad \|x_\ell\|_p \leq \|x\|_p$$

for every $x \in \ell^p$. *Define*

$$u_r : \ell^p \longrightarrow \ell^p\,, \quad x \longmapsto x_r \qquad (\textbf{the right shift on } \ell^p)\,,$$

$$u_\ell : \ell^p \longrightarrow \ell^p\,, \quad x \longmapsto x_\ell \qquad (\textbf{the left shift on } \ell^p)\,.$$

b) $u_r\,,\ u_\ell \in \mathcal{L}(\ell^p)$, $\|u_r\| = 1$, $\|u_\ell\| = 1$, u_r *is injective,* u_ℓ *is surjective.*

c) $u_\ell \circ u_r$ *is the identity operator on* ℓ^p.

d) *If* $\alpha \in \mathbb{K}\setminus\{0\}$ *and* $x, y \in \ell^p$ *such that*

$$(\alpha 1 - u_r)x = y$$

then

$$x(n) = \frac{1}{\alpha^{n+1}}\sum_{k=1}^{n} \alpha^k y(k)$$

for all $n \in \mathbb{N}$.

e) $\alpha 1 - u_r$ *is injective for all* $\alpha \in \mathbb{K}$.

All the above assertions hold for c *in place of* ℓ^p.

a), b), and c) are easy to see.

d) We prove the assertion by complete induction. The assertion is obvious for $n = 1$. Let $n \in \mathbb{N}$ and assume the assertion holds for n. Then

$$\alpha x(n+1) - x(n) = y(n+1),$$

so that

$$x(n+1) = \frac{1}{\alpha}x(n) + \frac{1}{\alpha}y(n+1) = \frac{1}{\alpha^{n+2}}\sum_{k=1}^{n}\alpha^k y(k) + \frac{1}{\alpha^{n+2}}\alpha^{n+1}y(n+1) =$$

$$= \frac{1}{\alpha^{n+2}}\sum_{k=1}^{n+1}\alpha^k y(k).$$

e) follows from d). ■

Example 1.2.2.10 (0) *Let* T *be a locally compact space and let* $\mathcal{C}_0(T)$ *be the set consisting of those elements of* $\mathcal{C}(T)$ *which vanish at the infinity (if* T *is compact then* $\mathcal{C}_0(T) := \mathcal{C}(T)$*). Then* $\mathcal{C}_0(T)$ *is a closed vector subspace of* $\mathcal{C}(T)$ *and* $\mathcal{C}_0(T)'$ *can be identified with the Banach space* $\mathcal{M}_b(T)$ *of bounded Radon measures on* T.

The proof can be found, for example, in N. Bourbaki, Integration, Ch. II (1952), §2.6. ■

Remark. If T is endowed with the discrete topology then

$$c_0(T) = \mathcal{C}_0(T).$$

Example 1.2.2.11 (1) (2) (3) *If* T *is a completely regular space, then* $\mathcal{C}(T)'$ *can be identified with the Banach space of Radon measures on the Stone-Čech compactification of* T.

This is an immediate consequence of Example 1.2.2.10. ■

Example 1.2.2.12 (0) $\ell^1(T)$, $c_0(T)'$, *and* $c(T)'$ *are isometric for any set* T.

By Example 1.2.2.3 e) $\ell^1(T)$ and $c_0(T)'$ are isometric. To show that $\ell^1(T)$ and $c(T)'$ are isometric, we may assume T to be infinite. Consider T whith the discrete topology and let T^* be the Alexandroff compactification of T. Then $c(T)$ can be identified with $\mathcal{C}(T^*)$. Hence $c(T)'$ can be identified with the Banach space $M_b(T^*)$ of bounded Radon measures on T^* (Example 1.2.2.10). But $M_b(T^*)$ is isometric to $\ell^1(T^*)$ and $\ell^1(T^*)$ is isometric to $\ell^1(T)$, since T and T^* have the same cardinality. Therefore $c(T)$ and $\ell^1(T)$ are isometric. ■

Remark. $c_0(T)'$ and $c(T)'$ are always isometric even though $c_0(T)$ and $c(T)$ are not isometric when T is infinite (Example 1.2.2.4 d)).

Proposition 1.2.2.13 (0) *Let $(E_t)_{t\in T}$ be a family of normed spaces. Take p and q weakly conjugate exponents. Define*

$$E := \left\{x \in \prod_{t\in T} E_t \mid (\|x_t\|)_{t\in T} \in \ell^p(T)\right\},$$

$$F := \left\{x' \in \prod_{t\in T} E'_t \mid (\|x'_t\|)_{t\in T} \in \ell^q(T)\right\},$$

and endow E and F with the norms

$$E \longrightarrow \mathbb{R}_+, \quad x \longmapsto \|(\|x_t\|)_{t\in T}\|_p,$$

$$F \longrightarrow \mathbb{R}_+, \quad x' \longmapsto \|(\|x'_t\|)_{t\in T}\|_q,$$

(Proposition 1.1.2.7 a)). Given $x' \in F$, define

$$\widetilde{x'} : E \longrightarrow \mathbb{K}, \quad x \longmapsto \sum_{t\in T}\langle x_t, x'_t\rangle .$$

a) *$\widetilde{x'} \in E'$ and $\|\widetilde{x'}\| = \|x'\|$ for every $x' \in F$.*

b) *If $p \neq \infty$ and $q \neq 0$ or if T is finite, then the map*

$$F \longrightarrow E', \quad x' \longmapsto \widetilde{x'}$$

is an isometry.

a) Given $x \in E$ and $x' \in F$,

$$\sum_{t\in T}|\langle x_t, x'_t\rangle| \leq \sum_{t\in T}\|x_t\|\,\|x'_t\| \leq \|x\|\,\|x'\|$$

(Proposition 1.2.1.4 a), Example 1.2.2.3 a)), so that $\widetilde{x'} \in E'$ and

$$\|\widetilde{x'}\| \le \|x'\|.$$

Take $x' \in F$ and $\alpha \in]0, \|x'\|[$. By Proposition 1.2.2.3 b),c), there is a $z \in \ell^p(T)^\#$ such that $z \ge 0$ and

$$\sum_{t\in T} \|x'_t\| z(t) > \alpha.$$

Let $S \in \mathfrak{P}_f(T)$ with

$$\sum_{t\in S} \|x'_t\| z(t) > \alpha.$$

By Proposition 1.2.1.4 b), there is an

$$(x_t)_{t\in T} \in \prod_{t\in T} E_t^\#$$

such that

$$x_t = 0$$

for every $t \in T \backslash S$ and

$$\sum_{t\in S} \langle x_t, x'_t \rangle z(t) > \alpha.$$

Put

$$y := (z(t)x_t)_{t\in T} \in E.$$

Then

$$\|y\| = \Big(\sum_{t\in T} z(t)^p \|x_t\|^p\Big)^{\frac{1}{p}} \le \Big(\sum_{t\in T} z(t)^p\Big)^{\frac{1}{p}} \le 1$$

if $p \in [1, \infty[$ and

$$\|y\| = \sup_{t\in T} z(t)\|x_t\| \le 1$$

if $p \in \{0, \infty\}$. It follows (Proposition 1.2.1.4 b))

$$\|\widetilde{x'}\| \ge \widetilde{x'}(y) = \sum_{t\in T} \langle y_t, x'_t \rangle =$$

$$= \sum_{t \in T} \langle x_t, x'_t \rangle z(t) > \alpha .$$

Since α is arbitrary,

$$\|\widetilde{x'}\| \geq \|x'\| , \qquad \|\widetilde{x'}\| = \|x'\| .$$

b) Take $\vartheta \in E'$. Let $s \in T$. Given $x \in E_s$, let $\widetilde{x}$ denote the element of E given by

$$t \longmapsto \begin{cases} x & \text{if } t = s \\ 0 & \text{if } t \neq s \end{cases}$$

and define

$$x'_s : E_s \longrightarrow \mathbb{K} , \quad x \longmapsto \vartheta(\widetilde{x}) .$$

Then $x'_s \in E'_s$. Put

$$x' := (x'_t)_{t \in T} \in \prod_{t \in T} E'_t .$$

Take $y \in \mathbb{K}^{(T)}$. Then

$$\left| \sum_{t \in T} y(t) x'_t(x_t) \right| = \left| \sum_{t \in T} y(t) \vartheta(\widetilde{x}_t) \right| = |\vartheta(yx)| \leq \|\vartheta\| \, \|yx\| \leq \|\vartheta\| \, \|y\|_p$$

for every $x \in E$ with

$$\|x_t\| \leq 1$$

for every $t \in T$. Hence

$$\left| \sum_{t \in T} y(t) \|x'_t\| \right| \leq \|\vartheta\| \, \|y\|_p$$

(Proposition 1.2.1.4 b)). Since y is arbitrary, $x' \in F$ and

$$\|x'\| \leq \|\vartheta\|$$

(Proposition 1.2.2.2), and $\widetilde{x'} = \vartheta$ on

$$G := \{ x \in E \mid \{ t \in T \mid x_t \neq 0 \} \text{ is finite} \} .$$

Since G is dense in E (Proposition 1.1.2.7 c)), we deduce that $\widetilde{x'} = \vartheta$ and

$$\|x'\| \leq \|\vartheta\| = \|\widetilde{x'}\| \leq \|x'\| , \qquad \|\widetilde{x'}\| = \|x'\| .$$

Hence the map

$$F \longrightarrow E' , \quad x' \longmapsto \widetilde{x'}$$

is an isometry. ∎

Corollary 1.2.2.14 *Let T be an infinite set endowed with the discrete topology and let βT denote the Stone–Čech compactification of T. Put*

$$E := c_0(T) \times \mathcal{C}(\beta T \backslash T),$$

$$F := \ell^1(T) \times \mathcal{M}_b(\beta T \backslash T)$$

and endow E and F with the norms

$$E \longrightarrow \mathbb{R}_+, \quad (x,y) \longmapsto \sup\{\|x\|_\infty, \|y\|\},$$
$$F \longrightarrow \mathbb{R}_+, \quad (z,\mu) \longmapsto \|z\|_1 + \|\mu\|$$

(Proposition 1.2.2.13 a)).

a) *For $(z,\mu) \in F$, define*

$$\widetilde{(z,\mu)} : E \longrightarrow \mathbb{K}, \quad (x,y) \longmapsto \sum_{t\in T} x(t)z(t) + \int y d\mu.$$

Then $\widetilde{(z,\mu)} \in E'$ for every $(z,\mu) \in F$ and the map

$$u : F \longrightarrow E', \quad (z,\mu) \longmapsto \widetilde{(z,\mu)}$$

is an isometry.

b) *For $(z,\mu) \in F$, define*

$$\widetilde{(z,\mu)} : \mathcal{C}(\beta T) \longrightarrow \mathbb{K}, \quad x \longmapsto \sum_{t\in T} x(t)z(t) + \int x \mid (\beta T\backslash T) d\mu.$$

Then $\widetilde{(z,\mu)} \in \mathcal{M}_b(\beta T)$ for every $(z,\mu) \in F$ (Example 1.2.2.10) and the map

$$v : F \longrightarrow \mathcal{M}_b(\beta T), \quad (z,\mu) \longmapsto \widetilde{(z,\mu)}$$

is an isometry.

c) *For $y \in \ell^\infty(T)$, denote by $\widetilde{y}$ its continuous extension to βT and for $\mu \in \mathcal{M}_b(\beta T)$, put*

$$\widetilde{\mu} : \ell^\infty(T) \longrightarrow \mathbb{K}, \quad y \longmapsto \int \widetilde{y} d\mu.$$

Then $\widetilde{\mu} \in (\ell^\infty(T))'$ for every $\mu \in \mathcal{M}_b(\beta T)$ and the map

$$w : \mathcal{M}_b(\beta T) \longrightarrow (\ell^\infty(T))', \quad \mu \longmapsto \widetilde{\mu}$$

is an isometry.

d) $w \circ v \circ u^{-1} : E' \to (\ell^\infty(T))'$ *is an isometry.*

a) follows from Example 1.2.2.3 e), Example 1.2.2.10, and Proposition 1.2.2.13.

b) is easy to see.

c) follows from Example 1.2.2.11.

d) follows from a), b), and c). ■

1.2.3 Infinite Matrices

Definition 1.2.3.1 (0) *An **infinite matrix** is a function $S \times T \to \mathbb{K}$, where S and T are sets. Let $k : S \times T \to \mathbb{K}$ be an infinite matrix and take $p \in [1, \infty] \cup \{0\}$. Let q be the conjugate exponent of p. If $k(s, \cdot) \in \ell^p(T)$ for every $s \in S$, then define*

$$\overset{\cap}{k}x : S \longrightarrow \mathbb{K}, \quad s \longmapsto \sum_{t \in T} k(s,t)x(t)$$

for $x \in \ell^q(T)$ (Example 1.2.2.3 a)). If $k(\cdot, t) \in \ell^p(S)$ for every $t \in T$, then we define

$$\overset{\cup}{k}x : T \longrightarrow \mathbb{K}, \quad t \longmapsto \sum_{s \in S} k(s,t)x(s)$$

for $x \in \ell^q(S)$ (Example 1.2.2.3 a)).

If S and T are finite, then k is a matrix and $\overset{\cap}{k}x$ is the usual multiplication of the matrix with a vector.

Definition 1.2.3.2 (0) *Let S, T be sets and take*

$$p, q \in [1, \infty] \cup \{0\}.$$

Let $\ell^{p,q}(S,T)$ denote the set of functions $k : S \times T \to \mathbb{K}$ such that $k(s, \cdot) \in \ell^q(T)$ for every $s \in S$ and that

$$\left(\|k(s,\cdot)\|_q\right)_{s \in S} \in \ell^p(S)$$

endowed with the norm

$$\ell^{p,q}(S,T) \longrightarrow \mathbb{R}_+, \quad k \longmapsto \|(\|k(s,\cdot)\|_q)_{s \in S}\|_p$$

(Proposition 1.1.2.7 a)). We define

$$\ell_0^{\infty,q}(S,T) := \left\{k \in \ell^{\infty,q}(S,T) \mid t \in T \Longrightarrow k(\cdot,t) \in c_0(S)\right\}$$

and endow $\ell_0^{\infty,q}(S,T)$ with the restriction of the norm of $\ell^{\infty,q}(S,T)$.

By Proposition 1.1.2.7, $\ell^{p,q}(S,T)$ is a Banach space. $\mathbb{K}^{(S \times T)}$ is dense in $\ell^{p,q}(S,T)$ whenever $p \neq \infty$ and $q \neq \infty$. It is easy to see that $\ell_0^{\infty,q}(S,T)$ is a closed subspace of $\ell^{\infty,q}(S,T)$, so that it is a Banach space. $\ell^{0,q}(S,T)$ is a closed subspace of $\ell_0^{\infty,q}(S,T)$.

Let $k \in \mathbb{K}^{S \times T}$. Then $k \in \ell^{2,2}(S,T)$ iff

$$\sum_{(s,t) \in S \times T} |k(s,t)|^2 < \infty$$

and in this case the above sum is $\|k\|^2$.

Proposition 1.2.3.3 *Let S,T be sets and take $p,p',q,q'\in[1,\infty[$ with*

$$p\le p',\quad q\le q'.$$

Then

$$\ell^{p,q}(S,T)\subset\ell^{p',q'}(S,T)\cap\ell^{p,0}(S,T)\cap\ell^{0,q}(S,T),$$

$$\ell^{p,0}(S,T)\cup\ell^{0,q}(S,T)\subset\ell^{0,0}(S,T),$$

and the various inclusion maps above have norm at most 1.

The assertion follows from Proposition 1.1.2.6 b), e). ■

Proposition 1.2.3.4 (0) *Let S,T be sets. Take*

$$p,r\in[1,\infty]\cup\{0\}$$

and let q be the conjugate exponent of p.

a) $\overset{\frown}{k}x\in\ell^r(S)$ *and*

$$\|\overset{\frown}{k}x\|_r\le\|k\|\,\|x\|_p$$

whenever $k\in\ell^{r,q}(S,T)$ and $x\in\ell^p(T)$.

b) *The map*

$$\overset{\frown}{k}:\ell^p(T)\longrightarrow\ell^r(S),\quad x\longmapsto\overset{\frown}{k}x$$

belongs to $\mathcal{L}(\ell^p(T),\ell^r(S))$ and

$$\|\overset{\frown}{k}\|\le\|k\|$$

whenever $k\in\ell^{r,q}(S,T)$. If $r\in\{0,\infty\}$ (resp. if $r\in\{0,\infty\}$ and $p=\infty$) then

$$\|\overset{\frown}{k}\|=\|k\|\qquad(\text{resp. }\|\overset{\frown}{k}\mid c_0(T)\|=\|\overset{\frown}{k}\|=\|k\|)$$

whenever $k\in\ell^{r,q}(S,T)$.

c) *For $S\neq\emptyset$ and $r=\infty$, the map*

$$\ell^{\infty,q}(S,T)\longrightarrow\mathcal{L}(\ell^p(T),\ell^\infty(S)),\quad k\longmapsto\overset{\frown}{k}$$

is an isometry iff p or T is finite.

d) *If* $p \neq \infty$, *then the following are equivalent for every* $k \in \ell^{r,q}(S,T)$:

d$_1$) $\overset{\cap}{k}(\ell^p(T)) \subset c_0(S)$ *(resp.* $c(S)$ *);*

d$_2$) $k(\cdot,t) \in c_0(S)$ *(resp.* $c(S)$ *) for every* $t \in T$.

e) *The map*

$$\ell_0^{\infty,1}(S,T) \longrightarrow \mathcal{L}(c_0(T), c_0(S)), \quad k \longmapsto \overset{\cap}{k}$$

is an isometry.

f) (Toeplitz, 1911) *Given* $k \in \ell_0^{\infty,1}(S,T)$, *the following are equivalent:*

f$_1$) $\overset{\cap}{k}x \in c(S)$ *for every* $x \in c(T)$;

f$_2$) $\sum_{t\in T} k(\cdot,t) \in c(S)$.

g) *If* R *is a set,* $u \in \mathcal{L}(\ell^2(S), \ell^2(R))$, *and* $k \in \ell^{2,2}(S,T)$, *then there is a unique* $h \in \ell^{2,2}(R,T)$ *such that*

$$\overset{\cap}{h} = u \circ \overset{\cap}{k};$$

we have

$$\|h\| \leq \|u\| \, \|k\|.$$

a) Given $s \in S$,

$$|(\overset{\cap}{k}x)(s)| = \left|\sum_{t\in T} k(s,t)x(t)\right| \leq \|k(s,\cdot)\|_q \|x\|_p$$

(Example 1.2.2.3 b), Proposition 1.2.1.4 a)). Thus $\overset{\cap}{k}x \in \ell^r(S)$ and

$$\|\overset{\cap}{k}x\|_r \leq \|k\| \, \|x\|_p.$$

b) By a), Proposition 1.1.6.11, and Proposition 1.2.1.1 d $\Rightarrow$ a,

$$\overset{\cap}{k} \in \mathcal{L}(\ell^p(T), \ell^r(S))$$

and

$$\|\overset{\cap}{k}\| \leq \|k\|.$$

Assume now that $r \in \{0, \infty\}$. By Example 1.2.2.3 b), c) and Proposition 1.2.1.4 b),

$$\|k(s,\cdot)\|_q = \sup\left\{ |(\overset{\cap}{k}x)(s)| \,\middle|\, x \in \ell^p(T)^{\#} \right\} \leq$$

$$\leq \sup\left\{ \|\overset{\cap}{k}x\|_\infty \,\middle|\, x \in \ell^p(T)^{\#} \right\} = \|\overset{\cap}{k}\|$$

for every $s \in S$, and so

$$\|k\| \leq \|\overset{\cap}{k}\|, \quad \|\overset{\cap}{k}\| = \|k\|.$$

Finally, assume that $r \in \{0, \infty\}$ and $p = \infty$. Then

$$\|k(s,\cdot)\|_1 = \sup\left\{ |(\overset{\cap}{k}x)(s)| \,\middle|\, x \in c_0(T)^{\#} \right\} \leq$$

$$\leq \sup\left\{ \|\overset{\cap}{k}x\|_\infty \,\middle|\, x \in c_0(T)^{\#} \right\} = \|\overset{\cap}{k} \mid c_0(T)\|$$

for every $s \in S$ (Example 1.2.2.3 c), Proposition 1.2.1.4 b)). Thus

$$\|k\| \leq \|\overset{\cap}{k} \mid c_0(T)\| \leq \|\overset{\cap}{k}\| = \|k\|$$

(Proposition 1.2.1.4 c)) and

$$\|\overset{\cap}{k} \mid c_0(T)\| = \|\overset{\cap}{k}\| = \|k\|.$$

c) Assume that either p or T is finite. Take $u \in \mathcal{L}(\ell^p(T), \ell^\infty(S))$. Put

$$k : S \times T \longrightarrow \mathbb{K}, \quad (s,t) \longmapsto (ue_t^T)(s).$$

Then

$$\left|\sum_{t\in T} k(s,t)x(t)\right| = \left|\sum_{t\in T} (ue_t^T)(s)x(t)\right| = \left|\sum_{t\in T} (ux(t)e_t^T)(s)\right| =$$

$$= |(ux)(s)| \leq \|ux\|_\infty \leq \|u\|\,\|x\|_p$$

for every $s \in S$, and $x \in \mathbb{K}^{(T)}$ (Proposition 1.2.1.4 a)). Thus

$$k(s,\cdot) \in \ell^q(T)$$

and

$$\|k(s,\cdot)\|_q \leq \|u\|$$

for every $s \in S$ (Example 1.2.2.2). Hence $k \in \ell^{\infty,q}(S,T)$. Since

$$(\overset{\cap}{k}e_t^T)(s) = k(s,t) = (ue_t^T)(s)$$

for every $(s,t) \in S \times T$, we have that

$$\overset{\cap}{k}e_t^T = ue_t^T$$

for every $t \in T$. So $\overset{\cap}{k} = u$ on $\mathbb{K}^{(T)}$. Since $\mathbb{K}^{(T)}$ is dense in $\ell^p(T)$ (Proposition 1.1.2.6 c)) we deduce that $\overset{\cap}{k} = u$. Hence the map

$$\ell^{\infty,q}(S,T) \longrightarrow \mathcal{L}(\ell^p(T)\,,\,\ell^\infty(S))\,, \quad k \longmapsto \overset{\cap}{k}$$

is surjective. By b) it is an isometry.

Assume now that $p = \infty$ and that T is infinite. Take $s \in S$ and let $\mathfrak{F}$ be a free ultrafilter (i.e. an ultrafilter containing no singleton set) on T. Define

$$u : \ell^\infty(T) \longrightarrow \ell^\infty(S)\,, \quad x \longmapsto (\lim x(\mathfrak{F}))e_s^S\,.$$

Then $u \in \mathcal{L}(\ell^\infty(T)\,,\,\ell^\infty(S))$, $u(e_T^T) = e_s^S$, and $u = 0$ on $\mathbb{K}^{(T)}$. Assume there is a $k \in \ell^{\infty,1}(S,T)$ with $\overset{\cap}{k} = u$. Then

$$k(s,t) = (\overset{\cap}{k}e_t^T)(s) = (ue_t^T)(s) = 0$$

for every $t \in T$ and we obtain the contradiction

$$1 = (ue_T^T)(s) = (\overset{\cap}{k}e_T^T)(s) = \sum_{t \in T} k(s,t)e_T^T(t) = 0\,.$$

Hence the map

$$\ell^{\infty,1}(S,T) \longrightarrow \mathfrak{L}\,(\ell^\infty(T)\,,\,\ell^\infty(S))\,, \quad k \longmapsto \overset{\cap}{k}$$

is not surjective.

$\mathrm{d}_1 \Rightarrow \mathrm{d}_2$ holds since $e_t^T \in \ell^p(T)$ and

$$k(\cdot,t) = \overset{\cap}{k}e_t^T \in c_0(S) \qquad (\text{resp. } c(T)\,)\,.$$

$\mathrm{d}_2 \Rightarrow \mathrm{d}_1$. By d_2), $\overset{\cap}{k}(\mathbb{K}^{(T)}) \subset c_0(S)$ (resp. $c(T)$). Since $\mathbb{K}^{(T)}$ is dense in $\ell^p(T)$ (Proposition 1.1.2.6 c)), d_1) follows from b) and Example 1.1.2.3.

e) By b) and $\mathrm{d}_2 \Rightarrow \mathrm{d}_1$, given $k \in \ell_0^{\infty,1}(S,T)$, $\overset{\cap}{k}$ may be taken to be an operator from $c_0(T)$ to $c_0(S)$. By c) and $\mathrm{d}_1 \Rightarrow \mathrm{d}_2$), the map

$$\ell_0^{\infty,1}(S,T) \longrightarrow \mathfrak{L}\,(c_0(T)\,,\,c_0(S))\,, \quad k \longmapsto \overset{\cap}{k}$$

is an isometry.

$f_1 \Rightarrow f_2$. We have that

$$\sum_{t\in T} k(\cdot,t) = \overset{\cap}{k} e_T^T \in c(S) .$$

$f_2 \Rightarrow f_1$. We may assume that T is infinite. Let $\mathfrak{F}$ denote the filter of cofinite subsets of T, i.e.

$$\mathfrak{F} := \{T \backslash B \mid \in \mathfrak{P}_f(T)\} ,$$

and put

$$\alpha := \lim x(\mathfrak{F}) .$$

Then $x - \alpha e_T^T \in c_0(T)$, so that by b) and $d_2 \Rightarrow d_1$,

$$\overset{\cap}{k}(x - \alpha e_T^T) \in c_0(S) .$$

Hence

$$\overset{\cap}{k} x = \overset{\cap}{k}(x - \alpha e_T^T) + \alpha \overset{\cap}{k} e_T^T = \overset{\cap}{k}(x - \alpha e_T^T) + \alpha \sum_{t\in T} k(\cdot,t) \in c(S) .$$

g) We set

$$h : R \times T \longrightarrow \mathbb{K} , \quad (r,t) \longmapsto (u\overset{\cap}{k} e_t^T)(r) .$$

Then

$$\sum_{(r,t)\in \mathbb{R}\times T} \left|h(r,t)\right|^2 = \sum_{t\in T} \sum_{r\in R} \left|(u\overset{\cap}{k} e_t^T)(r)\right|^2 = \sum_{t\in T} \left\|u\overset{\cap}{k} e_t^T\right\|_2^2 \leq$$

$$\leq \|u\|^2 \sum_{t\in T} \left\|\overset{\cap}{k} e_t^T\right\|_2^2 = \|u\|^2 \sum_{t\in T} \sum_{s\in S} |k(s,t)|^2 = \|u\|^2 \|k\|^2$$

(Proposition 1.2.1.4 a)), so that $h \in \ell^{2,2}(R,T)$ and

$$\|h\| \leq \|u\| \, \|k\| .$$

We have

$$(\overset{\cap}{h} e_t^T)(r) = h(r,t) = (u\overset{\cap}{k} e_t^T)(r)$$

for all $(r,t) \in R \times T$, so that

$$\overset{\cap}{h} e_t^T = u \overset{\cap}{k} e_t^T$$

for all $t \in T$. By Proposition 1.1.2.6 c),

$$\overset{\cap}{h} = u \circ \overset{\cap}{k} .$$

The uniqueness of h is obvious. ■

Remark. Some of the results in Proposition 1.2.3.4 are actual special cases of the theory of integral operators, which will be treated in Section 3.1.6.

Proposition 1.2.3.5 (1) (2) *Let R, S, T be sets and p, p' weakly conjugate exponents. Take $n, q \in [1, \infty] \cup \{0\}$. Let n' be the conjugate exponent of n. Take $k \in \ell^{p,n'}(S,T)$, $h \in \ell^{q,p'}(R,S)$, and let*

$$h \circ k : R \times T \longrightarrow \mathbb{K} , \quad (r,t) \longmapsto \sum_{s \in S} h(r,s)k(s,t)$$

(Example 1.2.2.3 a)). Then $h \circ k \in \ell^{q,n'}(R,T)$,

$$\|h \circ k\| \leq \|h\| \, \|k\| ,$$

and

$$\overset{\cap}{\overbrace{h \circ k}} x = \overset{\cap}{h} (\overset{\cap}{k} x)$$

whenever $x \in \ell^n(T)$. If $k \in \ell_0^{\infty,1}(S,T)$ and $h \in \ell_0^{\infty,1}(R,S)$, then

$$h \circ k \in \ell_0^{\infty,1}(R,S) .$$

Take $x \in \ell^n(T)$ and $r \in R$. Then

$$\sum_{(s,t) \in S \times T} |h(r,s)| \, |k(s,t)| \, |x(t)| = \sum_{s \in S} |h(r,s)| \Big(\sum_{t \in T} |k(s,t)| \, |x(t)| \Big) \leq$$

$$\leq \sum_{s \in S} |h(r,s)| \, \|k(s,\cdot)\|_{n'} \|x\|_n \leq \|h(r,\cdot)\|_{p'} \|k\| \, \|x\|_n$$

(Example 1.2.2.3 a)). It follows that

$$\sum_{t \in T} h \circ k(r,t)x(t) = \sum_{t \in T} \Big(\sum_{s \in S} h(r,s)k(s,t) \Big) x(t) =$$

$$= \sum_{s\in S} h(r,s)\Big(\sum_{t\in T} k(s,t)x(t)\Big) = \sum_{s\in S} h(r,s)(\overset{\cap}{k}x)(s) = \overset{\cap}{h}(\overset{\cap}{k}x)(r)\,,$$

$$\Big|\sum_{t\in T} h\circ k(r,t)x(t)\Big| \leq \sum_{(s,t)\in S\times T} |h(r,s)|\,|k(s,t)|\,|x(t)| \leq$$

$$\leq \|h(r,\cdot)\|_{p'}\,\|k\|\,\|x\|_n\,.$$

By Proposition 1.2.2.2,

$$h\circ k(r,\cdot) \in \ell^{n'}(T)\,, \quad \|h\circ k(r,\cdot)\|_{n'} \leq \|h(r,\cdot)\|_{p'}\|k\|\,.$$

Furthermore

$$\Big(\|h\circ k(r,\cdot)\|_{n'}\Big)_{r\in R} \in \ell^q(R)\,, \quad h\circ k \in \ell^{q,n'}(R,T)\,, \quad \|h\circ k\| \leq \|h\|\,\|k\|\,.$$

By the above result,

$$\overset{\cap}{\overbrace{h\circ k}}(x) = \overset{\cap}{h}(\overset{\cap}{k}x)\,.$$

The last assertion now follows from Proposition 1.2.3.4 $d_2 \Rightarrow d_1$. ■

Remark. If R, S and T are finite, then $h\circ k$ is the usual product of the matrices h and k.

Proposition 1.2.3.6 (1) (2) (3) *Let S, T be sets. Let p, p' and q, q' be weakly conjugate exponents.*

a) *$(k(s,t)k'(s,t))_{(s,t)\in S\times T}$ is summable whenever $k\in \ell^{p,q}(S,T)$ and*

$$k' \in \ell^{p',q'}(S,T)\,.$$

Given $k'\in \ell^{p',q'}(S,T)$, define

$$\widetilde{k'} : \ell^{p,q}(S,T) \longrightarrow \mathbb{K}\,, \quad k \longmapsto \sum_{(s,t)\in S\times T} k(s,t)k'(s,t)\,.$$

b) *$\widetilde{k'} \in \ell^{p,q}(S,T)'$ and $\|\widetilde{k'}\| = \|k'\|$.*

c) *If $p\neq\infty$ and $p'\neq 0$, then the map*

$$\ell^{p',q'}(S,T) \longrightarrow \ell^{p,q}(S,T)'\,, \quad k' \longmapsto \widetilde{k'}$$

is an isometry.

Given $s \in S$, $k(s,\cdot) \in \ell^q(T)$ and $k'(s,\cdot) \in \ell^{q'}(T) \subset \ell^q(T)'$ (Example 1.2.2.3 b)). The assertions follow from Proposition 1.2.2.13 (and Example 1.2.2.3 d), e)). ■

Proposition 1.2.3.7 (1) (2) *Let S,T be sets, βT the Stone-Čech compactification of T with respect to the discrete topology on T, and $\mathcal{M}$ the Banach space of Radon measures on βT. Take $p \in [1,\infty[\,\cup\{0\}$. Let q be the conjugate exponent of p and E the set of maps $\mu : S \to \mathcal{M}$ with*

$$(\|\mu_s\|)_{s\in S} \in \ell^q(S)\,.$$

Extend each $y \in \ell^\infty(T)$ to a continuous function on βT, which we denote with the same symbol. Given $\mu \in E$, define

$$\widetilde{\mu} : \ell^{p,\infty}(S,T) \longrightarrow \mathbb{K}\,, \quad k \longmapsto \sum_{s\in S} \int k(s,\cdot)d\mu_s$$

(Example 1.2.2.3 a)).

a) *E is a vector subspace of $\mathcal{M}^S$ and the map*

$$E \longrightarrow \mathbb{R}_+\,, \quad \mu \longmapsto \|(\|\mu_s\|)_{s\in S}\|_q$$

is a complete norm. Endow E with this norm.

b) *$\widetilde{\mu} \in \ell^{p,\infty}(S,T)'$ whenever $\mu \in E$ and the map*

$$E \longrightarrow \ell^{p,\infty}(S,T)'\,, \quad \mu \longmapsto \widetilde{\mu}$$

is an isometry.

The assertions follow from Proposition 1.1.2.7 and Proposition 1.2.2.13 (and Example 1.2.2.11). ■

Proposition 1.2.3.8 (1) (3) *Let S,T be sets, p,q weakly conjugate exponents, and take $r \in [1,\infty]$.*

a) *$\overset{\cup}{k}x \in \ell^r(T)$ and*

$$\|\overset{\cup}{k}x\|_r \leq \|k\|\,\|x\|_p$$

whenever $k \in \ell^{q,r}(S,T)$ and $x \in \ell^p(S)$.

b) *Given* $k \in \ell^{q,r}(S,T)$, *the map*

$$\overset{\cup}{k} : \ell^p(S) \longrightarrow \ell^r(T)\,, \quad x \longmapsto \overset{\cup}{k}x$$

belongs to $\mathcal{L}(\ell^p(S)\,,\,\ell^r(T))$ *and*

$$\|\overset{\cup}{k}\| \leq \|k\|\,.$$

c) *If* $p = 1$ *and* $q = \infty$, *then the map*

$$\ell^{\infty,r}(S,T) \longrightarrow \mathcal{L}\left(\ell^1(S), \ell^r(T)\right), \quad k \longmapsto \overset{\cup}{k}$$

is an isometry.

d) *If* $r = \infty$ *and* $p \neq \infty$ *then the following are equivalent whenever* $k \in \ell^{q,\infty}(S,T)$:

d₁) $\overset{\cup}{k}(\ell^p(S)) \subset c_0(T)$ (*resp.* $\overset{\cup}{k}(\ell^p(S)) \subset c(T)$);

d₂) $k(s,\cdot) \in c_0(T)$ (*resp.* $k(s,\cdot) \in c(T)$) *for every* $s \in S$.

a) Let r' be the conjugate exponent of r. Then

$$\left|\sum_{t\in T}(\overset{\cup}{k}x)(t)y(t)\right| = \left|\sum_{t\in T} y(t)\Big(\sum_{s\in S} k(s,t)x(s)\Big)\right| =$$

$$= \left|\sum_{s\in S} x(s)\Big(\sum_{t\in T} k(s,t)y(t)\Big)\right| \leq \sum_{s\in S}\left|x(s)\right|\left|\sum_{t\in T} k(s,t)y(t)\right| \leq$$

$$\leq \sum_{s\in S} |x(s)|\,\|k(s,\cdot)\|_r \|y\|_{r'} \leq \|y\|_{r'}\|k\|\,\|x\|_p$$

whenever $y \in \mathbb{K}^{(T)}$ (Proposition 1.1.6.11, Example 1.2.2.3 a)). Thus $\overset{\cup}{k}x \in \ell^r(T)$ and

$$\|\overset{\cup}{k}x\|_r \leq \|k\|\,\|x\|_p$$

(Proposition 1.2.2.2).

b) follows from a) and Proposition 1.2.1.1 d $\Rightarrow$ a (and Proposition 1.1.6.11).

c) The map

$$\ell^{\infty,r}(S,T) \longrightarrow \mathcal{L}(\ell^1(S)\,,\,\ell^r(T))\,, \quad k \longmapsto \overset{\cup}{k}$$

is linear (b) and Proposition 1.1.6.11).

Take $k \in \ell^{\infty,r}(S,T)$. Then, for every $s \in S$,

$$\|\overset{\cup}{k}\| = \|\overset{\cup}{k}\|\,\|e_s^S\|_1 \geq \|\overset{\cup}{k}(s,\cdot)\|_r$$

(Proposition 1.2.1.4 a)) and so

$$\|\overset{\cup}{k}\| \geq \sup_{s\in S} \|k(s,\cdot)\|_r = \|k\|\,.$$

By b), $\|\overset{\cup}{k}\| = \|k\|$.

Take $u \in \mathcal{L}(\ell^1(S), \ell^r(T))$. Put

$$k : S \times T \longrightarrow \mathbb{K}\,, \quad (s,t) \longmapsto (ue_s^S)(t)\,.$$

Then

$$k(s,\cdot) = ue_s^S \in \ell^r(T)$$

and

$$\|k(s,\cdot)\|_r = \|ue_s^S\|_r \leq \|u\|$$

whenever $s \in S$. Thus $k \in \ell^{\infty,r}(S,T)$. Given $s \in S$,

$$\overset{\cup}{k}e_s^S = k(s,\cdot) = ue_s^S\,,$$

and so $\overset{\cup}{k} = u$ on $\mathbb{K}^{(S)}$. By b), $\overset{\cup}{k} = u$, since $\mathbb{K}^{(S)}$ is dense in $\ell^1(S)$ (Proposition 1.1.2.6 c)).

$d_1 \Rightarrow d_2$ follows from $e_s^S \in \ell^p(S)$ and the fact that

$$k(s,\cdot) = \overset{\cup}{k}e_s^S \in c_0(T) \quad (\in c(T))$$

for every $s \in S$.

$d_2 \Rightarrow d_1$. Given $x \in \mathbb{K}^{(S)}$, $\overset{\cup}{k}x \in c_0(T) \quad (\in c(T))$ and d_1) follows from b) and Proposition 1.1.2.6 c) (and Example 1.1.2.3). ■

Remark. If $p \neq 1$, then the map

$$\ell^{q,r}(S,T) \longrightarrow \mathcal{L}(\ell^p(S), \ell^r(T))\,, \quad k \longmapsto \overset{\cup}{k}$$

need not be surjective. Indeed, if $S = T$ and $1 < p \leq r$, then the embedding $\ell^p(S) \subset \ell^r(S)$ is not of the form $\overset{\cup}{k}$.

Definition 1.2.3.9 (1) *A family* $(A_\iota)_{\iota\in I}$ *of sets is called* ***disjoint*** *if*

$$A_\iota \cap A_\lambda = \emptyset$$

for any distinct $\iota, \lambda \in I$.

Proposition 1.2.3.10 (1) *Let* S, T *be sets and* k *a function* $S \times T \to \mathbb{K}$ *such that* $k(s,\cdot) \in \ell^1(T)$ *for every* $s \in S$. *Then the following are equivalent:*

a) $\overset{\cap}{k}x \in \ell^\infty(S)$ *whenever* $x \in \ell^\infty(T)$.

b) $\overset{\cap}{k}x \in \ell^\infty(S)$ *for every* $x \in c_0(T)$.

c) $\overset{\cap}{k}e_B^T \in \ell^\infty(S)$ *for every countable subset* B *of* T .

d) *There is a sequence* $(\varepsilon_n)_{n\in\mathbb{N}}$ *in* $\mathbb{R}_+\setminus\{0\}$ *with the property that for every disjoint sequence* $(B_n)_{n\in\mathbb{N}}$ *in* $\mathfrak{P}_f(T)$ *there is a sequence* $(\alpha_n)_{n\in\mathbb{N}}$ *in* $\mathbb{R}$ *such that* $\alpha_n \geq \varepsilon_n$ *for every* $n \in \mathbb{N}$ *and that* $\overset{\cap}{k}x \in \ell^\infty(S)$, *where* x *is the map given by*

$$x : T \longrightarrow \mathbb{K}, \quad t \longmapsto \begin{cases} 0 & \text{if } t \in T \setminus \bigcup_{n\in\mathbb{N}} B_n \\ \alpha_n & \text{if } t \in B_n . \end{cases}$$

e) $k \in \ell^{\infty,1}(S,T)$.

If these conditions are fulfilled, then the map

$$\overset{\cap}{k} : \ell^\infty(T) \longrightarrow \ell^\infty(S), \quad x \longmapsto \overset{\cap}{k}x$$

is in $\mathcal{L}(\ell^\infty(T), \ell^\infty(S))$ *and*

$$\|\overset{\cap}{k}\| = \|\overset{\cap}{k} \mid c_0(T)\| = \sup_{s\in S} \|k(s,\cdot)\|_1 .$$

a $\Rightarrow$ b and a $\Rightarrow$ c are trivial.

b $\Rightarrow$ d. Given $n \in \mathbb{N}$, put $\varepsilon_n := \frac{1}{n}$.

c $\Rightarrow$ d. Given $n \in \mathbb{N}$, put $\varepsilon_n = 1$.

d $\Rightarrow$ e. Given $t \in T$,

$$k(\cdot, t) = \overset{\cap}{k}e_t^T \in \ell^\infty(S) .$$

Define

$$k_1 : S \times T \longrightarrow \mathbb{R}, \quad (s,t) \longmapsto \sup\{\operatorname{re} k(s,t), 0\}$$

$$k_2 : S \times T \longrightarrow \mathbb{R}, \quad (s,t) \longmapsto \sup\{-\operatorname{re} k(s,t), 0\}$$

$$k_3 : S \times T \longrightarrow \mathbb{R}, \quad (s,t) \longmapsto \sup\{\operatorname{im} k(s,t), 0\}$$

$$k_4 : S \times T \longrightarrow \mathbb{R}, \quad (s,t) \longmapsto \sup\{-\operatorname{im} k(s,t), 0\}$$

Assume that $(\|k(s,\cdot)\|_1)_{s\in S}$ is not bounded. Then there is a $j \in \{1,2,3,4\}$ such that $(\|k_j(s,\cdot)\|_1)_{s\in S}$ is not bounded. Put $B_0 := \emptyset$ and construct a sequence $(s_n)_{n\in\mathbb{N}}$ in S and an increasing sequence $(B_n)_{n\in\mathbb{N}}$ in $\mathfrak{P}_f(T)$ inductively such that

$$s_n \notin \{s_k \mid k \in \mathbb{N}_{n-1}\}, \qquad \sum_{t\in B_n\setminus B_{n-1}} k_j(s_n,t) > \frac{n}{\varepsilon_n}$$

for every $n \in \mathbb{N}$. Take $n \in \mathbb{N}$ and assume that the sequences have been constructed up to $n-1$. Then there is an

$$s_n \in S\setminus\{s_k \mid k \in \mathbb{N}_{n-1}\}$$

such that

$$\|k_j(s_n,\cdot)\|_1 > \frac{n}{\varepsilon_n} + 1 + (\operatorname{Card} B_{n-1}) \sup_{t\in B_{n-1}} \|k(\cdot,t)\|_\infty .$$

There is a finite subset B_n of T with $B_{n-1} \subset B_n$ and

$$\sum_{t\in T\setminus B_n} k_j(s_n,t) < 1 .$$

Then

$$\sum_{t\in B_n\setminus B_{n-1}} k_j(s_n,t) = \|k_j(s_n,\cdot)\|_1 - \sum_{t\in B_{n-1}} k_j(s_n,t) - \sum_{t\in T\setminus B_n} k_j(s_n,t) > \frac{n}{\varepsilon_n} .$$

This completes the inductive construction.

We put

$$C_n := \{t \in B_n\setminus B_{n-1} \mid k_j(s_n,t) > 0\} .$$

Then $(C_n)_{n\in\mathbb{N}}$ is a disjoint sequence in $\mathfrak{P}_f(T)$. By d), there is a sequence $(\alpha_n)_{n\in\mathbb{N}}$ in $\mathbb{R}$ such that $\alpha_n \geq \varepsilon_n$ for every $n \in \mathbb{N}$ and $\overset{\frown}{k}x \in \ell^\infty(S)$, where

$$x : T \longrightarrow \mathbb{K}, \quad t \longmapsto \begin{cases} 0 & \text{if } t \in T\setminus \bigcup_{n\in\mathbb{N}} C_n \\ \alpha_n & \text{if } t \in C_n ; \end{cases}$$

Given $n \in \mathbb{N}$,

$$|(\overset{\cap}{k}x)(s_n)| = \left|\sum_{t\in T} k(s_n,t)x(t)\right| \geq \sum_{t\in T} k_j(s_n,t)x(t) \geq$$

$$\geq \alpha_n \sum_{t\in B_n\setminus B_{n-1}} k_j(s_n,t) > \alpha_n \frac{n}{\varepsilon_n} \geq n\,,$$

which contradicts $\overset{\cap}{k}x \in \ell^\infty(S)$.

e $\Rightarrow$ a, and the last assertion follow from Proposition 1.2.3.4 b). ∎

Theorem 1.2.3.11 (1) (Kojima–Schur) *Let S, T be sets and $k : S \times T \to \mathbb{K}$ a function with $k(s,\cdot) \in \ell^1(T)$ for every $s \in S$. Then the following are equivalent:*

a) $\overset{\cap}{k}x \in c(S)$ *for every* $x \in c(T)$.

b) $k(\cdot,t) \in c(S)$ *for every* $t \in T$, $\sum_{t\in T} k(\cdot,t) \in c(S)$, *and* $k \in \ell^{\infty,1}(S,T)$.

If these conditions are fulfilled, then the map

$$c(T) \longrightarrow c(S)\,, \quad x \longmapsto \overset{\cap}{k}x$$

is in $\mathcal{L}(c(T)\,, c(S))$ and has norm $\|k\|$.

a $\Rightarrow$ b. Given $t \in T$, e_t^T and e_T^T belong to $c(T)$ and

$$k(\cdot,t) = \overset{\cap}{k}e_t^T \in c(S)\,, \quad \sum_{t\in T} k(\cdot,t) = \overset{\cap}{k}e_T^T \in c(S)\,.$$

By Proposition 1.2.3.10 b $\Rightarrow$ e, $k \in \ell^{\infty,1}(S,T)$.

b $\Rightarrow$ a. By Proposition 1.2.3.4 a),b), $\overset{\cap}{k}x \in \ell^\infty(S)$ whenever $x \in \ell^\infty(T)$ and the map

$$\ell^\infty(T) \longrightarrow \ell^\infty(S)\,, \quad x \longmapsto \overset{\cap}{k}x$$

is continuous. It is obvious that

$$\overset{\cap}{k}(\mathbb{K}^{(T)}) \subset c(S)\,.$$

By continuity,

$$\overset{\cap}{k}(c_0(T)) \subset c(S)\,.$$

There is an $\alpha \in \mathbb{K}$ with $x - \alpha e_T^T \in c_0(T)$. Since

$$k e_T^T = \sum_{t \in T} k(\cdot, t) \in c(S),$$

$\overset{\cap}{k} x \in c(S)$.

The final assertion follows from the last assertion of Proposition 1.2.3.10.

■

Remark. The above theorem still holds for $c_0(S)$ in place of $c(S)$.

Theorem 1.2.3.12 (1) (I. Schur) *Let S, T be sets and $k : S \times T \to \mathbb{K}$ a function with $k(s, \cdot) \in \ell^1(T)$ for every $s \in S$. Then the following are equivalent:*

a) $\overset{\cap}{k} x \in c(S)$ *for every* $x \in \ell^\infty(T)$.

b) $\overset{\cap}{k} e_B^T \in c(S)$ *for every countable subset* B *of* T.

c) $k(\cdot, t) \in c(S)$ *for every* $t \in T$*, and for any* $\varepsilon > 0$ *there is a finite subset* B *of* T *such that*

$$\sum_{t \in T \setminus B} |k(s, t)| < \varepsilon$$

for every $s \in S$.

Assume that $k(\cdot, t) \in c(S)$ *for every* $t \in T$. *Define*

$$\mathfrak{F} := \{S \setminus A \mid A \in \mathfrak{P}_f(S)\},$$

$$a : T \longrightarrow \mathbb{K}, \quad t \longmapsto \lim_{\mathfrak{F}} k(\cdot, t).$$

If the above conditions are fulfilled, then $a \in \ell^1(T)$,

$$\lim_{s, \mathfrak{F}} \|k(s, \cdot) - a\|_1 = 0,$$

the map

$$\overset{\cap}{k} : \ell^\infty(T) \longrightarrow c(S), \quad x \longmapsto \overset{\cap}{k} x$$

is in $\mathcal{L}(\ell^\infty(T), c(S))$,

$$\|\overset{\cap}{k}\| = \|\overset{\cap}{k} \mid c_0(T)\| = \sup_{s \in S} \|k(s, \cdot)\|_1$$

and

$$\lim_{\mathfrak{F}} \overset{\cap}{k} x = \sum_{t \in T} a(t)x(t) \,,$$

for every $x \in \ell^\infty(T)$.

a $\Rightarrow$ b is trivial.

b $\Rightarrow$ a. Let B be a subset of T. Assume that $\overset{\cap}{k} e_B^T \notin c(S)$. Then there is a sequence $(s_n)_{n\in\mathbb{N}}$ in S with

$$\liminf_{n\to\infty} \left| (\overset{\cap}{k} e_B^T)(s_n) - (\overset{\cap}{k} e_B^T)(s_{n+1}) \right| > 0 \,.$$

Given $n \in \mathbb{N}$, put

$$C_n := \{k(s_n, \cdot) \neq 0\} \cap B$$

and

$$C := \bigcup_{n\in\mathbb{N}} C_n \,.$$

Then C is countable and

$$(\overset{\cap}{k} e_B^T)(s_n) = \left(\overset{\cap}{k} e_C^T\right)(s_n)$$

for every $n \in \mathbb{N}$. Thus

$$\liminf_{n\to\infty} \left| (\overset{\cap}{k} e_C^T)(s_n) - (\overset{\cap}{k} e_C^T)(s_{n+1}) \right| > 0$$

which contradicts the fact that $\overset{\cap}{k} e_C^T \in c(S)$. Hence $\overset{\cap}{k} e_B^T \in c(S)$.

By Proposition 1.2.3.10, $\overset{\cap}{k} x \in \ell^\infty(S)$ whenever $x \in \ell^\infty(T)$ and the map

$$\ell^\infty(T) \longrightarrow \ell^\infty(S) \,, \quad x \longmapsto \overset{\cap}{k} x$$

is continuous. Since the vector subspace of $\ell^\infty(T)$ generated by $\{e_B^T \mid B \subset T\}$ is dense in $\ell^\infty(T)$ (Proposition 1.1.2.6 d)), $\overset{\cap}{k} x \in c(S)$ for every $x \in \ell^\infty(T)$ (Example 1.1.2.3).

a $\Rightarrow$ c. Given $t \in T$, $e_t^T \in \ell^\infty(T)$ and

$$k(\cdot, t) = \overset{\cap}{k} e_t^T \in c(S) \,.$$

Take $\varepsilon > 0$. Assume that for every finite subset B of T there is an $s \in S$ with

$$\sum_{t\in T\backslash B} |k(s,t)| \geq \varepsilon\,.$$

Put $B_0 := \emptyset$ and construct a sequence $(s_n)_{n\in\mathbb{N}}$ in S and an increasing sequence $(B_n)_{n\in\mathbb{N}}$ in $\mathfrak{P}_f(T)$ inductively such that

$$s_n \notin \{s_k \mid k \in \mathbb{N}_{n-1}\}\,, \qquad \sum_{t\in B_{n-1}} |k(s_n,t) - a(t)| < \frac{\varepsilon}{6}\,,$$

$$\sum_{t\in B_n\backslash B_{n-1}} |k(s_n,t)| > \frac{5\varepsilon}{6}\,, \qquad \sum_{t\in T\backslash B_n} |k(s_n,t)| < \frac{\varepsilon}{6}\,,$$

for every $n \in \mathbb{N}$. Take $n \in \mathbb{N}$ and assume that the sequences have been constructed up to $n-1$. Since

$$\lim_{\mathfrak{F}} k(\cdot,t) = a(t)$$

for every $t \in B_{n-1}$, there is a finite subset A of S with

$$\sum_{t\in B_{n-1}} |k(s,t) - a(t)| < \frac{\varepsilon}{6}$$

whenever $s \in S\backslash A$. Put

$$C := A \cup \{s_k \mid k \in \mathbb{N}_{n-1}\}.$$

Given $s \in C$, there is a finite subset D_s of T with

$$\sum_{t\in T\backslash D_s} |k(s,t)| < \varepsilon\,.$$

Define

$$D := B_{n-1} \cup \Big(\bigcup_{s\in C} D_s\Big)\,.$$

Then D is a finite subset of T and

$$\sum_{t\in T\backslash D} |k(s,t)| < \varepsilon$$

for every $s \in C$. By the above assumption, there is an $s_n \in S\backslash C$ with

$$\sum_{t\in T\backslash D} |k(s_n,t)| \geq \varepsilon\,.$$

Let B_n be a finite subset of T with $D \subset B_n$ and

$$\sum_{t\in T\setminus B_n} |k(s_n,t)| < \frac{\varepsilon}{6}.$$

Then

$$\sum_{t\in B_n\setminus B_{n-1}} |k(s_n,t)| = \sum_{t\in T\setminus B_{n-1}} |k(s_n,t)| - \sum_{t\in T\setminus B_n} |k(s_n,t)| \geq$$

$$\geq \sum_{t\in T\setminus D} |k(s_n,t)| - \sum_{t\in T\setminus B_n} |k(s_n,t)| > \varepsilon - \frac{\varepsilon}{6} = \frac{5\varepsilon}{6}.$$

This completes the inductive construction.

We now define a map $x : T \to \mathbb{K}$. We set $x = 0$ on

$$\Big(T\setminus \bigcup_{n\in\mathbb{N}} B_n\Big) \cup \Big(\bigcup_{n\in\mathbb{N}} (B_{2n-1}\setminus B_{2n-2})\Big).$$

Given $n \in \mathbb{N}$ and $t \in B_{2n}\setminus B_{2n-1}$, we set

$$x(t) := \frac{\overline{k(s_{2n},t)}}{|k(s_{2n},t)|}$$

if $k(s_{2n},t) \neq 0$, and $x(t) = 0$ otherwise. Then $x \in \ell^\infty(T)$ and

$$\left|(\overset{\frown}{k}x)(s_{2n}) - (\overset{\frown}{k}x)(s_{2n-1})\right| = \left|\sum_{t\in T} k(s_{2n},t)x(t) - \sum_{t\in T} k(s_{2n-1},t)x(t)\right| \geq$$

$$\geq \left|\sum_{t\in B_{2n}\setminus B_{2n-1}} k(s_{2n},t)x(t)\right| - \left|\sum_{t\in B_{2n-2}} \Big(k(s_{2n},t) - k(s_{2n-1},t)\Big)x(t)\right| -$$

$$-\left|\sum_{t\in T\setminus B_{2n}} k(s_{2n},t)x(t)\right| - \left|\sum_{t\in B_{2n}\setminus B_{2n-1}} k(s_{2n-1},t)x(t)\right| -$$

$$-\left|\sum_{t\in T\setminus B_{2n}} k(s_{2n-1},t)x(t)\right| \geq$$

$$\geq \sum_{t\in B_{2n}\setminus B_{2n-1}} \left|k(s_{2n},t)\right| - \sum_{t\in B_{2n-2}} \left|k(s_{2n},t) - a(t)\right| - \sum_{t\in B_{2n-2}} \left|k(s_{2n-1},t) - a(t)\right| -$$

$$- \sum_{t\in T\setminus B_{2n}} \left|k(s_{2n},t)\right| - \sum_{t\in T\setminus B_{2n-1}} \left|k(s_{2n-1},t)\right| >$$

$$> \frac{5\varepsilon}{6} - 2\frac{\varepsilon}{6} - 2\frac{\varepsilon}{6} = \frac{\varepsilon}{6}$$

for every $n \in \mathbb{N}$, which contradicts the fact that $\overset{\cap}{k}x \in c(S)$.

c $\Rightarrow$ a & the last assertion. Take $\varepsilon > 0$. By c) there is a finite subset B of T with

$$\sum_{t \in T \setminus B} |k(s,t)| < \varepsilon$$

whenever $s \in S$. We obtain successively that

$$\sum_{t \in T \setminus B} |a(t)| \leq \liminf_{s,\mathfrak{F}} \sum_{t \in T \setminus B} |k(s,t)| \leq \varepsilon\,, \quad a \in \ell^1(T)\,,$$

$$\|k(s,\cdot) - a\|_1 = \sum_{t \in T} |k(s,t) - a(t)| \leq$$

$$\leq \sum_{t \in B} |k(s,t) - a(t)| + \sum_{t \in T \setminus B} |k(s,t)| + \sum_{t \in T \setminus B} |a(t)| < \sum_{t \in B} |k(s,t) - a(t)| + 2\varepsilon$$

for every $s \in S$,

$$\limsup_{s,\mathfrak{F}} \|k(s,\cdot) - a\|_1 \leq 2\varepsilon\,,$$

and

$$\limsup_{s,\mathfrak{F}} \|k(s,\cdot) - a\|_1 = 0$$

since ε is arbitrary.

Take $x \in \ell^\infty(T)$. Then

$$\left|(\overset{\cap}{k}x)(s) - \sum_{t \in T} a(t)x(t)\right| = \left|\sum_{t \in T} \Big(k(s,t) - a(t)\Big)x(t)\right| \leq \|k(s,\cdot) - a\|_1 \|x\|_\infty$$

whenever $s \in S$. Thus

$$\lim_{s,\mathfrak{F}} \left|(\overset{\cap}{k}x)(s) - \sum_{t \in T} a(t)x(t)\right| = 0\,, \quad \lim_{s,\mathfrak{F}} (\overset{\cap}{k}x)(s) = \sum_{t \in T} a(t)x(t)\,,$$

$\overset{\cap}{k}x \in c(S)$.

By the last assertion of Proposition 1.2.3.10, the map

$$\overset{\cap}{k} : \ell^\infty(T) \longrightarrow c(S)\,, \quad x \longmapsto \overset{\cap}{k}x$$

is in $\mathcal{L}(\ell^\infty(T), c(S))$ and

$$\|\overset{\cap}{k}\| = \|\overset{\cap}{k} \mid c_0(T)\| = \sup_{s \in S} \|k(s,\cdot)\|_1\,.$$

■

Corollary 1.2.3.13 (1) *Let T be an arbitrary and S an infinite set. Let*

$$\mathfrak{F} := \{S \backslash A \mid A \in \mathfrak{P}_f(S)\},$$

and let $(x_s)_{s\in S}$ be a family in $\ell^1(T)$ such that

$$\lim_{s,\mathfrak{F}} \sum_{t\in B} x_s(t)$$

exists whenever B is a countable subset of T. Define

$$x : T \longrightarrow \mathbb{K}, \quad t \longmapsto \lim_{s,\mathfrak{F}} x_s(t).$$

Then $x \in \ell^1(T)$ and

$$\lim_{s,\mathfrak{F}} \|x_s - x\|_1 = 0.$$

We define

$$k : S \times T \longrightarrow \mathbb{K}, \quad (s,t) \longmapsto x_s(t).$$

By hypothesis, $\overset{\cap}{k} e_B^T \in c(S)$ for every countable subset B of T and the conclusion follows from Theorem 1.2.3.12. ∎

Corollary 1.2.3.14 (1) *Let S, T be sets and k a function $S \times T \to \mathbb{K}$ for which $k(s,\cdot) \in \ell^1(T)$ for every $s \in S$. Then the following are equivalent:*

a) $k \in \ell^{0,1}(S,T)$.

b) $\overset{\cap}{k} x \in c_0(S)$ *for every* $x \in \ell^\infty(T)$.

c) $k(\cdot,t) \in c_0(S)$ *for every* $t \in T$, *and given* $\varepsilon > 0$, *there is a finite subset* B *of* T *such that*

$$\sum_{t\in T\backslash B} |k(s,t)| < \varepsilon$$

whenever $s \in S$.

a $\Rightarrow$ b follows from Proposition 1.2.3.4 a).
b $\Rightarrow$ c. Given $t \in T$, $e_t^T \in \ell^\infty(T)$ so that

$$k(\cdot,t) = \overset{\cap}{k} e_t^T \in c_0(S).$$

By Theorem 1.2.3.12 a $\Rightarrow$ c, given $\varepsilon > 0$, there is a finite subset B of T such that

$$\sum_{t\in T\setminus B} |k(s,t)| < \varepsilon$$

whenever $s \in S$.

c $\Rightarrow$ a. Let $\varepsilon > 0$. By c) there is a finite subset B of T with

$$\sum_{t\in T\setminus B} |k(s,t)| < \frac{\varepsilon}{2}$$

whenever $s \in S$. Let n be the number of elements of B. Given $t \in B$, there is a finite subset A_t of S with

$$|k(s,t)| < \frac{\varepsilon}{2n+1}$$

whenever $s \in T\setminus A_t$. Put

$$A := \bigcup_{t\in B} A_t \,.$$

Then A is a finite subset of S and

$$\|k(s,\cdot)\|_1 = \sum_{t\in T} |k(s,t)| = \sum_{t\in B} |k(s,t)| + \sum_{t\in T\setminus B} |k(s,t)| < n\frac{\varepsilon}{2n+1} + \frac{\varepsilon}{2} < \varepsilon$$

for every $s \in T\setminus A$. Hence $\left(\|k(s,\cdot)\|_1\right)_{s\in S} \in c_0(S)$ and $k \in \ell^{0,1}(S,T)$. ∎

1.2.4 Quotient Spaces

Definition 1.2.4.1 (0) *Let E be a vector space. Take $A, B \in \mathfrak{P}(E)$, $z \in E$, and $\alpha \in \mathbb{K}$. Define*

$$A + B := \{x + y \mid (x, y) \in A \times B\},$$

$$\alpha A := \{\alpha x \mid x \in A\},$$

$$z + A := A + z := A + \{z\}.$$

Let F be a vector subspace of the vector space E. Given $x, y \in E$, define

$$x \sim y : \Longleftrightarrow x - y \in F.$$

Let E/F denote the set of equivalence classes of $\sim$. Then given $X, Y \in E/F$ and $\alpha \in \mathbb{K}$, $X + Y$ and αX belong to E/F. E/F is a vector space with respect to these operations, with F as the null element. The dimension of E/F is called **the codimension of F in E**. *The map $E \to F$ which maps each point of E into its equivalence class is linear and it is called* **the quotient map**.

Unlike the dimension of F, which is intrinsic to F, the codimension depends on the vector space in which F is embedded, as well as on F.

Theorem 1.2.4.2 (0) *Let F be a closed vector subspace of the normed space E and $q : E \to E/F$ the quotient map. Given $X \in E/F$, define*

$$\|X\| := \inf_{x \in X} \|x\|.$$

a) *The map*

$$E/F \longrightarrow \mathbb{R}_+, \quad X \longmapsto \|X\|$$

is a norm, called **the quotient norm of E/F**. *E/F endowed with this norm is called* **the quotient space of E with respect to F**.

b) *q is continuous and open (i.e. maps open sets into open sets) and $\|q\| = 1$ if $E \neq F$.*

c) *A subset U of E/F is open iff $\overset{-1}{q}(U)$ is open.*

d) *A map f of E/F in a topological space T is continuous iff $f \circ q$ is continuous.*

e) *E/F is complete whenever E is complete.*

f) *If F and E/F are complete, then E is complete.*

a) Take $X, Y \in E/F$ and $\alpha \in \mathbb{K}$. Take $x \in X$ and $y \in Y$. Then $x + y \in X + Y$, $\alpha x \in \alpha X$, so that

$$\|X+Y\| \leq \|x+y\| \leq \|x\| + \|y\|, \quad \|\alpha X\| \leq \|\alpha x\| = |\alpha|\,\|x\|.$$

Since x and y are arbitrary,

$$\|X+Y\| \leq \|X\| + \|Y\|, \quad \|\alpha X\| \leq |\alpha|\,\|X\|.$$

For $\alpha \neq 0$,

$$\|X\| = \|\frac{1}{\alpha}(\alpha X)\| \leq \frac{1}{|\alpha|}\|\alpha X\|, \quad |\alpha|\,\|X\| \leq \|\alpha X\|, \quad \|\alpha X\| = |\alpha|\,\|X\|.$$

Now suppose $\|X\| = 0$. Then there is a sequence $(x_n)_{n\in\mathbb{N}}$ in X converging to 0. Since F is closed, so is X. Hence $0 \in X$, i.e. $X = F$ and $X = 0$ in E/F.

b) Given $x \in E$,

$$\|qx\| \leq \|x\|.$$

Hence q is continuous (Proposition 1.2.1.1 d $\Rightarrow$ a) and $\|q\| \leq 1$. Assume $E \neq F$. Then we can find $X \in E/F$, $X \neq F$, and we have that

$$\|X\| = \|qx\| \leq \|q\|\,\|x\|$$

for all $x \in X$. Thus

$$\|X\| \leq \|q\|\,\|X\|, \quad 1 \leq \|q\|, \quad \|q\| = 1.$$

Now take an open set, U, of E and take $x \in U$. For some $\varepsilon > 0$, $U_\varepsilon^E(x) \subset U$. Take $Y \in E/F$ with $\|Y\| < \varepsilon$. There is a $y \in Y$ with $\|y\| < \varepsilon$. Thus

$$x + y \in U_\varepsilon^E(x) \subset U, \quad qx + Y = q(x+y) \in q(U).$$

Hence

$$U_\varepsilon^{E/F}(qx) \subset q(U),$$

and $q(U)$ is open.

c) If U is open, then $\overset{-1}{q}(U)$ is open as well since q is continuous by b). Now assume that $\overset{-1}{q}(U)$ is open. Then

$$U = q\left(\overset{-1}{q}(U)\right)$$

is open, since, by b), q is an open mapping.

d) If f is continuous, then $f \circ q$ is continuous as well. Now assume that $f \circ q$ is continuous and let U be an open set of T. Then

$$\overset{-1}{q}\left(\overset{-1}{f}(U)\right) = \overset{-1}{\overbrace{f \circ q}}(U)$$

is open. By c), $\overset{-1}{f}(U)$ is open. Hence f is continuous.

e) Let $(X_\iota)_{\iota \in I}$ be an absolutely summable family in E/F. For $\iota \in I$ take $x_\iota \in X_\iota$ with

$$\|x_\iota\| \leq 2\|X_\iota\|.$$

Then $(x_\iota)_{\iota \in I}$ is an absolutely summable family in E. By Corollary 1.1.6.10 a $\Rightarrow$ b, $(x_\iota)_{\iota \in I}$ is summable. By Proposition 1.2.1.16 $(X_\iota)_{\iota \in I}$ is summable. Thus by Corollary1.1.6.10 b $\Rightarrow$ a, E/F is complete.

f) Let $(x_n)_{n \in \mathbb{N}}$ be a Cauchy sequence in E. By b), $(qx_n)_{n \in \mathbb{N}}$ is a Cauchy sequence in E/F. Since E/F is complete there is an $x \in E$ such that $(qx_n)_{n \in \mathbb{N}}$ converges to qx. Hence $(q(x_n - x))_{n \in \mathbb{N}}$ converges to 0. For every $n \in \mathbb{N}$ let $y_n \in F$ such that

$$\|x_n - x - y_n\| \leq \|q(x_n - x)\| + \frac{1}{n}.$$

Then

$$\|y_m - y_n\| = \|(x_n - x - y_n) - (x_m - x - y_m) + (x_m - x_n)\| \leq$$

$$\leq \|x_n - x - y_n\| + \|x_m - x - y_m\| + \|x_m - x_n\| \leq$$

$$\leq \|q(x_n - x)\| + \frac{1}{n} + \|q(x_m - x)\| + \frac{1}{m} + \|x_m - x_n\|$$

for all $m, n \in \mathbb{N}$, so that $(y_n)_{n \in \mathbb{N}}$ is a Cauchy sequence in F. Since F is complete $(y_n)_{n \in \mathbb{N}}$ converges to a $y \in F$. It follows that $(x_n)_{n \in \mathbb{N}}$ converges to $x + y$. Hence E is complete. ■

Remark. If G is a closed vector subspace of E it does not follow that $q(G)$ is closed even if E is complete (Example 1.2.4.15 d)).

Corollary 1.2.4.3 (0) *Let F be a closed vector subspace of the normed space E and G a finite-dimensional vector subspace of E. Then $F+G$ is a closed vector subspace of E.*

Let $q : E \to E/F$ be the quotient map. Then $q(G)$ is a finite–dimensional vector subspace and hence a closed vector subspace of E/F (Corollary 1.2.3.6). The conclusion now follows from

$$F + G = \overset{-1}{q}(q(G))$$

and Theorem 1.2.4.2 b). ■

Remark. In the above corollary we may not replace the above hypothesis "G is finite–dimensional" by the weaker one "G is closed". A counterexample may be found in Example 1.2.4.15 e).

Corollary 1.2.4.4 *Let F be a closed subspace of the normed space E. E is separable iff F and E/F are both separable.*

It is clear that if E is separable, then both F and E/F are also separable. So assume that F and E/F are separable. Let A and B be countable dense sets of F and E/F respectively. Let $q : E \to E/F$ be the quotient map and let f be a map of B in E with

$$q(f(X)) = X$$

for every $X \in B$. Take $x \in E$ and $\varepsilon > 0$. There is a $Y \in B$ such that

$$\|qx - Y\| < \varepsilon\,.$$

Then

$$\|q(x - f(Y))\| = \|qx - q(f(Y))\| = \|qx - Y\| < \varepsilon\,.$$

$x - f(Y) - A$ is a dense set of $\overset{-1}{q}(q(x - f(Y)))$. Hence there is a $y \in A$ with

$$\|x - f(Y) - y\| < \varepsilon\,.$$

Since x and ε are arbitrary, $A + f(B)$ is a dense set of E. But $A + f(B)$ is countable, and so E is separable. ■

Definition 1.2.4.5 (0) *Let E, F be vector spaces and $u : E \to F$ be a linear map. Define*

$$\operatorname{Ker} u := \overset{-1}{u}(0) \qquad (\textbf{\textit{the kernel of }} u),$$

$$\operatorname{Im} u := u(E) \qquad (\textbf{\textit{the image of }} u),$$

$$\operatorname{Coker} u := F/\operatorname{Im} u \qquad (\textbf{\textit{the cokernel of }} u),$$

u is a vector subspace of E and $\operatorname{Im} u$ is a vector subspace of F.
$p \in \{0\} \cup [1, \infty]$ and u_r, u_ℓ are the right and the left shift of ℓ^p, ively, then

$$\operatorname{Ker} u_r = \{0\}, \qquad \operatorname{Im} u_r = \{x \in \ell^p \mid x_1 = 0\},$$

$$\operatorname{Ker} u_\ell = \{x \in \ell^p \mid n \in \mathbb{N}\backslash\{1\} \Longrightarrow x_n = 0\}, \qquad \operatorname{Im} u_\ell = \ell^p.$$

a 1.2.4.6 (0) *Let E, F be vector spaces, G a vector subspace of $E \to E/G$ the quotient map, and $u : E \to F$ a linear map which is 0 Then there is a unique map $v : E/G \to F$ such that $u = v \circ q$. v is* ***he factorization of*** u ***through*** E/G. *v is linear and it is injective iff $= G$. In this case the map*

$$E/G \longrightarrow \operatorname{Im} u, \quad x \longmapsto vx$$

tive and is called ***the algebraic isomorphism associated to*** u**.**

uniqueness of v follows from the surjectivity of q. Take $X \in E/G$ $y \in X$. Then $x - y \in G$, and so

$$ux - uy = u(x - y) = 0, \quad ux = uy.$$

remark, the map $v : E/G \to F$, defined by putting

$$vX := ux$$

$\in E/G$, where x is any element of X, is well-defined. Clearly,

$$v \circ q = u.$$

$X, Y \in E/G$ and $\alpha, \beta \in \mathbb{K}$. Take $x \in X$ and $y \in Y$. Then

$$\alpha x + \beta y \in \alpha X + \beta Y,$$

$$v(\alpha X + \beta Y) = u(\alpha x + \beta y) = \alpha ux + \beta uy = \alpha vX + \beta vY,$$

v is linear.
ume that $\operatorname{Ker} u = G$. Take $X \in E/G$ with $vX = 0$. Then

$$ux = vX = 0$$

for every $x \in X$. Hence $X \subset \operatorname{Ker} u = G$, so that $X = G$. Thus v is injective.

Now suppose that v is injective. Then given $x \in \operatorname{Ker} u$,

$$v(qx) = v \circ q(x) = ux = 0,$$

so that

$$qx = 0, \quad \text{i.e. } x \in G.$$

Hence $\operatorname{Ker} u = G$. ■

Proposition 1.2.4.7 (0) *Let E, F be normed spaces, G a closed vector subspace of E, $q : E \to E/G$ the quotient map, and put*

$$\mathcal{F} := \{u \in \mathcal{L}(E,F) \mid u \mid G = 0\}.$$

$\mathcal{F}$ is a closed vector subspace of $\mathcal{L}(E,F)$ and the map

$$\mathcal{L}(E/G, F) \longrightarrow \mathcal{F}, \quad v \longmapsto v \circ q$$

is an isometry.

It is obvious that $\mathcal{F}$ is a vector subspace of $\mathcal{L}(E,F)$. Take $u \in \overline{\mathcal{F}}$ and $x \in G$. There is a sequence $(u_n)_{n\in\mathbb{N}}$ in $\mathcal{F}$ converging to u. Given $n \in \mathbb{N}$,

$$\|ux - u_n x\| = \|(u - u_n)x\| \le \|u - u_n\| \, \|x\|$$

(Proposition 1.2.1.4 a)). Hence

$$ux = \lim_{n\to\infty} u_n x = 0.$$

Thus u restricts to 0 on G and so $u \in \mathcal{F}$. Hence $\mathcal{F}$ is closed.

The map

$$\mathcal{L}(E/G, F) \longrightarrow \mathcal{F}, \quad v \longmapsto v \circ q$$

is obviously linear and

$$\|v \circ q\| \le \|v\| \, \|q\| \le \|v\|$$

whenever $v \in \mathcal{L}(E/G, F)$ (Corollary 1.2.1.5, Theorem 1.2.4.2 b)). Take $u \in \mathcal{F}$ and let v be the factorization of u through E/G (Lemma 1.2.4.6). Take $X \in E/G$ and $x \in X$. Then

$$\|vX\| = \|v(qx)\| = \|ux\| \le \|u\|\,\|x\|\,,$$

so that

$$\|vX\| \le \|u\|\,\|X\|\,, \quad v \in \mathcal{L}(E/G, F) \text{ and } \|v\| \le \|u\|\,, \quad \|v\| = \|v \circ q\|$$

(Proposition 1.2.1.1 d $\Rightarrow$ a). The above map is therefore an isometry. ■

Proposition 1.2.4.8 $\left(0 \right)$ *Let E, F be normed spaces and $u : E \to F$ a linear map. If E is finite–dimensional, then u is continuous.*

First assume that u is injective. We put

$$p : E \longrightarrow \mathbb{R}_+\,, \quad x \longmapsto \|ux\|\,.$$

Then

$$p(x+y) = \|u(x+y)\| = \|ux + uy\| \le \|ux\| + \|uy\| = p(x) + p(y)\,,$$

$$p(\alpha x) = \|u(\alpha x)\| = \|\alpha ux\| = |\alpha|\,\|ux\| = |\alpha| p(x)\,,$$

$$(p(x) = 0) \Longrightarrow (\|ux\| = 0) \Longrightarrow (ux = 0) \Longrightarrow (x = 0)$$

whenever $x, y \in E$ and $\alpha \in \mathbb{K}$. Thus, p is a norm. By Minkowski's Theorem (Theorem 1.1.3.4), there is a $\beta > 0$ such that

$$p(x) \le \beta \|x\|$$

for every $x \in E$. Thus

$$\|ux\| = p(x) \le \beta \|x\|$$

for every $x \in E$, i.e. u is continuous (Proposition 1.2.1.1 d $\Rightarrow$ a).

Now let u be arbitrary. $\operatorname{Ker} u$ is a finite–dimensional vector subspace of E and hence closed (Corollary 1.1.3.6). Let

$$q : E \longrightarrow E/\operatorname{Ker} u$$

be the quotient map and v the factorization of u through $E/\operatorname{Ker} u$. Then v is injective (Lemma 1.2.4.6). Since $E/\operatorname{Ker} u$ is finite-dimensional, the above considerations show that v is continuous. Thus $u = v \circ q$ is also continuous. ■

Corollary 1.2.4.9 (0) *If E, F are finite–dimensional normed spaces, then every bijective linear map $E \to F$ is an isomorphism.* ■

Corollary 1.2.4.10 (0) *The dual and the algebraic dual of the finite–dimensional normed space E coincide. In particular, the dimension of E and E' coincide.* ■

Corollary 1.2.4.11 (0) *Let E, F be normed spaces and $u : E \to F$ be a linear map. If $\operatorname{Ker} u$ is closed (e.g. finite-dimensional) and $\operatorname{Im} u$ is finite-dimensional, then u is continuous.*

Let $q : E \to E/\operatorname{Ker} u$ be the quotient map and v the factorization of u through $E/\operatorname{Ker} u$. Then v is injective (Lemma 1.2.4.6). Since

$$v(E/\operatorname{Ker} u) = \operatorname{Im} u\,,$$

$E/\operatorname{Ker} u$ is finite–dimensional. Hence v is continuous (Proposition 1.2.4.8) and so $u = v \circ q$ is also continuous.

If $\operatorname{Ker} u$ is finite–dimensional, then it is closed by Corollary 1.1.3.6. ■

Corollary 1.2.4.12 (0) *Let x' be a linear form on the normed space E. The following are equivalent:*

a) *x' is continuous.*

b) *$\operatorname{Ker} x'$ is closed.*

c) *$x' = 0$ or $\operatorname{Ker} x'$ is not dense.*

a ⇒ b ⇒ c is trivial.

b ⇒ a follows from Corollary 1.2.4.11.

c ⇒ b. Assume that $\operatorname{Ker} x'$ is not closed. Then $\operatorname{Ker} x'$ has a point of adherence x which does not belong to $\operatorname{Ker} x'$. Then given $y \in E$,

$$x'\left(y - \frac{x'(y)}{x'(x)}x\right) = x'(y) - \frac{x'(y)}{x'(x)}x'(x) = 0\,,$$

$$y - \frac{x'(y)}{x'(x)}x \in \operatorname{Ker} x' \subset \overline{\operatorname{Ker} x'}\,,$$

$$y = \left(y - \frac{x'(y)}{x'(x)}x\right) + \frac{x'(y)}{x'(x)}x \in \overline{\operatorname{Ker} x'}$$

(Corollary 1.1.5.4),

$$E \subset \overline{\operatorname{Ker} x'}\,,$$

which is a contradiction. Hence $\operatorname{Ker} x'$ is closed. ■

Corollary 1.2.4.13 *Every infinite–dimensional normed space contains a dense proper vector subspace.*

Let E be an infinite–dimensional normed space. By Corollary 1.2.1.2, E admits a discontinuous linear form x'. By Corollary 1.2.4.12 c $\Rightarrow$ a, $\operatorname{Ker} x'$ is a dense proper vector subspace of E. ■

Example 1.2.4.14 *Let βT be the Stone-Čech compactification of the discrete space T. We put $\Delta := \beta T \backslash T$ and consider every $x \in \ell^\infty(T)$ extended continuously on βT. Put*

$$u : \ell^\infty(T) \longrightarrow \mathcal{C}(\Delta)\,, \quad x \longmapsto x|\Delta\,.$$

Then the factorization

$$\ell^\infty(T)/c_0(T) \longrightarrow \mathcal{C}(\Delta)$$

of u through $\ell^\infty(T)/c_0(t)$ is an isometry.

By Lemma 1.2.4.6 and Tietze's theorem, the factorization

$$\ell^\infty(T)/c_0(T) \longrightarrow \mathcal{C}(\Delta)$$

of u is bijective and it is easy to see that it is also norm– preserving. ■

Example 1.2.4.15 *Take*

$$E := \ell^1 \times c_0\,,$$

$$F := \{(x,x) \mid x \in \ell^1\}\,,$$

$$G := \{(x,0) \mid x \in \ell^1\}\,,$$

$$u : E \longrightarrow c_0\,, \quad (x,y) \longmapsto y - x\,,$$

and endow E with the 1–norm.

a) $u \in \mathcal{L}(E, c_0)$, $\|u\| = 1$.

b) $F = \operatorname{Ker} u$.

c) *The factorization of u through E/F is an isometry.*

d) *G is a closed vector subspace of E but $q(G)$ is not closed in E/F. (Here $q : E \to E/F$ is the quotient map.)*

e) $F + G = \overset{-1}{q}(q(G)) = \{(x, y) \mid x, y \in \ell^1\}$ *is not closed.*

a) u is linear and

$$\|u(x, y)\| = \|y - x\|_0 \le \|y\|_0 + \|x\|_0 \le \|x\|_1 + \|y\|_0 = \|(x, y)\|_1 .$$

b) is obvious.

c) By Lemma 1.2.4.6, the factorization of u is bijective. Take $z \in c_0$ and $(x, y) \in E$ with

$$z = u(x, y) = y - x .$$

By a),

$$\|z\|_0 \le \|(y, x)\|_1 .$$

It follows from

$$z = u(0, z) \text{ and } \|z\|_0 = \|(0, z)\|_1$$

that the factorization is an isometry.

d) By c), we may identify E/F with c_0. With this identification,

$$q(G) = \ell^1 .$$

e) is easy to see. ■

1.2.5 Complemented Subspaces

Proposition 1.2.5.1 (0) *Let F and G be subspaces of the normed space E. Then the map*

$$F \times G \longrightarrow E\,, \quad (x,y) \longmapsto x+y$$

is linear and continuous.

It is obvious that the above map is linear. If we endow $F \times G$ with the 1–norm of the product then

$$\|x+y\| \le \|x\| + \|y\| = \|(x,y)\|_1$$

proving continuity (Proposition 1.2.1.1 d $\Rightarrow$ a). ∎

Proposition 1.2.5.2 (0) *Let E be a normed space, F, G subspaces of E, $q : E \to E/F$ the quotient map, and put*

$$u : F \times G \longrightarrow E\,, \quad (x,y) \longmapsto x+y\,,$$

$$v : G \longrightarrow E/F\,, \quad y \longmapsto qy\,.$$

Then the following are equivalent:

a) *u is an isomorphism.*

b) *F and G are closed in E and v is an isomorphism.*

c) *F is closed in E and v is an isomorphism.*

a $\Rightarrow$ b. $F \times \{0\}$ and $\{0\} \times G$ are closed sets of $F \times G$. Since u is a homeomorphism, F and G are closed sets of E. q is continuous (Theorem 1.2.4.2 b)) and so v is also continuous. Take $Z \in E/F$ and $z \in Z$. There is a pair $(x,y) \in F \times G$ with

$$u(x,y) = z\,.$$

It follows that

$$vy = qy = q(x+y) = q(u(x,y)) = qz = Z\,.$$

Hence v is surjective.

We show that v is lower bounded. Take $y \in G$. Then

$$\|y\| \le \|x\| + \|y\| = \|(x,y)\|_1 = \|u^{-1}(x+y)\| \le \|u^{-1}\|\,\|x+y\|$$

for every $x \in F$ (Proposition 1.2.1.4 a)), so that

$$\|y\| \le \|u^{-1}\| \inf_{x\in F} \|x+y\| = \|u^{-1}\|\,\|qy\|, \quad \|vy\| = \|qy\| \ge \frac{1}{\|u^{-1}\|}\|y\|.$$

By Proposition 1.2.1.18 a) , b) , v is bijective and v^{-1} is continuous, i.e. v is an isomorphism.

b $\Rightarrow$ c is trivial.

c $\Rightarrow$ a. Let $z \in E$. There is a $y \in G$ with

$$qy = qz.$$

Hence $z - y \in F$ and

$$u(z-y,y) = z - y + y = z,$$

i.e. u is surjective.

We show that u is lower bounded. Take $(x,y) \in F \times G$. Then

$$q(x+y) = qx + qy = qy,$$

so that (Proposition 1.2.1.4 a), Theorem 1.2.4.2 b))

$$\|y\| = \|v^{-1}q(x+y)\| \le \|v^{-1}\|\,\|q(x+y)\| \le \|v^{-1}\|\,\|x+y\|,$$

$$\|x\| = \|x+y-y\| \le \|x+y\| + \|y\| \le (1+\|v^{-1}\|)\|x+y\|,$$

$$\|(x,y)\|_\infty = \sup\{\|x\|, \|y\|) \le (1+\|v^{-1}\|)\|x+y\|$$

$$\|u(x,y)\| = \|x+y\| \ge \frac{1}{1+\|v^{-1}\|}\|(x,y)\|_\infty.$$

By Proposition 1.2.1.18 a) , b) , u is injective and u^{-1} is continuous. Since u is continuous (Proposition 1.2.5.1), it is an isomorphism. ∎

Definition 1.2.5.3 (0) *Let E be a normed space and F,G be subspaces of E. We say that E is* ***the direct sum of*** *F* ***and*** *G, and we denote this by*

$$E = F \oplus G$$

if the map

$$F \times G \longrightarrow E\,, \quad (x,y) \longmapsto x+y$$

is an isomorphism.

Let E be a normed space. A ***complemented subspace of*** *E is a subspace F of E for which there is a subspace G of E with*

$$E = F \oplus G\,.$$

In this case G is also a complemented subspace of E; it is called a ***complement of F in E.***

By Proposition 1.2.5.2 a $\Rightarrow$ b, complemented subspaces of a normed space are closed. But there are closed subspaces of Banach spaces which are not complemented subspaces (see Corollary 1.2.5.14).

Let $(E_\iota)_{\iota\in I}$ be a finite family of normed space. For each $\lambda \in I$, the space E_λ (canonically identified with a subspace of $\prod\limits_{\iota\in I} E_\iota$) is a complemented subspace of $\prod\limits_{\iota\in I} E_\iota$.

Corollary 1.2.5.4 *Let E be a normed space, F a complemented subspace of E, and G,H complements of F in E. Then G and H are isomorphic.*

By Proposition 1.2.5.2 a $\Rightarrow$ c, F is closed and G and H are isomorphic to E/F. Hence F and G are isomorphic to each other. ∎

Corollary 1.2.5.5 *Let E be a normed space and F be a complemented subspace of E. E is complete iff F and E/F are both complete.*

By Proposition 1.2.5.2 a $\Rightarrow$ c, F is closed. Thus the assertion follows from Theorem 1.2.4.2 e),f). ∎

Corollary 1.2.5.6 (0) *Let F be a closed subspace of the normed space E which is of finite codimension in E. Then E has a finite-dimensional subspace G such that*

$$E = F \oplus G\,.$$

As G we may take any algebraic complement of F in E. In particular, E is complete iff F is complete.

Let $q : E \to E/F$ be the quotient map and $(X_\iota)_{\iota \in I}$ an algebraic basis for E/F. Given $\iota \in I$, take $x_\iota \in X_\iota$ and let G be the vector subspace of E generated by $(x_\iota)_{\iota \in I}$. Then G is finite–dimensional and $q|G$ is an algebraic isomorphism. By Corollary 1.2.4.9, $q|G$ is an isomorphism, so by Proposition 1.2.5.2 c $\Rightarrow$ a

$$E = F \oplus G.$$

If H is an algebraic complement of F in E, then there is an isomorphism $u : G \to H$. The map

$$F \times G \longrightarrow F \times H, \quad (x, y) \longmapsto (x, uy)$$

is then an isomorphism, so

$$E = F \oplus H.$$

The last assertion follows from Proposition 1.1.5.1, since every finite–dimensional normed space is complete (Corollary 1.1.3.5). ∎

Definition 1.2.5.7 (0) *Let F be a vector subspace of the normed space E. A* ***projection in*** *E is an operator p on E such that $p \circ p = p$. If $F = \operatorname{Im} p$ then we say p is a* ***projection of*** *E* ***onto*** *F. In this case*

$$x \in F \Longleftrightarrow px = x$$

for every $x \in E$, so F is closed.

For every $u \in \mathcal{L}(E)$ with $u(E) \subset F$, u is a projection of E on F iff $ux = x$ for every $x \in F$.

Theorem 1.2.5.8 (0) (Murray, 1937) *Let F and G be subspaces of the space E. Then the following are equivalent:*

a) $E = F \oplus G$.

b) *there is a projection p of E onto F such that $G = \operatorname{Ker} p$.*

If these assertions hold, then $1-p$ is a projection of E onto G, $F = \operatorname{Ker}(1-p)$, and E/F is isomorphic to G.

a $\Rightarrow$ b. Let $q : E \to E/G$ be the quotient map. By Proposition 1.2.5.2 a $\Rightarrow$ c, G is closed and the map

$$u : F \longrightarrow E/G\,, \quad x \longmapsto qx$$

is an isomorphism. Put

$$j : F \longrightarrow E\,, \quad x \longmapsto x\,,$$

$$p := j \circ u^{-1} \circ q\,.$$

Then p is an operator on E and $\operatorname{Im} p \subset F$. Given $x \in F$,

$$px = u^{-1}(qx) = x\,.$$

Thus $\operatorname{Im} p = F$ and $p \circ p = p$. That $G = \operatorname{Ker} p$ is obvious.

b $\Rightarrow$ a & the last assertion. Since

$$(1-p) \circ (1-p) = 1 - p - p + p = 1 - p\,,$$

$1-p$ is a projection in E. Take $x \in E$. Then

$$(x \in F) \Longleftrightarrow (x = px) \Longleftrightarrow ((1-p)x = 0) \Longleftrightarrow (x \in \operatorname{Ker}(1-p))$$

$$(x \in G) \Longleftrightarrow (px = 0) \Longleftrightarrow ((1-p)x = x) \Longleftrightarrow (x \in \operatorname{Im}(1-p))\,.$$

Thus

$$F = \operatorname{Ker}(1-p)\,, \quad G = \operatorname{Im}(1-p)\,.$$

Define

$$u : F \times G \longrightarrow E\,, \quad (x,y) \longmapsto x + y\,,$$

$$v : E \longrightarrow F \times G\,, \quad z \longmapsto (pz, (1-p)z)\,.$$

u and v are linear and continuous (Proposition 1.2.5.1, Proposition 1.1.5.1). Since

$$uvz = pz + (1-p)z = z\,,$$

for every $z \in E$ and

$$vu(x,y) = (p(x+y)\,, (1-p)(x+y)) = (x,y)$$

for every $(x,y) \in F \times G$, it follows that u is an isomorphism. Hence

$$E = F \oplus G\,.$$

By Proposition 1.2.5.2 a $\Rightarrow$ c, E/F and G are isomorphic. ∎

Corollary 1.2.5.9 (0) *Let E be a normed space, F a subspace of E, and G a subspace of F.*

a) *If F is a complemented subspace of E and G is a complemented subspace of F, then G is a complemented subspace of E.*

b) *If G is a complemented subspace of E, then it is a complemented subspace of F.*

a) By Theorem 1.2.5.8 a $\Rightarrow$ b, there are projections u of E onto F and v of F onto G. Then

$$E \longrightarrow E, \quad x \longmapsto vux$$

is a projection of E onto G. By Theorem 1.2.5.8 b $\Rightarrow$ a, G is a complemented subspace of E.

b) By Theorem 1.2.5.8 a $\Rightarrow$ b, there is a projection u of E onto G. Then

$$F \longrightarrow F, \quad x \longmapsto ux$$

is a projection of F onto G. By Theorem 1.2.5.8 b $\Rightarrow$ a, G is a complemented subspace of F. ■

Corollary 1.2.5.10 (0) *Let E, F be normed spaces. Take $u \in \mathcal{L}(E, F)$ such that the map*

$$E \longrightarrow \operatorname{Im} u, \quad x \longmapsto ux$$

is an isomorphism. Let G, H be vector subspaces of E and F respectively, such that G is not a complemented subspace of E. If $u(G) \subset H$ and there is a projection of H onto $u(G)$, then H is not a complemented subspace of F.

Assume H to be a complemented subspace of F. By Murray's Theorem, $u(G)$ is a complemented subspace of H. By Corollary 1.2.5.9 a), $u(G)$ is a complemented subspace of F and by the same corollary b), it is a complemented subspace of $\operatorname{Im} u$. It follows that G is a complemented subspace of E and this is a contradiction. ■

Example 1.2.5.11 (0) *If T is an infinite set, then there exists a projection of $c(T)$ on $c_0(T)$ whose norm is 2 and every such projection has a norm at least 2. $c_0(T)$ is a complemented subspace of $c(T)$ with codimension 1.*

Let $\mathfrak{F}$ denote

$$\{A \subset T \mid T \backslash A \text{ is finite}\},$$

the filter on T consisting of the cofinite subsets of T. It is easy to see that

$$c(T) \longrightarrow c(T), \quad x \longmapsto x - (\lim x(\mathfrak{F}))e_T$$

is a projection of $c(T)$ onto $c_0(T)$ with norm 2 and with one-dimensional kernel. By Murray's theorem, $c_0(T)$ is a complemented subspace of $c(T)$ with codimension 1.

Let u be a projection of $c(T)$ onto $c_0(T)$. Set

$$x := ue_T .$$

Then, given $t \in T$,

$$e_t \in c_0(T), \quad \|e_T - 2e_t\| = 1,$$

$$\|u\| \geq \|u(e_T - 2e_t)\| = \|x - 2e_t\| \geq |x(t) - 2| \geq 2 - |x(t)|,$$

and so

$$\|u\| \geq \lim_{t,\mathfrak{F}}(2 - |x(t)|) = 2 .$$

■

Remark. This result will be generalized in Proposition 4.2.8.23.

Example 1.2.5.12 *Let E be a normed space, T a topological space, S a closed set of T, and*

$$u : \mathcal{C}(S, E) \longrightarrow \mathcal{C}(T, E)$$

an operator with

$$(ux) \mid S = x$$

for every $x \in \mathcal{C}(S,E)$. Then $\{x \in \mathcal{C}(T,E) \mid x = 0 \text{ on } S\}$ is a complemented subspace of $\mathcal{C}(T,E)$.

The map

$$\mathcal{C}(T,E) \longrightarrow \mathcal{C}(T,E), \quad x \longmapsto x - u(x|S)$$

is a projection of $\mathcal{C}(T,E)$ onto $\{x \in \mathcal{C}(T,E) \mid x = 0 \text{ on } S\}$ and the conclusion follows from Murray's Theorem. ■

Remark. Exercises 1.2.14, 1.2.16 (resp. Corollary 1.2.5.15) present examples where such an operator u exists (resp. does not exist).

Example 1.2.5.13 (0) *If a complemented subspace of ℓ^∞ contains c_0, then it is not separable.*

Let E be a complemented subspace of ℓ^∞ containing c_0 and let u be a projection in ℓ^∞ with $E = \operatorname{Ker} u$ (Theorem 1.2.5.8 a $\Rightarrow$ b). Put

$$F := \operatorname{Im} u$$

and define

$$x'_n : F \longrightarrow \mathbb{K}, \quad x \longmapsto x(n)$$

for every $n \in \mathbb{N}$. By Lemma 1.1.2.17, there is an uncountable set $\mathfrak{A}$ of infinite subsets of $\mathbb{N}$ such that $A \cap B$ is finite for distinct $A, B \in \mathfrak{A}$.

Let $\mathfrak{B}$ be a finite subset of $\mathfrak{A}$. Given $A \in \mathfrak{B}$, put

$$C(A) := A \backslash \bigcup_{\substack{B \in \mathfrak{B} \\ B \neq A}} B$$

and

$$x := \sum_{A \in \mathfrak{B}} e_{C(A)} \,.$$

Then $\|x\| \leq 1$. Since E contains c_0,

$$ue_A = ue_{C(A)}$$

for every $A \in \mathfrak{A}$. Hence

$$ux = \sum_{A \in \mathfrak{B}} ue_A \,.$$

Take $x' \in F'$. By the above,

$$\sup_{\mathfrak{B} \in \mathfrak{P}_f(\mathfrak{A})} \left| \sum_{A \in \mathfrak{B}} x'(ue_A) \right| = \sup_{\mathfrak{B} \in \mathfrak{P}_f(\mathfrak{A})} \left| x'\Big(\sum_{A \in \mathfrak{B}} (ue_A) \Big) \right| \leq$$

$$\leq \sup_{x \in E^{\#}} |x'(ux)| \leq \|x' \circ u\| \leq \|x'\| \, \|u\|$$

(Proposition 1.2.1.4 b), Corollary 1.2.1.5). Hence $(x'(ue_A))_{A \in \mathfrak{A}}$ is a summable family (Proposition 1.1.6.14 d $\Rightarrow$ a) and

$$\Big\{ A \in \mathfrak{A} \mid x'(ue_A) \neq 0 \Big\}$$

is a countable set (Corollary 1.1.6.7). Thus

$$\mathfrak{A}_0 := \bigcup_{n\in\mathbb{N}} \left\{A \in \mathfrak{A} \mid x_n'(ue_A) \neq 0\right\}$$

is also countable.

Take $A \in \mathfrak{A}\backslash\mathfrak{A}_0$. Then

$$(ue_A)(n) = x_n'(ue_A) = 0$$

for every $n \in \mathbb{N}$, so that

$$ue_A = 0\,.$$

Thus

$$e_A \in \operatorname{Ker} u = E\,.$$

$\{e_A \mid A \in \mathfrak{A}\backslash\mathfrak{A}_0\}$ is thus an uncountable subset of E such that

$$\|e_A - e_B\| = 1$$

for distinct $A, B \in \mathfrak{A}\backslash\mathfrak{A}_0$. Hence E is not separable. ∎

Remark. Lindenstrauss (1967) proved that every infinite-dimensional complemented subspace of ℓ^∞ is isomorphic to ℓ^∞, and therefore not separable (Example 1.1.2.2).

Corollary 1.2.5.14 (0) (Phillips, 1940) *For no infinite set T are $c(T)$ and $c_0(T)$ complemented subspaces of $\ell^\infty(T)$.*

Let S be a countable infinite subset of T. Given $x \in \ell^\infty(S)$, define

$$\widetilde{x} : T \longrightarrow \mathbb{K}\,, \quad t \longmapsto \begin{cases} x(t) & \text{if } t \in S \\ 0 & \text{if } t \in T\backslash S \end{cases}\,.$$

Define

$$u : \ell^\infty(S) \longrightarrow \ell^\infty(T)\,, \quad x \longmapsto \widetilde{x}\,.$$

Then

$$\ell^\infty(S) \longrightarrow \operatorname{Im} u\,, \quad x \longmapsto ux$$

is an isometry, $u(c_0(S)) \subset c_0(T)$, and

$$c_0(T) \longrightarrow c_0(T)\,, \quad y \longmapsto e_S y$$

is a projection of $c_0(T)$ onto $u(c_0(S))$. By Example 1.1.2.3 and Example 1.2.5.13, $c_0(S)$ is not a complemented subspace of $\ell^\infty(S)$, so that by Corollary 1.2.5.10, $c_0(T)$ is not a complemented subspace of $\ell^\infty(T)$.

By Example 1.2.5.11, $c_0(T)$ is a complemented subspace of $c(T)$ so that by the above considerations and Corollary 1.2.5.9 a), $c(T)$ is not a complemented subspace of $\ell^\infty(T)$. ∎

Corollary 1.2.5.15 *Let T be an infinite set endowed with the discrete topology and let T^* be its Stone-Čech compactification. Then there is no operator*

$$u : \mathcal{C}(T^*\backslash T) \longrightarrow \mathcal{C}(T^*)$$

with the property that

$$x = ux|(T^*\backslash T)$$

for every $x \in \mathcal{C}(T^*\backslash T)$.

By Example 1.2.5.12, the existence of such an operator would imply that $\{x \in \mathcal{C}(T^*) \mid x = 0 \text{ on } T^*\backslash T\}$ is a complemented subspace of $\mathcal{C}(T^*)$, i.e. that $c_0(T)$ is a complemented subspace of $\ell^\infty(T)$, contradicting Corollary 1.2.5.14. ∎

Proposition 1.2.5.16 (0) *Let T be a set and take $p \in [1,\infty] \cup \{0\}$. Given $x \in \ell^\infty(T)$, define*

$$\widetilde{x} : \ell^p(T) \longrightarrow \ell^p(T)\,, \quad y \longmapsto xy\,,$$

and for $u \in \mathcal{L}(\ell^\infty(T))$, define

$$\dot{u} : T \longrightarrow \mathbb{K}\,, \quad t \longmapsto (ue_t)(t)\,.$$

a) *Take $x \in \ell^\infty(T)$. Then $\widetilde{x} \in \mathcal{L}(\ell^p(T))$, $\|\widetilde{x}\| = \|x\|_\infty$. We define*

$$\varphi : \ell^\infty(T) \longrightarrow \mathcal{L}(\ell^p(T))\,, \quad x \longmapsto \widetilde{x}\,.$$

b) *φ is linear and the map*

$$\ell^\infty(T) \longrightarrow \operatorname{Im}\varphi\,, \quad x \longmapsto \varphi x$$

is an isometry.

c) $u \in \mathcal{L}(\ell^p(T)) \Rightarrow \dot{u} \in \ell^\infty(T)\,, \quad \|\dot{u}\|_\infty \le \|u\|\,.$

d) $x \in \ell^\infty(T) \Rightarrow \dot{\tilde{x}} = x\,.$

e) *The map*

$$\mathcal{L}(\ell^p(T)) \longrightarrow \mathcal{L}(\ell^p(T))\,, \quad u \longmapsto \tilde{\dot{u}}$$

is a projection of $\mathcal{L}(\ell^p(T))$ *onto* $\operatorname{Im}\varphi$ *of norm* 1.

f) $\operatorname{Im}\varphi$ *is a complemented subspace of* $\mathcal{L}(l^p(T))$.

a), b), c), and d) are easy to see.

e) By c) and a) the map is a well-defined operator of norm at most 1. By d), it is a projection of $\mathcal{L}(\ell^p(T))$ onto $\operatorname{Im}\varphi$. Hence it has norm 1.

f) follows from e) and Murray's Theorem. ∎

1.2.6 The Topology of Pointwise Convergence

Definition 1.2.6.1 (0) *Let S be a set, A a subset of S, T a topological space, and $\mathcal{F}$ a set of maps of S into T. $\mathcal{F}_A$ denotes the set $\mathcal{F}$ endowed with the topology of pointwise convergence in A.*

$\mathcal{F}_S$ denotes the set $\mathcal{F}$ endowed with the topology of pointwise convergence, i.e. with the topology on $\mathcal{F}$ induced by the product topology on T^S.

Let E, F be normed spaces. By Proposition 1.2.1.4 a), the topology of pointwise convergence on $\mathcal{L}(E,F)$ is coarser than the norm topology of $\mathcal{L}(E,F)$. In particular, the topology of E'_E is coarser than the norm topology of E'.

Proposition 1.2.6.2 (0) *Let E be a vector space, F a vector space of linear forms on E, and V' a 0–neighbourhood in F_E. Then there is a finite subset A of E such that*

$$\{x' \in F \mid x \in A \Longrightarrow |x'(x)| \leq 1\} \subset V'.$$

Given $x' \in F$, there is an $\alpha \in \mathbb{R}_+$ with $x' \in \alpha V'$.

By the definition of the topology of pointwise convergence, there is a finite family $(x_\iota)_{\iota\in I}$ in E and a family $(\varepsilon_\iota)_{\iota\in I}$ in $\mathbb{R}_+ \setminus \{0\}$ such that

$$\{x' \in E' \mid \iota \in I \Longrightarrow |x'(x_\iota)| \leq \varepsilon_\iota\} \subset V',$$

The set

$$A := \left\{ \frac{1}{\varepsilon_\iota} x_\iota \,\middle|\, \iota \in I \right\}$$

has the desired properties.

Now we prove the last assertion. We may assume that $A \neq \emptyset$. We put

$$\alpha := 1 + \sup_{x\in A} |x'(x)| \in \mathbb{R}_+ .$$

Then, for each $x \in A$,

$$\left|\frac{1}{\alpha} x'(x)\right| \leq 1,$$

so that $\frac{1}{\alpha} x' \in V$, i.e. $x' \in \alpha V$. ∎

Lemma 1.2.6.3 (0) *Let x' be a linear form on the vector space E, and $(x'_\iota)_{\iota \in I}$ a finite, nonempty family of linear forms on E with*

$$\bigcap_{\iota \in K} \operatorname{Ker} x'_\iota \subset \operatorname{Ker} x' .$$

Then x' is a linear combination of the x'_ι $(\iota \in I)$.

We may assume that the family $(x'_\iota)_{\iota \in I}$ is linearly independent. We prove the lemma by complete induction on the cardinality of I. Take $\lambda \in I$ and put $J := I \backslash \{\lambda\}$. By the hypothesis of the induction, there is an

$$x \in \Big(\bigcap_{\iota \in J} \operatorname{Ker} x'_\iota\Big) \backslash \operatorname{Ker} x'_\lambda$$

(if $J = \emptyset$, replace the intersection by E). Put

$$y' := x' - \frac{x'(x)}{x'_\lambda(x)} x'_\lambda .$$

Take

$$y \in \bigcap_{\iota \in J} \operatorname{Ker} x'_\iota .$$

Then

$$x'_\lambda \left(y - \frac{x'_\lambda(y)}{x'_\lambda(x)} x \right) = 0 ,$$

$$y - \frac{x'_\lambda(y)}{x'_\lambda(x)} x \in \bigcap_{\iota \in I} \operatorname{Ker} x'_\iota ,$$

$$0 = x'(y) - \frac{x'_\lambda(y)}{x'_\lambda(x)} x'(x) = y'(y) ,$$

$$\bigcap_{\iota \in J} \operatorname{Ker} x'_\iota \subset \operatorname{Ker} y' .$$

By the hypothesis of the induction, y' is a linear combination of the x'_ι $(\iota \in J)$. Hence x' is a linear combination of the x'_ι $(\iota \in I)$. ■

Lemma 1.2.6.4 (0) *Let A be a finite subset of the vector space E. Let F be a vector space of linear forms on E and x'' a linear form on F which is bounded on*

$$\{x' \in F' \mid x \in A \Longrightarrow |x'(x)| < 1\}.$$

Given $x \in E$, *define*

$$\widetilde{x} : F \longrightarrow \mathbb{K}, \quad x' \longmapsto x'(x).$$

Then x'' *is a linear combination of* $(\widetilde{x})_{x \in A}$. *In particular, there is an* $x \in E$ *with* $\widetilde{x} = x''$.

We may assume that A is not empty. Then

$$\bigcap_{x \in A} \operatorname{Ker} \widetilde{x} \subset \operatorname{Ker} x'',$$

and the assertion follows from Lemma 1.2.6.3. ■

Corollary 1.2.6.5 (0) *Let* E *be a normed space,* F *a vector subspace of* E', *and* x'' *a continuous linear form on* F_E. *Then there is an* $x \in E$ *such that*

$$\langle x, \cdot \rangle | F = x''.$$

By Proposition 1.2.6.2, there is a finite subset A of E such that x'' is bounded on

$$\{x' \in E' \mid x \in A \Longrightarrow |x'(x)| \leq 1\}.$$

The assertion now follows from Lemma 1.2.6.4. ■

Proposition 1.2.6.6 (0) *If* E *is a normed space, then the map*

$$E'_E \longrightarrow \mathbb{R}_+, \quad x' \longmapsto \|x'\|$$

is lower semicontinuous and $E'^{\#}$ *is a closed set of* E'_E.

Given $x \in E$, the map

$$E'_E \longrightarrow \mathbb{R}_+, \quad x' \longmapsto |x'(x)|$$

is continuous. The assertion now follows from Proposition 1.2.1.4 b). ■

Remark. $E'^{\#}$ is even a compact set of E'_E. (See Theorem 1.2.8.1).

Proposition 1.2.6.7 *If E is a normed space then the map*

$$\mathcal{L}(E)^{\#}_{E} \times \mathcal{L}(E)_{E} \longrightarrow \mathcal{L}(E)_{E}\,, \qquad (u,v) \longmapsto uv$$

is continuous.

Take $(u_0, v_0) \in \mathcal{L}(E)^{\#} \times \mathcal{L}(E)$, A a finite subset of E, and $\varepsilon > 0$. There are neighbourhoods $\mathcal{U}$, $\mathcal{V}$ of u_0 and v_0 in $\mathcal{L}(E)^{\#}_{E}$ and $\mathcal{L}(E)_{E}$, respectively, such that

$$\|uv_0x - u_0v_0x\| < \frac{\varepsilon}{2}\,, \qquad \|vx - v_0x\| < \frac{\varepsilon}{2}$$

for all $u \in \mathcal{U}$, $v \in \mathcal{V}$, and $x \in A$. It follows

$$\|uvx - u_0v_0x\| \leq \|uvx - uv_0x\| + \|uv_0x - u_0v_0x\| <$$

$$< \|vx - v_0x\| + \frac{\varepsilon}{2} < \frac{\varepsilon}{2} + \frac{\varepsilon}{2} = \varepsilon$$

for all $u \in \mathcal{U}$, $v \in \mathcal{V}$, and $x \in A$. Hence the above map is continuous. ■

1.2.7 Convex Sets

Definition 1.2.7.1 (0) *The subset A of the vector space E is called* ***convex*** *(resp.* ***absolutely convex****) if*

$$\alpha A + \beta A \subset A$$

for every $\alpha, \beta \in \mathbb{R}_+$ (resp. $\alpha, \beta \in \mathbb{K}$) with

$$\alpha + \beta = 1 \quad (\textit{resp.}\ |\alpha| + |\beta| \leq 1)\,.$$

It is easy to see that a subset A of a vector space is absolutely convex iff it is convex and

$$\alpha A \subset A$$

for every $\alpha \in \mathbb{K}$ with $|\alpha| \leq 1$.

Proposition 1.2.7.2 (0) *If E is a normed space, then*

$$\{x \in E \mid \|x\| < \alpha\} \quad \textit{and} \quad \{x \in E \mid \|x\| \leq \alpha\}$$

are absolutely convex for every $\alpha \in \mathbb{R}_+$.

Let $x, y \in E$ and $\alpha, \beta \in \mathbb{K}$ with

$$|\alpha| + |\beta| \leq 1\,.$$

Then

$$\|\alpha x + \beta x\| \leq |\alpha|\, \|x\| + |\beta|\, \|y\| \leq \sup\{\|x\|, \|y\|\}\,.$$

The assertion now follows. ■

Proposition 1.2.7.3 (0) *Let E be a vector space and A, B be convex (resp. absolutely convex) sets of E. Then $\alpha A + \beta B$ is a convex (resp. absolutely convex) set of E for every $\alpha, \beta \in \mathbb{K}$.*

Take $\gamma, \delta \in \mathbb{R}_+$ (resp. $\gamma, \delta \in \mathbb{K}$) with

$$\gamma + \delta = 1 \quad (\text{resp. } |\gamma| + |\delta| \leq 1)\,.$$

Then

$$\gamma(\alpha A + \beta B) + \delta(\alpha A + \beta B) = \alpha(\gamma A + \delta A) + \beta(\gamma B + \delta B) \subset \alpha A + \beta B\,.$$

Hence $\alpha A + \beta B$ is convex (resp. absolutey convex). ■

Proposition 1.2.7.4 (0) *Let E be a vector space and $(A_\iota)_{\iota\in I}$ a nonempty family of convex (resp. absolutely convex) sets of E. Then $\bigcap_{\iota\in I} A_\iota$ is convex (resp. absolutely convex). Given any subset A of E, there is a smallest convex (resp. absolutely convex) set of E containing A; it is equal to*

$$\Big\{\sum_{\iota\in I}\alpha_\iota x_\iota \mid \big((\alpha_\iota, x_\iota)\big)_{\iota\in I} \text{ finite family in } \mathbb{R}_+ \times A,\ \sum_{\iota\in I}\alpha_\iota = 1\Big\}$$

$$\Big(\text{resp. } \Big\{\sum_{\iota\in I}\alpha_\iota x_\iota \mid \big((\alpha_\iota, x_\iota)\big)_{\iota\in I} \text{ finite family in } \mathbb{K} \times A,\ \sum_{\iota\in I}|\alpha_\iota| \le 1\Big\}\Big),$$

*and it is called **the convex** (resp. **absolutely convex**) **hull of** A.*

Take $x, y \in \bigcap_{\iota\in I} A_\iota$ and $\alpha, \beta \in \mathbb{R}_+$ (resp. $\alpha, \beta \in \mathbb{K}$) with

$$\alpha + \beta = 1 \quad (\text{resp. } |\alpha| + |\beta| \le 1).$$

Then $x, y \in A_\iota$ and so

$$\alpha x + \beta y \in A_\iota$$

for every $\iota \in I$. Hence

$$\alpha x + \beta y \in \bigcap_{\iota\in I} A_\iota,$$

i.e. $\bigcap_{\iota\in I} A_\iota$ is convex (resp. absolutely convex).

The latter assertion is easy to see. ■

Proposition 1.2.7.5 (0) *The closure of a convex (resp. absolutely convex) set of a normed space is convex (resp. absolutely convex).*

Let E be a normed space and A a convex (resp. absolutely convex) set of E. Take $x, y \in \overline{A}$ and $\alpha, \beta \in \mathbb{R}_+$ (resp. $\alpha, \beta \in \mathbb{K}$) with

$$\alpha + \beta = 1 \quad (\text{resp. } |\alpha| + |\beta| \le 1).$$

There are sequences $(x_n)_{n\in\mathbb{N}}, (y_n)_{n\in\mathbb{N}}$ in A converging to x and y, respectively. Then $\alpha x_n + \beta y_n \in A$ for every $n \in \mathbb{N}$, so that

$$\alpha x + \beta y = \lim_{n\to\infty}(\alpha x_n + \beta y_n) \in \overline{A}.$$

Hence $\overline{A}$ is convex (resp. absolutely convex). ■

Corollary 1.2.7.6 (0) *Let B be the convex (resp. absolutely convex) hull of the subset A of the normed space E. Then $\overline{B}$ is the smallest convex (resp. absolutely convex) closed set of E containing A, and it is called* ***the convex** (resp.* ***absolutely convex) closed hull of** A.*

The corollary follows immediately from Proposition 1.2.7.5. ■

Proposition 1.2.7.7 (0) *Let E, F be vector spaces and $u : E \to F$ a linear map. Let A be a convex (resp. absolutely convex) set of E and B a convex (resp. absolutely convex) set of F. Then $u(A)$ is a convex (resp. absolutely convex) set of F and $\overset{-1}{u}(B)$ is a convex (resp. absolutely convex) set of E.*

Take $\alpha, \beta \in \mathbb{R}_+$ (resp. $\alpha, \beta \in \mathbb{K}$) with

$$\alpha + \beta = 1 \quad (\text{resp. } |\alpha| + |\beta| \leq 1).$$

Then

$$\alpha u(A) + \beta u(A) \subset u(\alpha A + \beta A) \subset u(A).$$

$$\alpha \overset{-1}{u}(B) + \beta \overset{-1}{u}(B) \subset \overset{-1}{u}(\alpha B + \beta B) \subset \overset{-1}{u}(B).$$

Hence $u(A)$ and $\overset{-1}{u}(B)$ are convex (resp. absolutely convex). ■

Proposition 1.2.7.8 (0) *Let E be a Banach space and $(x_n)_{n\in\mathbb{N}}$ a sequence in E converging to 0. Then the convex (resp. absolutely convex) closed hull of $\{x_n \mid n \in \mathbb{N}\}$ is compact.*

Given $x \in \ell^1$, $(x(n)x_n)_{n\in\mathbb{N}}$ is absolutely summable (and hence summable) in E (Corollary 1.1.6.10 a $\Rightarrow$ c)). Define

$$u : \ell^1 \longrightarrow E, \, x \longmapsto \sum_{n\in\mathbb{N}} x(n)x_n .$$

$u(\ell^{1\#})$ contains $\{x_n \mid n \in \mathbb{N}\}$. Since $\ell^{1\#}$ is absolutely convex (Proposition 1.2.7.2), $u(\ell^{1\#})$ is also absolutely convex (Proposition 1.2.7.7). We show that $u(\ell^{1\#})$ is compact. Let $\mathfrak{F}$ be an ultrafilter on $\ell^{1\#}$. Put

$$x : \mathbb{N} \longrightarrow \mathbb{K}, \quad n \longmapsto \lim_{y,\mathfrak{F}} y(n).$$

Then $x \in \ell^{1\#}$. Take $\varepsilon > 0$. There is an $m \in \mathbb{N}$ with $\|x_n\| < \frac{\varepsilon}{4}$ for every $n \in \mathbb{N}$, $n > m$. Given $n \in \mathbb{N}_m$, there is an $A_n \in \mathfrak{F}$ with

$$|x(n) - y(n)| < \frac{\varepsilon}{2m(\|x_n\| + 1)}$$

whenever $y \in A_n$. Then

$$\|ux - uy\| = \|\sum_{n\in\mathbb{N}}(x(n) - y(n))x_n\| \leq$$

$$\leq \sum_{n=1}^{m} |x(n) - y(n)| \, \|x_n\| + \sum_{n=m+1}^{\infty} |x(n) - y(n)| \, \|x_n\| < \frac{\varepsilon}{2} + \frac{\varepsilon}{2} = \varepsilon$$

for every $y \in \bigcap_{n=1}^{m} A_n$ (Corollary 1.1.6.10). Hence $u(\mathfrak{F})$ converges to ux, so that $u(\ell^{1\#})$ is compact.

Let A be an absolutely convex closed set of E containing $\{x_n \mid n \in \mathbb{N}\}$ and take $x \in \ell^{1\#}$. Then

$$\sum_{n=1}^{m} x(n)x_n \in A$$

for every $m \in \mathbb{N}$. Thus $ux \in A$. Hence $u(\ell^{1\#}) \subset A$ and $u(\ell^{1\#})$ is the absolutely convex closed hull of $\{x_n \mid n \in \mathbb{N}\}$.

The convex closed hull of $\{x_n \mid n \in \mathbb{N}\}$ is a closed subset of $u(\ell^{1\#})$ and is therefore compact. ■

Definition 1.2.7.9 (0) *Let E be a vector space and A a convex set of E. A* **face of** *A is a nonempty convex subset B of A with the property that*

$$x, y \in A, \; \alpha \in]0,1[, \; \alpha x + (1-\alpha)y \in B \Longrightarrow x, y \in B.$$

An **extreme point of** *A is an element x of A such that $\{x\}$ is a face of A.*

A is, of course, a face of A and the face of a face of A is a face of A. In particular, if B is a face of A, then $x \in B$ is an extreme point of B iff it is an extreme point of A.

Proposition 1.2.7.10 (0) *Let E, F be vector spaces and $u : E \to F$ a linear map. Let A be a convex set of E and B a face of $u(A)$. Then $A \cap \overset{-1}{u}(B)$ is a face of A.*

By Proposition 1.2.7.4 and Proposition 1.2.7.7, $u(A)$ and $A \cap \overset{-1}{u}(B)$ are convex. Take $x, y \in A$ and $\alpha \in]0,1[$, with

$$\alpha x + (1-\alpha)y \in A \cap \overset{-1}{u}(B).$$

Then $ux, uy \in u(A)$ and

$$\alpha ux + (1-\alpha)uy \in B\,.$$

Hence $ux, uy \in B$, i.e. $x, y \in \overset{-1}{u}(B)$. Thus $A \cap \overset{-1}{u}(B)$ is a face of A. ∎

Proposition 1.2.7.11 (0) *If $E \neq \{0\}$ is a normed space and x an extreme point of $E^{\#}$, then $\|x\| = 1$.*

If $\|x\| = 0$, then

$$x = \frac{1}{2}y + \frac{1}{2}(-y)$$

for every $y \in E^{\#}$, so that x cannot be an extreme point of $E^{\#}$. If $0 < \|x\| < 1$ then

$$\frac{1}{\|x\|}x\,,\ 0 \in E^{\#}\,,$$

$$\|x\|\left(\frac{1}{\|x\|}x\right) + (1-\|x\|)\,0 = x\,,$$

so that x is not an extreme point of $E^{\#}$. Hence $\|x\| = 1$. ∎

Example 1.2.7.12 *Let T be a nonempty set. $x \in \ell^1(T)^{\#}$ is an extreme point of $\ell^1(T)^{\#}$ iff $|x(t)| = 1$ for some $t \in T$.*

First assume that $|x(t)| = 1$ for some $t \in T$. Take $y, z \in \ell^1(T)^{\#}$ and $\alpha \in]0,1[$ with

$$\alpha y + (1-\alpha)z = x\,.$$

Then

$$\alpha y(t) + (1-\alpha)z(t) = x(t)$$

so that

$$y(t) = z(t) = x(t)\,.$$

Since x, y, and z all vanish on $T\backslash\{t\}$,

$$y = z = x\,.$$

Thus x is an extreme point of $\ell^1(T)^{\#}$.

Now assume that x is an extreme point of $\ell^1(T)^\#$. Assume further that $|x(t)| \neq 1$ for every $t \in T$. By Proposition 1.2.7.11, $\|x\|_1 = 1$. Hence there are distinct elements $s, t \in T$ with

$$x(s) \neq 0\,, \quad x(t) \neq 0\,.$$

Define

$$y : T \longrightarrow \mathbb{K}\,, \quad r \longmapsto \begin{cases} x(s) \quad \left(1 + \frac{|x(t)|}{|x(s)|}\right) & \text{if } r = s \\ 0 & \text{if } r = t \\ x(r) & \text{if } r = T \backslash \{s, t\}\,, \end{cases}$$

$$z : T \longrightarrow \mathbb{K}\,, \quad r \longmapsto \begin{cases} 0 & \text{if } r = s \\ x(t) \quad \left(1 + \frac{|x(s)|}{|x(t)|}\right) & \text{if } r = t \\ x(r) & \text{if } r = T \backslash \{s, t\}\,. \end{cases}$$

Then $y, z \in \ell^1(T)^\#$ and

$$\frac{|x(s)|}{|x(s)| + |x(t)|} y + \frac{|x(t)|}{|x(s)| + |x(t)|} z = x\,.$$

This contradicts the hypothesis that x is an extreme point of $\ell^1(T)^\#$. Hence $|x(t)| = 1$ for some $t \in T$. ■

The above result admits a generalization.

Example 1.2.7.13 *Let μ be a Radon measure on the Hausdorff space T. Then $x \in L^1(\mu)^\#$ is an extreme point of $L^1(\mu)^\#$ iff*

$$|x(t)\mu(\{t\})| = 1 \qquad \textit{for some} \quad t \in T\,.$$

First assume that

$$|x(t)\mu(\{t\})| = 1 \qquad \text{for some} \quad t \in T\,.$$

Take $y, z \in L^1(\mu)^\#$ and $\alpha \in]0, 1[$ with

$$\alpha y + (1 - \alpha) z = x\,.$$

Then

$$\alpha y(t) + (1 - \alpha) z(t) = x(t)\,,$$

so that

$$y(t) = z(t) = x(t) .$$

Since x, y, and z all vanish μ–a.e. on $T \backslash (t)$,

$$y = z = x .$$

Thus x is an extreme point of $L^1(\mu)^\#$.

Now assume that x is an extreme point of $L^1(\mu)^\#$. Assume further that

$$|x(t)\mu(\{t\})| \neq 1$$

for every $t \in T$. By Proposition 1.2.7.11, $\|x\| = 1$. Hence there are two disjoint compact sets K, L of T with

$$\alpha := \int_K |x|\, d|\mu| \neq 0 , \quad \beta := \int_L |x|\, d|\mu| \neq 0 .$$

Define

$$y := \frac{\alpha + \beta}{\alpha} x e_K^T + x e_{T\backslash(K \cup L)}^T ,$$

$$z := \frac{\alpha + \beta}{\beta} x e_L^T + x e_{T\backslash(K \cup L)}^T .$$

Then $y, z \in L^1(\mu)^\#$ and

$$\frac{\alpha}{\alpha + \beta} y + \frac{\beta}{\alpha + \beta} z = x .$$

This contradicts the hypothesis that x is an extreme point of $L^1(\mu)^\#$. Hence

$$|x(t)\mu(\{t\})| = 1$$

for some $t \in T$. ■

Example 1.2.7.14 *Let T be a Hausdorff space, $\mathcal{M}_b(T)$ the Banach space of bounded Radon measures on T (Example 1.1.2.26), and take $\mu \in \mathcal{M}_b(T)^\#$. Then μ is an extreme point of $\mathcal{M}_b(T)^\#$ iff there is a $t \in T$ such that*

$$\operatorname{Supp} \mu = \{t\} , \quad |\mu(\{t\})| = 1 .$$

$$\mathcal{M} := \{\nu \in \mathcal{M}_b(T) \mid \nu \textit{ is positive and } \nu(T) = 1\}$$

is a face of $\mathcal{M}_b(T)^\#$ and

$$\{\delta_t \mid t \in T\}$$

is the set of extreme points of $\mathcal{M}$ *, where* δ_t *denotes* **the Dirac measure on** T **at the point** $t \in T$ *(i.e.*

$$\delta_t(A) = e_A(t)$$

for every Borel set A *of* T *).*

Assume that μ is an extreme point $\mathcal{M}_b(T)^\#$. Assume further that the support of μ contains at least two points. Then there are disjoint Borel sets A and B of T such that

$$A \cup B = \operatorname{Supp} \mu, \quad |\mu|(A) \neq 0, \quad |\mu|(B) \neq 0.$$

Define

$$\mu_A := \frac{1}{|\mu|(A)} e_A \cdot \mu, \quad \mu_B := \frac{1}{|\mu|(B)} e_B \cdot \mu.$$

Then $\mu_A, \mu_B \in \mathcal{M}(T)^\#$,

$$|\mu|(A) + |\mu|(B) \leq 1, \quad \mu = |\mu|(A)\mu_A + |\mu|(B)\mu_B,$$

contradicting the hypothesis that μ is an extreme point. Hence there is a $t \in T$ with

$$\operatorname{Supp} \mu = \{t\}, \quad |\mu(\{t\})| = 1$$

(Proposition 1.2.7.11).

The converse is easy to see. The last assertion follows from the first one. ■

Example 1.2.7.15 (0) *Let* T *be a completely regular space. Then* $x \in \mathcal{C}(T)$ *is an extreme point of* $\mathcal{C}(T)^\#$ *iff* $|x| = e_T$.

Take $x \in \mathcal{C}(T)^\#$ and assume that there is a $t \in T$ with $|x(t)| \neq 1$. Let U be a neighbourhood of t such that

$$\alpha := \sup_{s \in U} |x(s)| < 1.$$

Take $y \in \mathcal{C}(T)$ with

$$\{y \neq 0\} \subset U, \quad \|y\|_\infty = 1 - \alpha.$$

Then

$$x \pm y \in \mathcal{C}(T)^{\#},$$

$$\frac{1}{2}(x+y) + \frac{1}{2}(x-y) = x.$$

Thus x is not an extremal element of $\mathcal{C}(T)^{\#}$.

Now assume that $|x| = e_T$. Take $y, z \in \mathcal{C}(T)^{\#}$ and $\alpha \in]0,1[$ with

$$\alpha y + (1-\alpha)z = x.$$

Then

$$\alpha y(t) + (1-\alpha)z(t) = x(t),$$

so that

$$y(t) = z(t) = x(t)$$

for every $t \in T$. Hence

$$y = z = x$$

and x is thus an extreme point of $\mathcal{C}(T)^{\#}$. ■

Example 1.2.7.16 *Let T be a set. Then $x \in \ell^\infty(T)$ is an extreme point of $\ell^\infty(T)$ iff $|x| = e_T$.*

This assertion follows immediately from Example 1.2.7.15. ■

Example 1.2.7.17 *Let T be a locally compact space. Then x of $\mathcal{C}_0(T)$ is an extreme point of $\mathcal{C}_0(T)^{\#}$ iff $|x| = e_T$. In particular, $\mathcal{C}_0(T)^{\#}$ admits extreme points iff T is compact.*

Take first $x \in \mathcal{C}_0(T)^{\#}$ and assume that there is a $t \in T$ with $|x(t)| \neq 1$. Let K be a compact neighbourhood of T, so that

$$\alpha := \sup_{s \in K} |x(s)| < 1.$$

Take $y \in \mathcal{C}_0(T)^{\#}$ with

$$\{y \neq 0\} \subset K, \quad \|y\|_\infty = 1 - \alpha.$$

Then

$$x \pm y \in \mathcal{C}_0(T)^{\#} ,$$

$$\frac{1}{2}(x+y) + \frac{1}{2}(x-y) = x .$$

Thus x is not an extreme point of $\mathcal{C}_0(T)^{\#}$.

Now suppose that $|x| = e_T$. Then T is compact and, by Example 1.2.7.15, x is an extreme point of $\mathcal{C}_0(T)^{\#}$. ■

Corollary 1.2.7.18 *Let T be a compact space and S a locally compact non–compact space. Then $\mathcal{C}(T)$ and $\mathcal{C}_0(S)$ are not isometric.* ■

1.2.8 The Alaoglu–Bourbaki Theorem

Theorem 1.2.8.1 (0) (Alaoglu 1940, Bourbaki 1938) *Let E be a normed space. Then $E_E'^{\#}$ (i.e. the unit ball of E' endowed with the topology of pointwise convergence) is compact. If E is separable then $E_E'^{\#}$ is metrizable.*

Let $\mathfrak{F}$ be an ultrafilter on $E'^{\#}$. Define

$$y' : E \longrightarrow \mathbb{K}, \quad x \longmapsto \lim_{x',\mathfrak{F}} x'(x);$$

y' is linear and

$$|y'(x)| = \lim_{x',\mathfrak{F}} |x'(x)| \le \|x\|$$

for every $x \in E$ (Proposition 1.2.1.4 a)). Hence y' is continuous and $\|y'\| \le 1$ (Proposition 1.2.1.1 d $\Rightarrow$ a), i.e. $y' \in E'^{\#}$. Since $\mathfrak{F}$ converges to y' in the topology of pointwise convergence, $E_E'^{\#}$ is compact.

Now suppose that E is separable. Let A be a countable dense set of E. Since $E'^{\#}$ is equicontinuous, the topology of pointwise convergence in A coincides with the topology of pointwise convergence in E, i.e. $E_E'^{\#} = E_A'^{\#}$ (Proposition 1.1.2.15). But $E_A'^{\#}$ is metrizable (since A is countable). Hence $E_E'^{\#}$ is metrizable. ■

Theorem 1.2.8.2 (0) (Banach–Dieudonné) *Let E be a normed space and $\mathfrak{A}$ the set of subsets A of E, such that*

$$\{x \in A \mid \|x\| \ge \varepsilon\}$$

is finite for every $\varepsilon > 0$. Then of the topologies on E', the topology of uniform convergence on the sets of $\mathfrak{A}$ is the finest one to induce the topology of pointwise convergence on the equicontinuous sets of E'.

Put

$$\mathfrak{T}' := \left\{ U' \subset E' \;\middle|\; \begin{array}{l} \text{given an equicontinuous set } A' \text{ of } E', \\ A' \cap U' \text{ is an open subset of } A'_E \end{array} \right\}.$$

It is easy to see that $\mathfrak{T}'$ is the finest tology on E' inducing the topology of pointwise convergence on the equicontinuous sets of E'. Let $\mathfrak{S}'$ be the topology (on E') of uniform convergence on the sets of $\mathfrak{A}$. Since the sets of $\mathfrak{A}$ are relatively compact, $\mathfrak{S}'$ induces the topology of pointwise convergence on the equicontinuous sets of E (Proposition 1.1.2.15). Hence $\mathfrak{S}' \subset \mathfrak{T}'$.

Take $U' \in \mathfrak{T}'$ and $x' \in U'$. Let $n \in \mathbb{N}$. Given $A \subset E$, put

$$\widetilde{A} := \{y' \in nE'^{\#} \mid x \in A \Longrightarrow |\langle x, y' - x'\rangle| \leq 1\}.$$

Then

$$\left\{\widetilde{A}\middle| A \in \mathfrak{P}_f\left(\frac{1}{n}E^{\#}\right)\right\}$$

is a downward directed set of closed sets of $nE'^{\#}_E$, the intersection of which is $\{x'\} \cap nE'^{\#}$. Since $U' \cap nE'^{\#}$ is an open set of $nE'^{\#}_E$ and since $nE'^{\#}_E$ is compact (Theorem 1.2.8.1), there is an $A_n \in \mathfrak{P}_f(\frac{1}{n}E^{\#})$ with

$$\widetilde{A}_n \subset U'.$$

Define

$$A := \bigcup_{n \in \mathbb{N}} A_n.$$

Then $A \in \mathfrak{A}$ and

$$x' \in \{y' \in E' \mid x \in A \Longrightarrow |\langle x, y' - x'\rangle| \leq 1\} \subset U'.$$

Hence U' is a neighbourhood of x' with respect to $\mathfrak{S}$. We deduce that $\mathfrak{T}' \subset \mathfrak{S}$. Thus $\mathfrak{S}' = \mathfrak{T}'$. ■

Corollary 1.2.8.3 (0) *Let E be a Banach space and $\mathfrak{K}$ the set of compact convex sets of E. Then the topology on E' of uniform convergence on the sets in $\mathfrak{K}$ is the finest one to induce the topology of pointwise convergence on the equicontinuous sets of E'.*

Endow E' with $\mathfrak{S}'$, the topology of uniform convergence on the sets of $\mathfrak{K}$ and let $\mathfrak{T}'$ be the finest topology on E' inducing the topology of pointwise convergence on the equicontinuous sets of E' (Theorem 1.2.8.2). We have to prove that $\mathfrak{S}' = \mathfrak{T}'$. $\mathfrak{S}' \subset \mathfrak{T}'$ follows from Proposition 1.1.2.15. By Theorem 1.2.8.2, to prove the reverse inclusion it suffices to show that, given a sequence $(x_n)_{n \in \mathbb{N}}$ in E converging to 0, there is a $K \in \mathfrak{K}$ containing $\{x_n \mid n \in \mathbb{N}\}$. But this was proved in Proposition 1.2.7.8. ■

1.2.9 Bilinear Maps

Definition 1.2.9.1 (0) *Let E, F, G be vector spaces. A map $u : E \times F \to G$ is called* ***bilinear*** *if it is linear in each variable (i.e. $u(x, \cdot)$ and $u(\cdot, y)$ are linear for every $x \in E$ and $y \in F$).*

Proposition 1.2.9.2 (0) *Let E, F, G be normed spaces and $u : E \times F \to G$ a bilinear map. Then the following are equivalent:*

a) *u is continuous.*

b) *u is continuous at $(0, 0)$.*

c) *There is an $\alpha \in \mathbb{R}_+$ such that*

$$\|u(x,y)\| \leq \alpha \|x\| \, \|y\|$$

for every $(x, y) \in E \times F$.

d) *The restriction of u to $E^\# \times F^\#$ is uniformly continuous.*

d $\Rightarrow$ a $\Rightarrow$ b is trivial.
b $\Rightarrow$ c . There is a $\delta > 0$ so that

$$\|u(x,y)\| \leq 1$$

whenever $(x, y) \in E \times F$ satisfies

$$\|(x,y)\|_\infty \leq \delta \, .$$

Put

$$\alpha := \frac{1}{\delta^2} \, .$$

Take $(x, y) \in E \times F$ with $x \neq 0$, $y \neq 0$. Then

$$\left\| \left(\frac{\delta}{\|x\|} x \, , \, \frac{\delta}{\|y\|} y \right) \right\|_\infty =$$

$$= \sup \left\{ \left\| \frac{\delta}{\|x\|} x \right\| , \left\| \frac{\delta}{\|y\|} y \right\| \right\} = \delta \, ,$$

so

$$\frac{\delta^2}{\|x\| \, \|y\|} \|u(x,y)\| = \left\| u \left(\frac{\delta}{\|x\|} x \, , \, \frac{\delta}{\|y\|} y \right) \right\| \leq 1 \, .$$

$$\|u(x,y)\| \le \frac{1}{\delta^2}\|x\|\,\|y\| = \alpha\|x\|\,\|y\|\,.$$

c $\Rightarrow$ d. Given $(x_1, y_1)\,,\ (x_2, y_2) \in E^\# \times F^\#$,

$$\|u(x_1,y_1) - u(x_2,y_2)\| \le \|u(x_1,y_1) - u(x_1,y_2)\| + \|u(x_1,y_2) - u(x_2,y_2)\| \le$$

$$\le \|u(x_1, y_1 - y_2)\| + \|u(x_1 - x_2, y_2)\| \le$$

$$\le \alpha\|x_1\|\,\|y_1 - y_2\| + \alpha\|x_1 - x_2\|\,\|y_2\| \le$$

$$\le \alpha\|y_1 - y_2\| + \alpha\|x_1 - x_2\| \le 2\|(x_1,y_1) - (x_2,y_2)\|_1\,.$$ ■

Corollary 1.2.9.3 (0) *Given normed spaces* E, F, G,

$$\mathcal{L}(E,F) \times \mathcal{L}(F,G) \longrightarrow \mathcal{L}(E,G)\,, \quad (u,v) \longmapsto v \circ u$$

is bilinear and continuous.

It is easy to see that the map is bilinear. By Corollary 1.2.1.5 and Proposition 1.2.9.2 c $\Rightarrow$ a, it is also continuous. ■

Corollary 1.2.9.4 (0) *If* E, F *are normed spaces, then the map*

$$E \times \mathcal{L}(E,F) \longrightarrow F\,, \quad (x,u) \longmapsto ux$$

is bilinear and continuous.

The assertion follows from Proposition 1.2.1.4 a) and Proposition 1.2.9.2 c $\Rightarrow$ a. ■

Corollary 1.2.9.5 (0) *Given a normed space* E,

$$E \times E' \longrightarrow \mathbb{K}\,, \quad (x,x') \longmapsto \langle x, x'\rangle$$

$$\mathbb{K} \times E \longrightarrow E\,, \quad (\alpha, x) \longmapsto \alpha x$$

are bilinear and continuous. ■

Proposition 1.2.9.6 (0) *Let* E, F, G *be Banach spaces, and* $u : E \times F \to G$ *a continuous bilinear map. Let* $(x_n)_{n\in\mathbb{N}\cup\{0\}}$, $(y_n)_{n\in\mathbb{N}\cup\{0\}}$ *be absolutely summable families in* E *and* F, *respectively. Given* $n \in \mathbb{N}$, *put*

$$z_n := \sum_{k=0}^{n} u(x_k, y_{n-k})$$

for every $n \in \mathbb{N}\cup\{0\}$. *Then* $(z_n)_{n\in\mathbb{N}\cup\{0\}}$ *is absolutely summable and*

$$\sum_{n\in\mathbb{N}\cup\{0\}} z_n = u\Big(\sum_{n\in\mathbb{N}\cup\{0\}} x_n\,,\ \sum_{n\in\mathbb{N}\cup\{0\}} y_n\Big)\,.$$

Given $p \in \mathbb{N}$, define

$$a_p := u\left(\sum_{n=0}^{p} x_n\,,\ \sum_{n=0}^{p} y_n\right),$$

$$b_p := \left(\sum_{n=0}^{p} \|x_n\|\right)\left(\sum_{n=0}^{p} \|y_n\|\right).$$

Then

$$\lim_{p\to\infty} a_p = u\Big(\sum_{n\in\mathbb{N}\cup\{0\}} x_n\,,\ \sum_{n\in\mathbb{N}\cup\{0\}} y_n\Big)$$

and $(b_p)_{p\in\mathbb{N}}$ is a convergent sequence.

By Proposition 1.2.9.2 a $\Rightarrow$ c, there is an $\alpha \in \mathbb{R}_+$ such that

$$\|u(x,y)\| \leq \alpha\|x\|\,\|y\|$$

for every $(x,y) \in E \times F$. Take $\varepsilon > 0$. There is an $m \in \mathbb{N}$ such that

$$b_p - b_m < \frac{\varepsilon}{1+\alpha}$$

for every $p \in \mathbb{N}$ with $p \geq m$. Take $p \in \mathbb{N}$, with $p \geq 2m$ and let

$$A := \Big\{(i,j) \in (\mathbb{N}_p \cup \{0\}) \times (\mathbb{N}_p \cup \{0\}) \mid m < \sup\{i,j\}\Big\},$$
$$B := \Big\{(i,j) \in A \mid i+j > p\Big\}.$$

Then

$$\left\|\sum_{n=0}^{p} z_n - a_p\right\| = \left\|\sum_{(i,j)\in B} u(x_i,y_j)\right\| \leq \sum_{(i,j)\in B} \alpha\|x_i\|\,\|y_j\| \leq$$

$$\leq \alpha \sum_{(i,j)\in A} \|x_i\|\,\|y_j\| = \alpha(b_p - b_m) < \varepsilon\,.$$

Thus

$$\lim_{p\to\infty}\Big(\sum_{n=0}^{p} z_n - a_p\Big) = 0\,,$$

$$u\Big(\sum_{n\in\mathbb{N}\cup\{0\}} x_n\,,\ \sum_{n\in\mathbb{N}\cup\{0\}} y_n\Big) = \lim_{p\to\infty} a_p = \lim_{p\to\infty}\sum_{n=0}^{p} z_n\,.$$

Since

$$\|z_n\| \leq \alpha\sum_{k=0}^{n} \|x_k\|\,\|y_{n-k}\|$$

for every $n \in \mathbb{N}$, it follows that $(z_n)_{n\in\mathbb{N}\cup\{0\}}$ is absolutely summable and that

$$\sum_{n\in\mathbb{N}\cup\{0\}} z_n = u\Big(\sum_{n\in\mathbb{N}\cup\{0\}} x_n\,,\ \sum_{n\in\mathbb{N}\cup\{0\}} y_n\Big)\,. \qquad \blacksquare$$

Exercises

E 1.2.1 Let E, F be normed spaces and $u : E \to F$ be a linear map such that, if $(x_n)_{n\in\mathbb{N}}$ is a null sequence in E (i.e. $\lim\limits_{n\to\infty} x_n = 0$), then $(ux_n)_{n\in\mathbb{N}}$ is bounded. Show that u is continuous.

E 1.2.2 Let E, F be normed spaces and $u : E \to F$ a map, satisfying

1) $\sup\limits_{x\in E^\#} \|ux\| < \infty$.

2) $x, y \in E \Rightarrow u(x+y) = ux + uy$.

3) If $\mathbb{K} = \mathbb{C}$, then $u(ix) = iux$ for every $x \in E$.

Show that u is linear and continuous.

E 1.2.3 Let E be a normed space. For $x \in E$ define

$$u_x : \mathbb{K} \longrightarrow E, \quad \alpha \longmapsto \alpha x.$$

Prove the following:

a) Given $x \in E$, $u_x \in \mathcal{L}(\mathbb{K}, E)$ and $\|u_x\| = \|x\|$.

b) The map

$$E \longrightarrow \mathcal{L}(\mathbb{K}, E), \quad x \longmapsto u_x$$

is an isometry.

E 1.2.4 Let E, F be normed spaces, A a dense set of E, and $(u_n)_{n\in\mathbb{N}}$ a bounded sequence in $\mathcal{L}(E, F)$ such that $(u_n x)_{n\in\mathbb{N}}$ is a Cauchy sequence (resp. a null sequence) for each $x \in A$. Show that $(u_n x)_{n\in\mathbb{N}}$ is a Cauchy sequence (resp. a null sequence) for each $x \in E$.

E 1.2.5 Take $p \in [1, \infty[$, $q \in]1, \infty]$ with $\frac{1}{p} + \frac{1}{q} = 1$. Show that

$$\lim_{s\to 0} \int_{-\infty}^{\infty} x(s+t)y(t)dt = \int_{-\infty}^{\infty} x(t)y(t)dt$$

whenever $x \in \mathcal{L}^p(\mathbb{R})$, $y \in \mathcal{L}^q(\mathbb{R})$.

E 1.2.6 Let E, F be normed spaces, $(x', y') \in E' \times F'$, and

$$z' : E \times F \longrightarrow \mathbb{K}, \quad (x, y) \longmapsto x'(x) + x'(y).$$

Show that $z' \in (E \times F)'$ and

$$\|z'\| = \left(\|x'\|^2 + \|y'\|^2\right)^{\frac{1}{2}},$$

where $E \times F$ is endowed with the 2–norm.

E 1.2.7 Let T be a set. Take $x \in \ell^\infty(T)$, and define

$$\widetilde{x} : \ell^1(T) \longrightarrow \mathbb{K}, \quad y \longmapsto \sum_{t \in T} x(t)y(t).$$

Show that the following are equivalent (Example 1.2.2.3 b)):

a) There is a $y \in \ell^1(T)$ with $\|y\| = 1$ and $\widetilde{x}(y) = \|\widetilde{x}\|$.

b) There is a $t \in T$ with $|x(t)| = \|x\|_\infty$.

E 1.2.8 Let T be a set. Take $x \in \ell^1(T)$, and define

$$\widetilde{x} : c_0(T) \quad (\text{resp. } c(T)) \longrightarrow \mathbb{K}, \quad y \longmapsto \sum_{t \in T} x(t)y(t).$$

Show that the following are equivalent (Example 1.2.2.3 b)):

a) There is a $y \in c_0(T)$ (resp. $y \in c(T)$) with $\|y\| = 1$ and $|\widetilde{x}(y)| = \|\widetilde{x}\|$.

b) $x \in \mathbb{K}^{(T)}$ (resp. the map

$$T \setminus \overset{-1}{x}(0) \longrightarrow \mathbb{K}, \quad s \longmapsto \frac{x(s)}{|x(s)|}$$

is in $c(S)$).

E 1.2.9 Take $p, q, r \in [1, \infty]$ with $\frac{1}{p} + \frac{1}{q} \geq 1$ and $k : \mathbb{N} \times \mathbb{N} \rightarrow \mathbb{K}$ such that $k(m, \cdot) \in \ell^p$ for every $m \in \mathbb{N}$. Given $x \in \ell^q$, define

$$\widetilde{x} : \mathbb{N} \longrightarrow \mathbb{K}, \quad m \longmapsto \sum_{n \in \mathbb{N}} k(m, n)x_n.$$

Show that if $\widetilde{x} \in \ell^r$ whenever $x \in \ell^q$, then the map

$$\ell^q \longrightarrow \ell^r, \quad x \longmapsto \widetilde{x}$$

is continuous.

E 1.2.10 Take $k \in \mathcal{C}([0,1] \times [0,1])$. Given $x \in \mathcal{C}([0,1])$, define

$$\widetilde{k}x : [0,1] \longrightarrow \mathbb{K}, \quad s \longmapsto \int_0^1 k(s,t)x(t)dt .$$

Show the following:

a) $\widetilde{k}x \in \mathcal{C}([0,1])$ whenever $x \in \mathcal{C}([0,1])$.

b) The map

$$\widetilde{k} : \mathcal{C}([0,1]) \longrightarrow \mathcal{C}([0,1]), \quad x \longmapsto \widetilde{k}x$$

is linear and continuous and

$$\|\widetilde{k}\| = \sup_{s\in[0,1]} \int_0^1 |k(s,t)|dt .$$

E 1.2.11 Take $p \in [1,\infty]$. Given $x \in \mathbb{K}^{\mathbb{Z}}$, define

$$x_r : \mathbb{Z} \longrightarrow \mathbb{K}, \quad n \longmapsto x(n-1),$$

$$x_\ell : \mathbb{Z} \longrightarrow \mathbb{K}, \quad n \longmapsto x(n+1).$$

Show that:

a) $x \in \ell^p(\mathbb{Z}) \Rightarrow x_r, x_\ell \in \ell^p(\mathbb{Z})$, $\|x_r\|_p = \|x_\ell\|_p = \|x\|_p$.

b) $x \in c(\mathbb{Z})$ (resp. $c_0(\mathbb{Z})$) $\Rightarrow x_r, x_\ell \in c(\mathbb{Z})$ (resp. $c_0(\mathbb{Z})$).

c) The maps $\ell^p(\mathbb{Z}) \to \ell^p(\mathbb{Z})$, $c(\mathbb{Z}) \to c(\mathbb{Z})$, $c_0(\mathbb{Z}) \to c_0(\mathbb{Z})$ defined by $x \mapsto x_r$ (resp. x_ℓ) are isometries. They are called **the right** (resp. **left**) **shift of** $\ell^p(\mathbb{Z})$, $c(\mathbb{Z})$, and $c_0(\mathbb{Z})$, respectively.

E 1.2.12 Let $(\alpha_n)_{n\in\mathbb{N}}$ be a sequence in $\mathbb{K}$. Take $p,q \in]1,\infty[$ with $\frac{1}{p}+\frac{1}{q}=1$, and put

$$F := \left\{ x \in \ell^p \,\middle|\, \lim_{n\to\infty} \sum_{k=1}^{n} \alpha_k x_k = 0 \right\}.$$

Show that the following are equivalent:

a) $\sum\limits_{n=1}^{\infty} |\alpha_n|^q = 0$ or ∞.

b) F is dense in ℓ^p.

E 1.2.13 Let F be a closed vector subspace of the Banach space E, $q : E \to E/F$ the quotient map, and T a set. Take $u \in \mathcal{L}(\ell^1(T), E/F)$ and $\varepsilon > 0$. Show that there is a $v \in \mathcal{L}(\ell^1(T), E)$ such that

$$u = q \circ v, \quad \|v\| \leq (1+\varepsilon)\|u\|.$$

E 1.2.14 Let T be a metrizable topological space, S a closed set of T, and E a normed space. Prove the following:

a) There exists an operator

$$u : \mathcal{C}(S, E) \longrightarrow \mathcal{C}(T, E)$$

of norm 1 such that

$$(ux) \mid S = x$$

for every $x \in \mathcal{C}(S, E)$ (see e.g. Jun–Iti Nagata, Modern general topology, Theorem VII.14).

b) $\{x \in \mathcal{C}(T, E) \mid x = 0 \text{ on } S\}$ is a complemented subspace of $\mathcal{C}(T, E)$.

E 1.2.15 Let E be a normed space, T a completely regular space, S a closed set of T, A a countable subset of T such that $A \backslash U$ is finite for every neighbourhood U of S and put

$$\mathcal{F} := \{x \in \mathcal{C}(T, E) \mid x = 0 \text{ on } S\}.$$

Show that there is an operator

$$u : c_0(A, E) \longrightarrow \mathcal{F}$$

of norm 1 such that

$$(ux) \mid A = x$$

for every $x \in c_0(A, E)$.

E 1.2.16 Let E be a normed space, T a completely regular space, and K a compact set of T. For each closed set S of T let S' denote the set of $t \in T$ such that t is not a point of adherence of $S \backslash \{t\}$, and define K_ξ for each ordinal number ξ by means of transfinite induction as follows:

$$K_0 := K\,, \quad K_\xi := \Big(\bigcap_{\eta\in\xi} K_\eta\Big)'\,.$$

Assume that there is a countable ordinal number ξ with $K_\xi = \emptyset$. Show that there is an operator

$$u : \mathcal{C}(K,E) \longrightarrow \mathcal{C}(T,E)$$

such that

$$(ux) \mid K = x$$

for every $x \in \mathcal{C}(K,E)$ and that $\{x \in \mathcal{C}(T,E) \mid x = 0 \text{ on } K\}$ is a complemented subspace of $\mathcal{C}(T,E)$. (Hint: Use the preceding exercise.)

E 1.2.17 Let E be a vector space, A a convex (resp. absolutely convex) set of E, $(x_\iota)_{\iota\in I}$ a finite nonempty family in A, and $(\alpha_\iota)_{\iota\in I}$ a family in $\mathbb{R}_+$ (resp. in $\mathbb{K}$), such that

$$\sum_{\iota\in I} \alpha_\iota = 1 \quad \Big(\text{resp. } \sum_{\iota\in I} |\alpha_\iota| \le 1\Big)\,.$$

Prove the following:

a) $\sum\limits_{\iota\in I} \alpha_\iota x_\iota \in A$.

b) There is an $x \in A$ such that

$$\sum_{\iota\in I} x_\iota = (\operatorname{Card} I)\, x\,.$$

E 1.2.18 Let E be a (separable) normed space. Show that there exists a (metrizable) compact space T such that E is isometric to a subspace of $\mathcal{C}(T)$.

E 1.2.19 Let E be a separable normed space. Show that E is isometric to a subspace of ℓ^∞.

E 1.2.20 Let E, F be normed space with F finite–dimensional. Show that the closed unit ball in $\mathcal{L}(E,F)$ endowed with the topology of pointwise convergence is compact (and metrizable if E is separable) (generalization of the Theorem of Alaoglu–Bourbaki).

E 1.2.21 Let E, F, G be normed spaces. Define

$$\mathcal{B}(E, F; G) := \{u : E \times F \longrightarrow G \mid u \text{ is bilinear and continuous}\}$$

and

$$\|u\| := \inf\{\alpha \in \mathbb{R}_+ \mid (x, y) \in E \times F \Longrightarrow \|u(x, y)\| \leq \alpha \|x\| \, \|y\|\},$$

$$\widetilde{u} : E \longrightarrow \mathcal{L}(F, G), \quad x \longmapsto u(x, \cdot)$$

for every $u \in \mathcal{B}(E, F; G)$. Prove the following:

a) $\mathcal{B}(E, F; G)$ is a vector subspace of $G^{E \times F}$ and the map

$$\mathcal{B}(E, F; G) \longrightarrow \mathbb{R}_+, \quad u \longmapsto \|u\|$$

is a norm.

b) $\widetilde{u}$ is linear and continuous and $\|\widetilde{u}\| = \|u\|$ for every $u \in \mathcal{B}(E, F; G)$.

c) The map

$$\mathcal{B}(E, F; G) \longrightarrow \mathcal{L}(E, \mathcal{L}(F, G)), \quad u \longmapsto \widetilde{u}$$

is an isometry.

E 1.2.22 Let F be a closed vector subspace of a Banach space. Show that the set of projections of E on F is a convex set of $\mathcal{L}(E)$, closed with respect to the topology of pointwise convergence.

1.3 The Hahn–Banach Theorem

The Hahn–Banach Theorem is the most important result in the theory of normed spaces, without which the theory would lose all interest. It ensures that the dual space of a normed space contains sufficiently many vectors to allow every normed space to be isometrically imbedded — by means of the evaluation map — into its bidual. It also allows us to associate to each operator its transpose and bitranspose. The evaluation map enables us to define the most important class of Banach spaces: the reflexive ones.

1.3.1 The Banach Theorem

Lemma 1.3.1.1 (0) *Let E be a vector space and F, G vector subspaces of E with $F \cap G = \{0\}$. Put*

$$H := F + G .$$

Let x' and y' be linear forms on F and G, respectively. Then there is a unique linear form z' on H with

$$z'|F = x' , \quad z'|G = y' .$$

The uniqueness is obvious. Take $(x_1, y_1), (x_2, y_2) \in F \times G$ with

$$x_1 + y_1 = x_2 + y_2 .$$

Then

$$x_1 - x_2 = y_2 - y_1 \in F \cap G ,$$

so that

$$x_1 = x_2 , \quad y_1 = y_2 .$$

It follows that the map

$$z' : H \longrightarrow \mathbb{K} , \quad x + y \longmapsto x'(x) + y'(y)$$

is well–defined. z' has the required properties. ■

Theorem 1.3.1.2 (0) (Banach, 1929) *Let F be a vector subspace of the real vector space E. Let p be a real function on E and y' be linear form on F such that*

a) *$p(x+y) \leq p(x)+p(y)$ for every $x, y \in E$.*

b) *$p(\alpha x) = \alpha p(x)$ for every $x \in E$ and $\alpha \in \mathrm{I\!R}_+$.*

c) *$y'(y) \leq p(y)$ for every $y \in F$.*

Then there is a linear form x' on E such that $x'|F = y'$ and $x'(x) \leq p(x)$ for every $x \in E$.

Let Ω be the set of pairs (G, z') satisfying the following conditions:

1) G is a vector subspace of E containing F;

2) z' is a linear form on G with $z'|F = y'$;

3) $z'(z) \leq p(z)$ for every $z \in G$.

Define an order on Ω by

$$(G_1, z_1') \leq (G_2, z_2') :\Longleftrightarrow G_1 \subset G_2\,,\ z_2'|G_1 = z_1'\,.$$

It is easy to see that this order is inductive. By Zorn's Lemma Ω has a maximal element (G, z'). We show that $G = E$.

So suppose that $G \neq E$. Take $x \in E \backslash G$ and put

$$H := \{\alpha x + y | \alpha \in \mathrm{I\!R}\,,\ y \in G\}\,.$$

Then H is a vector subspace of E, which strictly contains G. Then

$$z'(y) + z'(z) = z'(y+z) \leq p(y+z) = p(y-x+z+x) \leq p(y-x) + p(z+x)$$

for every $y, z \in G$. It follows that

$$\sup_{y \in G}\left(z'(y) - p(y-x)\right) \leq \inf_{z \in G}(p(z+x) - z'(z))\,.$$

Take

$$\beta \in \left[\sup_{y \in G}\left(z'(y) - p(y-x)\right),\quad \inf_{z \in G}\left(p(z+x) - z'(z)\right)\right],$$

and let z_0' be the linear form on H with

$$z_0'|G = z'\,,\quad z_0'(x) = \beta$$

(Lemma 1.3.1.1). Take $y \in G$ and $\alpha > 0$. Then

$$z_0'(\alpha x + y) = \alpha\beta + z'(y) = \alpha\Big(\beta + z'\big(\frac{1}{\alpha}y\big)\Big) \leq$$

$$\leq \alpha\Big(p\big(\frac{1}{\alpha}y + x\big) - z'\big(\frac{1}{\alpha}y\big) + z'\big(\frac{1}{\alpha}y\big)\Big) = \alpha p\big(\frac{1}{\alpha}y + x\big) = p(\alpha x + y)\,,$$

$$z_0'(-\alpha x + y) = -\alpha\beta + z'(y) = \alpha\big(-\beta + z'\big(\frac{1}{\alpha}y\big)\big) \leq$$

$$\leq \alpha\Big(p\big(\frac{1}{\alpha}y - x\big) - z'\big(\frac{1}{\alpha}x\big) + z'\big(\frac{1}{\alpha}y\big)\Big) = \alpha p\big(\frac{1}{\alpha}y - x\big) = p(-\alpha x + y)\,.$$

Hence $(H, z_0') \in \Omega$ and

$$(G, z') < (H, z_0')\,,$$

which contradicts the maximality of (G, z'). ■

Example 1.3.1.3 *Consider $\mathbb{K} = \mathbb{R}$. There is a continuous linear form x' on ℓ^∞ such that*

$$\liminf_{n\to\infty} \frac{1}{n}\sum_{m=1}^{n} x_m \leq x'(x) \leq \limsup_{n\to\infty} \frac{1}{n}\sum_{m=1}^{n} x_m$$

for every $x \in \ell^\infty$. Then

$$\|x'\| = 1\,, \quad x'|c_0 = 0\,,$$

$$x \in \ell^\infty \Longrightarrow \liminf_{n\to\infty} x_n \leq x'(x) \leq \limsup_{n\to\infty} x_n\,.$$

Moreover,

$$x' \circ u_r = x' \circ u_\ell = x'$$

for every such x', where u_r (resp. u_ℓ) denotes the right (resp. left) shift on ℓ^∞.

Define

$$p : \ell^\infty \longrightarrow \mathbb{R}\,, \quad x \longmapsto \limsup_{n\to\infty} \frac{1}{n}\sum_{m=1}^{n} x_m\,,$$

$$F := \left\{ x \in \ell^\infty \;\middle|\; \left(\frac{1}{n}\sum_{m=1}^{n} x_m\right)_{n\in\mathbb{N}} \text{ converges} \right\},$$

$$y' : F \longrightarrow \mathbb{R}, \quad x \longmapsto \lim_{n\to\infty} \frac{1}{n} \sum_{m=1}^{n} x_m .$$

Then F is a vector subspace of ℓ^∞, y' is a linear form on F with $y' \le p\,|\,F$, and

$$p(x+y) \le p(x) + p(y), \quad p(\alpha x) = \alpha p(x)$$

for every $x, y \in \ell^\infty$ and $\alpha \in \mathbb{R}_+$. By Banach's Theorem, there is a linear form x' on ℓ^∞ with

$$x'|F = y', \quad x' \le p .$$

Thus

$$x'(x) = -x'(-x) \ge -p(-x) = \liminf_{n\to\infty} \frac{1}{n} \sum_{m=1}^{n} x_m$$

for every $x \in \ell^\infty$. The inqualities

$$\liminf_{n\to\infty} x_n \le \liminf_{n\to\infty} \frac{1}{n} \sum_{m=1}^{n} x_m \le x'(x) \le \limsup_{n\to\infty} \frac{1}{n} \sum_{m=1}^{n} x_m \le \limsup_{n\to\infty} x_n$$

are trivial for $x \in \ell^\infty$. We deduce that x' is continuous with norm 1 and vanishes on c_0.

Take $x \in \ell^\infty$ and put

$$y := u_\ell x \quad (\text{resp. } y := u_r x) .$$

Then

$$\frac{1}{n} \sum_{m=1}^{n} (x_m - y_m) = \frac{x_n}{n} \qquad (\text{resp. } \frac{x_1 - x_{n+1}}{n})$$

for every $n \in \mathbb{N}$. Hence

$$\lim_{n\to\infty} \frac{1}{n} \sum_{m=1}^{n} (x_m - y_m) = 0 .$$

Thus $x - y \in F$ and

$$x'(x-y) = y'(x-y) = 0, \quad x'(x) = x'(y) . \qquad \blacksquare$$

Remark. Let $\mathfrak{F}$ be a free ultrafilter on $\mathbb{N}$. Then

$$x' : \ell^\infty \longrightarrow \mathbb{R}, \quad x \longmapsto \lim_{n,\mathfrak{F}} \frac{1}{n} \sum_{m=1}^{n} x_m$$

has the properties required in the above example. This is another proof for the existence of x' which does not use Banach's Theorem. Of course, not all x' are of the above form.

Proposition 1.3.1.4 (0) *Let E be a vector space and A a convex (resp. absolutely convex) set of E such that for every $x \in E$ there is an $\alpha \in \mathbb{R}_+$ with $x \in \alpha A$. Define*

$$p : E \longrightarrow \mathbb{R}, \quad x \longmapsto \inf\{\alpha \in \mathbb{R}_+ | x \in \alpha A\}.$$

Then

$$p(x+y) \le p(x) + p(y), \quad p(\alpha x) = |\alpha| p(x)$$

for every $x, y \in E$ and $\alpha \in \mathbb{R}_+$ (resp. $\alpha \in \mathbb{K}$).

Take $\beta, \gamma \in \mathbb{R}_+ \setminus \{0\}$ with

$$x \in \beta A, \quad y \in \gamma A.$$

Then $\frac{1}{\beta}x$, $\frac{1}{\gamma}y \in A$, so that

$$\frac{1}{\beta+\gamma}(x+y) = \frac{\beta}{\beta+\gamma}\frac{1}{\beta}x + \frac{\gamma}{\beta+\gamma}\frac{1}{\gamma}y \in A.$$

Hence

$$x + y \in (\beta+\gamma)A, \quad p(x+y) \le \beta + \gamma.$$

Since β and γ are arbitrary,

$$p(x+y) \le p(x) + p(y).$$

The other assertion is trivial. ■

Lemma 1.3.1.5 (0) *Let E, F be complex vector spaces and $u : E \to F$ an $\mathbb{R}$–linear map. Then*

$$E \longrightarrow F, \quad x \longmapsto ux - iu(ix)$$

is $\mathbb{C}$–linear.

Define

$$v : E \longrightarrow F, \quad x \longmapsto ux - iu(ix).$$

v is obviously $\mathbb{R}$–linear and

$$v(ix) = u(ix) - iu(-x) = i(ux - iu(ix)) = ivx$$

for every $x \in E$. Hence

$$v((\alpha + i\beta)x) = v(\alpha x) + v(i\beta x) = \alpha vx + i\beta vx = (\alpha + i\beta)vx$$

for every $x \in E$ and $\alpha, \beta \in \mathbb{R}$, i.e. v is $\mathbb{C}$–linear. ■

Proposition 1.3.1.6 (0) *Let A be a nonempty convex set of the vector space E. Let B be an absolutely convex set of E such that for each $y \in E$ there is an $\alpha \in \mathbb{R}_+$ with $y \in \alpha B$. Take $x \in E \backslash (A+B)$. Then there is a linear form x' on E, bounded on B, such that*

$$\sup_{y \in A} \operatorname{re} x'(y) < \operatorname{re} x'(x).$$

Now $0 \in B$. We may assume that $0 \in A$ (otherwise we replace x and A by $x - a$ and $A - a$ for an $a \in A$). We put

$$C := A + \frac{1}{2} B,$$

$$p : E \longrightarrow \mathbb{R}, \quad y \longmapsto \inf\{\alpha \in \mathbb{R}_+ | y \in \alpha C\},$$

$$F := \{\alpha x \mid \alpha \in \mathbb{R}\},$$

$$y' : F \longrightarrow \mathbb{R}, \quad \alpha x \longmapsto \alpha p(x).$$

By Proposition 1.2.7.3, C is convex and by Proposition 1.3.1.4,

$$p(y + z) \leq p(y) + p(z), \quad p(\alpha y) = \alpha p(y)$$

for every $y, z \in \mathbb{R}_+$ and $\alpha \in \mathbb{R}_+$.

Take $\alpha, \beta \in \mathbb{R}_+$ with

$$x \in (\alpha B) \cap (\beta C).$$

Then $\alpha > 1$, $\beta > 1$. There is a pair $(a, b) \in A \times B$, with

$$\frac{1}{\beta} x = a + \frac{1}{2} b.$$

Then

$$x - \left(\left(1 - \frac{1}{\beta}\right) x + \frac{1}{2} b\right) = \frac{1}{\beta} x - \frac{1}{2} b = a \in A,$$

$$\left(1 - \frac{1}{\beta}\right) x + \frac{1}{2} b \notin B,$$

$$\left(1 - \frac{1}{\beta}\right) x \notin \frac{1}{2} B, \quad x \notin \frac{\beta}{2(\beta - 1)} B.$$

$$\frac{\beta}{2(\beta-1)} < \alpha\,, \quad \beta > \frac{2\alpha}{2\alpha-1}\,.$$

Since $x \notin C$, it follows that

$$p(x) \geq \frac{2\alpha}{2\alpha-1} > 1\,.$$

Thus we have $y' \leq p|F$. By Theorem 1.3.1.2, there is an $\mathbb{R}$–linear map $z' : E \to \mathbb{R}$ such that $z' \leq p$ and

$$z'(x) = y'(x) = p(x) > 1\,.$$

Put

$$x' := z'$$

if $\mathbb{K} = \mathbb{R}$ and

$$x' : E \longrightarrow \mathbb{C}\,, \quad y \longmapsto z'(y) - iz'(iy)$$

if $\mathbb{K} = \mathbb{C}$. By Lemma 1.3.1.5, x' is a linear form on E. Since $p \leq 2$ on B and $-B = B$, x' is bounded on B. Then

$$\sup_{y\in A} \operatorname{re} x'(y) = \sup_{y\in A} z'(y) \leq \sup_{y\in A} p(y') \leq 1 < z'(x) = \operatorname{re} x'(x)\,. \qquad \blacksquare$$

Corollary 1.3.1.7 (0) *Let E be a normed space, A a nonempty closed (absolutely) convex set of E, and take $x \in E\backslash A$. Then there is an $x' \in E'$ such that*

$$\sup_{y\in A} \operatorname{re} x'(y) < \operatorname{re} x'(x)\,, \quad (\sup_{y\in A} |x'(y)| < x'(x))\,.$$

Take $\varepsilon > 0$ with

$$U_\varepsilon^E(x) \subset E\backslash A\,.$$

Then $x \in E\backslash\left(A + U_\varepsilon^E(0)\right)$. By Proposition 1.3.1.6 (and Proposition 1.2.7.2), there is a linear form x' on E, bounded on $U_\varepsilon^E(0)$, such that

$$\sup_{y\in A} \operatorname{re} x'(y) < \operatorname{re} x'(x)\,.$$

By Proposition 1.2.1.1 e $\Rightarrow$ a, $x' \in E'$.

Now suppose that A is absolutely convex. Then $x'(x) \neq 0$. Define

$$y' := \frac{\overline{x'(x)}}{|x'(x)|} x' .$$

Take $y \in A$ with $y'(y) \neq 0$. Then

$$\frac{\overline{x'(x)}}{|x'(x)|} \frac{\overline{y'(y)}}{|y'(y)|} y \in A ,$$

so that

$$|y'(y)| = \operatorname{re} x' \left(\frac{\overline{x'(x)}}{|x'(x)|} \frac{\overline{y'(y)}}{|y'(y)|} y \right) \leq \sup_{z \in A} \operatorname{re} x'(z) ,$$

$$\sup_{y \in A} |y'(y)| \leq \sup_{z \in A} \operatorname{re} x'(z) < \operatorname{re} x'(x) \leq |x'(x)| = \frac{\overline{x'(x)}}{|x'(x)|} x'(x) = y'(x) . \qquad \blacksquare$$

Proposition 1.3.1.8 (0) *Let E be a vector space, F a vector space of linear forms on E, A' a nonempty (absolutely) convex closed set of F_E, and take $x' \in F \backslash A'$. Then there is an $x \in E$ such that*

$$\sup_{y' \in A'} \operatorname{re} y'(x) < \operatorname{re} x'(x) ,$$

$$\left(\sup_{y' \in A'} |y'(x)| < x'(x) \right) .$$

There is a 0–neighbourhood U' in F_E with

$$x' + U' \subset F \backslash A' .$$

By Proposition 1.2.6.2, there is a finite subset A of E such that

$$V' := \{ x' \in F | x \in A \Longrightarrow |x'(x)| \leq 1 \} \subset U'$$

and for each $y' \in F$ there is an $\alpha \in \mathbb{R}_+$ with $y' \in \alpha V'$. V' is absolutely convex and

$$x' \in F \backslash (A' + V') .$$

By Proposition 1.3.1.6, there is a linear form x'' on F, bounded on V', such that

$$\sup_{y' \in A'} \operatorname{re} x''(y') < \operatorname{re} x''(x') .$$

By Lemma 1.2.6.4, there is an $x \in E$ with

$$y'(x) = x''(y')$$

for every $y' \in F$. Then

$$\sup_{y' \in A'} \operatorname{re} y'(x) < \operatorname{re} x'(x) .$$

Assume now A' absolutely convex. Then $x'(x) \neq 0$. We set

$$y := \frac{\overline{x'(x)}}{|x'(x)|} x .$$

Let $y' \in A'$ with $y'(y) \neq 0$. Then

$$\frac{\overline{x'(x)}}{|x'(x)|} \frac{\overline{y'(y)}}{|y'(x)|} y' \in A'$$

so

$$\sup_{z' \in A'} \operatorname{re} z'(x) \geq \operatorname{re} \frac{\overline{x'(x)}}{|x'(x)|} \frac{\overline{y'(y)}}{|y'(y)|} y'(x) =$$

$$= \operatorname{re} \frac{\overline{y'(y)}}{|y'(y)|} y'(y) = |y'(y)| .$$

Hence

$$\sup_{y' \in A'} |y'(y)| \leq \sup_{z' \in A'} z'(x) < \operatorname{re} x'(x) \leq |x'(x)| = x'(y) .$$ ■

Corollary 1.3.1.9 (0) *Let E be a normed space and A' an absolutely convex closed set of E'_E such that*

$$\|x\| = \sup_{x' \in A'} |x'(x)|$$

for every $x \in E$. Then $A' = E'^{\#}$.

Let $x' \in E' \backslash A'$. By Proposition 1.3.1.8, there is an $x \in E$ such that

$$\sup_{y' \in A'} |y'(x)| < x'(x) .$$

We get

$$\|x\| < x'(x) \leq \|x'\| \, \|x\| ,$$

so $x' \in E^{\#}$. Hence $E'^{\#} \subset A'$. The converse inclusion is trivial. ■

Theorem 1.3.1.10 (0) (Krein–Milman) *Let* E *be a normed space,* K' *a convex compact set of* E'_E , L' *the set of extreme points of* K' , *and* $\mathfrak{A}'$ *the set of closed faces of* K'_E .

a) *The intersection of every nonempty family in* $\mathfrak{A}'$ *belongs to* $\mathfrak{A}'$ *if it is nonempty.*

b) *Every* $A' \in \mathfrak{A}'$ *contains an extreme point of* K' .

c) K' *is the smallest closed convex set of* E'_E *containing* L' .

a) Let $(A'_\iota)_{\iota\in I}$ be a nonempty family in $\mathfrak{A}'$. Take $x, y \in K'$ and $\alpha \in]0,1[$ with

$$\alpha x + (1-\alpha)y \in \bigcap_{\iota\in I} A'_\iota .$$

Then $x, y \in A'_\iota$ for every $\iota \in I$, so that $\bigcap_{\iota\in I} A'_\iota$ is a closed face of K'_E (Proposition 1.2.7.4).

b) Order $\mathfrak{A}'$ by reverse inclusion. By a) and Zorn's Lemma, there is a maximal element B' of $\mathfrak{A}'$ contained in A' . Take $x \in E$. The set

$$\{x'(x) | x' \in B'\}$$

is a convex compact set of $\mathbb{K}$ (Proposition 1.2.7.7), and so it contains an extreme point α . By Proposition 1.2.7.10,

$$\{x' \in B' | x'(x) = \alpha\}$$

is a face of B' and hence a face of K' . It is obviously closed in E'_E , so that it belongs to $\mathfrak{A}'$. Since B' is minimal, it coincides with B' . Thus

$$x'(x) = y'(x)$$

for every $x', y' \in B'$. Since x is arbitrary, B' must be a one point set, and this point is our extreme point of K' .

c) Let M' be a convex closed set of E'_E containing L' . Assume there is an $x' \in K' \setminus M'$. By Proposition 1.3.1.8, there is an $x \in E$ such that

$$\sup_{y'\in M'} \operatorname{re} y'(x) < \operatorname{re} x'(x) .$$

The set

$$\{y'(x) | y' \in K'\}$$

is a convex compact set of $\mathbb{K}$ (Proposition 1.2.7.7), so that it contains an extreme point α with

$$\sup_{y' \in M'} \operatorname{re} y'(x) < \operatorname{re} \alpha .$$

By Proposition 1.2.7.10,

$$\{y' \in K' | y'(x) = \alpha\}$$

is a face of K', which is obviously closed in E'_E. By b) it contains an extreme point of K'. This extreme point does not belong to M', which is a contradiction. Hence $K' \subset M'$ and K' is the smallest convex closed set of E'_E containing L'. ■

Definition 1.3.1.11 (0) *A normed space E is called a **dual space** if there is a Banach space F such that E is isometric to F'. F is called a **predual of** E.*

Every dual space is complete (Corollary 1.2.1.10, Proposition 1.2.1.14).

The notion "dual space" is a geometric and not a topological one. More precisely, there are isomorphic Banach spaces such that the one is a dual space whereas the other is not (see Example 1.3.1.16 c)). The predual of a Banach space is not unique. Indeed, $c_0 \times (\ell^\infty / c_0)$ is a predual of $(\ell^\infty)'$ (Corollary 1.2.2.14), but $c_0 \times (\ell^\infty / c_0)$ is not isomorphic to ℓ^∞ (Remark of Example 1.2.5.13).

Corollary 1.3.1.12 (0) *If E is a dual space then $E^\#$ has extreme points.*

Let F be a Banach space and $u : F' \to E$ an isometry. By Proposition 1.2.7.2 and by the Alaoglu–Bourbaki Theorem, $F'^{\#}_F$ is convex and compact. By the Krein–Milman Theorem, it has extreme points. It follows that $E^\#$ also has extreme points. ■

Example 1.3.1.13 *If T is a locally compact non-compact space, then $\mathcal{C}_0(T)$ is not a dual space.*

By Example 1.2.7.17, $\mathcal{C}_o(T)^\#$ has no extreme points. Hence, by Corollary 1.3.1.12, $\mathcal{C}_0(T)$ is not a dual space. ■

Example 1.3.1.14 *If T is a set, then $c_0(T)$ is a dual space iff T is finite.*

If T is finite, then $c_0(T)$ is isometric to $\ell^1(T)'$ (Example 1.2.2.3 g)), so that $c_0(T)$ is a dual space. If T is infinite, then $c_0(T)$ is not a dual space by Example 1.3.1.13. ■

Example 1.3.1.15 *Let T be a Hausdorff space. If $\mu \neq 0$ is an atomless Radon measure on T, then $L^1(\mu)$ is not a dual space.*

By Example 1.2.7.13, $L^1(\mu)^\#$ has no extreme points, so that by Corollary 1.3.1.12, $L^1(\mu)$ is not a dual space. ■

Example 1.3.1.16 *Let $\mathfrak{F}$ be a free ultrafilter on $\mathbb{N}$ and take*

$$p : \ell^\infty \longrightarrow \mathbb{R}_+ , \quad x \longmapsto \|x\|_\infty + \lim_{n,\mathfrak{F}} |x(n)| .$$

a) *p is a norm and*

$$\|\cdot\|_\infty \le p \le 2\|\cdot\|_\infty .$$

b) *$\{x \in \ell^\infty | p(x) \le 1\}$ has no extreme points.*

c) *ℓ^∞ endowed with the norm p is not a dual space but it is isomorphic to ℓ^∞ endowed with the usual norm $\|\cdot\|_\infty$, which is a dual space.*

a) is easy to see.

b) Let x be an extreme point of $\{x \in \ell^\infty | p(x) \le 1\}$ and take $n \in \mathbb{N}$. Assume that $|x(n)| \neq 1$. Define

$$y_\pm : \mathbb{N} \longrightarrow \mathbb{K} , \quad k \longmapsto \begin{cases} x(k) & \text{if } k \neq n \\ x(n) \pm (1 - |x(n)|) & \text{if } k = n . \end{cases}$$

Then $p(y_\pm) \le 1$ and

$$x = \frac{1}{2} y_+ + \frac{1}{2} y_- .$$

This contradicts the assumption that x is an extreme point of $\{x \in \ell^\infty | p(x) \le 1\}$. Hence $|x(n)| = 1$ for every $n \in \mathbb{N}$. This leads to the contradiction that $p(x) = 2$.

c) By b) and Corollary 1.3.1.12, ℓ^∞ endowed with the norm p is not a dual space. By a), it is isomorphic with ℓ^∞ endowed with the usual norm $\|\cdot\|_\infty$, which is the dual of ℓ^1. ■

1.3.2 Examples in Measure Theory

(S. Banach, 1923)

Lemma 1.3.2.1 *Let $\mathbb{K} = \mathbb{R}$. Let T be an additive group, Ω the class of all finite nonempty families in T, and define*

$$\varphi : \Omega \longrightarrow \mathbb{N}, \quad (t_\iota)_{\iota \in I} \longmapsto \operatorname{Card} I .$$

Put

$$u_t x : T \longrightarrow \mathbb{R}, \quad s \longmapsto x(s+t),$$

$$ux : T \longrightarrow \mathbb{R}, \quad s \longmapsto x(-s),$$

$$p_\omega(x) := \frac{1}{2\varphi(\omega)} \sup_{s\in T} \Big(\sum_{\iota \in I} u_{t_\iota}(x + ux) \Big)(s)$$

for every $x \in \ell^\infty(T)$, $t \in T$, and $\omega := (t_\iota)_{\iota\in I} \in \Omega$. Define further

$$p(x) := \inf_{\omega \in \Omega} p_\omega(x)$$

for every $x \in \ell^\infty(T)$. Then for every $x, y \in \ell^\infty(T)$, $\alpha \in \mathbb{R}_+$, and $t \in T$:

a) $\inf\limits_{t\in T} x(t) \le p(x) \le \sup\limits_{t \in T} x(t)$.

b) $p(x+y) \le p(x) + p(y)$.

c) $p(\alpha x) = \alpha p(x)$.

d) $p(x - u_t x) \le 0$.

e) $p(x - ux) \le 0$.

f) *If T is a compact additive group and λ is its Haar measure normed by $\lambda(T) = 1$, then*

$$\int_* x d\lambda \le p(x).$$

g) *If T is a topological additive group with the Baire property, then*

$$\sum_{\lambda\in L} \alpha_\lambda \le p\Big(\sum_{\lambda \in L} \alpha_\lambda e_{A_\lambda}\Big)$$

for every finite family $(\alpha_\lambda)_{\lambda\in L}$ in $\mathbb{R}$ and every family $(A_\lambda)_{\lambda \in L}$ of dense G_δ–sets of T.

a) is obvious.

b) Take $\omega_1 := (s_\iota)_{\iota \in I} \in \Omega$ and $\omega_2 := (t_\lambda)_{\lambda \in L} \in \Omega$. Put

$$\omega := (s_\iota + t_\lambda)_{(\iota,\lambda) \in I \times L} \in \Omega .$$

Then

$$p(x+y) \leq p_\omega(x+y) = \frac{1}{2\varphi(\omega)} \sup_{s \in T} \left(\sum_{(\iota,\lambda) \in I \times L} u_{s_\iota + t_\lambda}(x+y+ux+uy) \right)(s) =$$

$$= \frac{1}{2\varphi(\omega)} \sup_{s \in T} \Big(\sum_{\lambda \in L} \Big(\sum_{\iota \in I} u_{s_\iota}(x+ux) \Big)(s+t_\lambda) +$$

$$+ \sum_{\iota \in I} \Big(\sum_{\lambda \in L} u_{t_\lambda}(y+uy) \Big)(s+s_\iota) \Big) \leq$$

$$\leq \frac{1}{\varphi(\omega_2)} \sum_{\lambda \in L} \frac{1}{2\varphi(\omega_1)} \sup_{s \in T} \Big(\sum_{\iota \in I} u_{s_\iota}(x+ux) \Big)(s+t_\lambda) +$$

$$+ \frac{1}{\varphi(\omega_1)} \sum_{\iota \in I} \frac{1}{2\varphi(\omega_2)} \sup_{s \in T} \Big(\sum_{\lambda \in L} u_{t_\lambda}(y+uy) \Big)(s+s_\iota) =$$

$$= \frac{1}{\varphi(\omega_2)} \sum_{\lambda \in L} p_{\omega_1}(x) + \frac{1}{\varphi(\omega_1)} \sum_{\iota \in I} p_{\omega_2}(y) = p_{\omega_1}(x) + p_{\omega_2}(y) .$$

Since ω_1 and ω_2 are arbitrary,

$$p(x+y) \leq p(x) + p(y) .$$

c) is trivial.

d) Take $n \in \mathbb{N}$. Put

$$\omega := (kt)_{k \in \mathbb{N}_n} \in \Omega .$$

Then

$$p(x - u_t x) \leq p_\omega(x - u_t x) =$$

$$= \frac{1}{2n} \sup_{s \in T} \left(\sum_{k=1}^{n} (x - u_t x + ux - uu_t x)(s+kt) \right) =$$

$$= \frac{1}{2n} \sup_{s\in T} \begin{pmatrix} \sum_{k=1}^{n} (x(s+kt) - x(s+(k+1)t)) + \\ \sum_{k=1}^{n} (x(-x-kt) - x(-s-(k-1)t)) \end{pmatrix} =$$

$$= \frac{1}{2n} \sup_{s\in T} (x(s+t) - x(s+(n+1)t) + x(-s-nt) - x(-s)) \leq \frac{2\|x\|_\infty}{n}.$$

Since n is arbitrary,

$$p(x - u_t x) \leq 0.$$

e) $p(x - ux) \leq \frac{1}{2} \sup_{s\in T} (x - ux + ux - uux)(s) = 0$.

f) Take $\omega := (t_\iota)_{\iota\in I} \in \Omega$. Then

$$p_\omega(s) = \frac{1}{2\varphi(\omega)} \sup_{s\in T} \Big(\sum_{\iota\in I} u_{t_\iota}(x+ux)\Big)(s) \geq \frac{1}{2\varphi(\omega)} \int_* \Big(\sum_{\iota\in I} u_{t_\iota}(x+ux)\Big) d\lambda \geq$$

$$\geq \frac{1}{2\varphi(\omega)} \sum_{\iota\in I} \Big(\int_* u_{t_\iota} x d\lambda + \int_* u_{t_\iota} ux d\lambda\Big) = \int_* x d\lambda.$$

g) Take $\omega := (t_\iota)_{\iota\in I} \in \Omega$. The sets $-t_\iota \pm A_\lambda$ are dense G_δ–sets of T for every $(\iota, \lambda) \in I \times L$. Since T has the Baire property, we can choose

$$s_0 \in \Big(\bigcap_{\substack{\iota\in I\\ \lambda\in L}} (-t_\iota + A_\lambda)\Big) \cap \Big(\bigcap_{\substack{\iota\in I\\ \lambda\in L}} (-t_\iota - A_\lambda)\Big).$$

Then

$$p_\omega\Big(\sum_{\lambda\in L} \alpha_\lambda e_{A_\lambda}\Big) = \frac{1}{2\varphi(\omega)} \sup_{s\in T} \Big(\sum_{\iota\in I} u_{t_\iota}\Big(\sum_{\lambda\in L} \alpha_\lambda \Big(e_{A_\lambda} + ue_{A_\lambda}\Big)\Big)\Big)(s) =$$

$$= \frac{1}{2\varphi(\omega)} \sup_{s\in T} \Big(\sum_{\iota\in I} \sum_{\lambda\in L} \alpha_\lambda \Big(e_{-t_\iota+A_\lambda} + e_{-t_\iota-A_\lambda}\Big)\Big)(s) \geq$$

$$\geq \frac{1}{2\varphi(\omega)} \sum_{\iota\in I} \sum_{\lambda\in L} \alpha_\lambda \Big(e_{-t_\iota+A_\lambda} + e_{-t_\iota-A_\lambda}\Big)(s_0) \geq \frac{1}{2\varphi(\omega)} \Big(\sum_{\iota\in I} \sum_{\lambda\in L} 2\alpha_\lambda\Big) = \sum_{\lambda\in L} \alpha_\lambda.$$

Since ω is arbitrary,

$$p\Big(\sum_{\lambda\in L} \alpha_\lambda e_{A_\lambda}\Big) \geq \sum_{\lambda\in L} \alpha_\lambda. \qquad ■$$

Example 1.3.2.2 *Using the notation of Lemma 1.3.2.1, let* $\mathcal{F}$ *be a vector subspace of* $\ell^\infty(T)$ *and* y' *a linear form on* $\mathcal{F}$ *with* $y' \le p\,|\,\mathcal{F}$. *Then there is an* $x' \in \ell^\infty(T)'$ *such that*

1) $\|x'\| \le 1$;

2) $x' \circ u_t = x' \circ u = x'$ *for all* $t \in T$;

3) $x'(x) \in \mathbb{R}_+$ *for every positive function* x *of* $\ell^\infty(T)$;

4) $x' \mid \mathcal{F} = y'$.

By Lemma 1.3.2.1 b), c), and the Banach Theorem (Theorem 1.3.1.2), there is a linear extention x' of y' on $\ell^\infty(T)$ such that $x' \le p$. Then

$$x'(x) \le p(x) \le \|x\|, \quad -x'(x) = x'(-x) \le p(-x) \le \|-x\| = \|x\|$$

(Lemma 1.3.2.1 a)). Hence

$$|x'(x)| \le \|x\|$$

for every $x \in \ell^\infty(T)$. Therefore x' is continuous and $\|x'\| \le 1$ (Proposition 1.2.1.1 d $\Rightarrow$ a).

Take $t \in T$ and $x \in \ell^\infty(T)$. Then

$$x'(x - u_t x) \le p(x - u_t x) \le 0, \quad x'(x - ux) \le p(x - ux) \le 0$$

(Lemma 1.3.2.1 d), e)). Since x is arbitrary,

$$-x'(x - u_t x) = x'((-x) - u_t(-x)) \le 0, \quad -x'(x - ux) = x'((-x) - u(-x)) \le 0.$$

Hence

$$x'(x) - x' \circ u_t(x) = x'(x - u_t x) = 0, \quad x'(x) - x' \circ u(x) = x'(x - ux) = 0,$$

and

$$x' \circ u_t = x' \circ u = x'.$$

Thus

$$-x'(x) = x'(-x) \le p(-x) \le 0$$

(Lemma 1.3.2.1 a)), so that

$$x'(x) \ge 0$$

for every positive function x of $\ell^\infty(T)$. ∎

Lemma 1.3.2.3 *Let S be an infinite subgroup of the additive group T, $\mathfrak{R}$ a ring of subsets of T. Take $C \in \mathfrak{R}$ and let μ be a bounded positive real function on $\mathfrak{R}$ such that for every $A, B \in \mathfrak{R}$ with $A \cap B = \emptyset$ and every $s \in S$:*

1) $\mu(A \cup B) = \mu(A) + \mu(B)$,

2) $s + A \in \mathfrak{R}, \quad \mu(s + A) = \mu(A)$,

3) $\mathfrak{P}(C) \subset \mathfrak{R}$,

4) *for every $t \in T$ there is an $r \in C$ with $t - r \in S$.*

Then there is a subset B of C such that $(s + B)_{s \in S}$ is a partition of T and

$$\mu(s + B) = 0$$

for every $s \in S$.

We define

$$s \sim t : \Longleftrightarrow s - t \in S$$

for $s, t \in T$. $\sim$ is an equivalence relation in T. By 4), there is a subset B of C containing exactly one point in each equivalence class of $\sim$. Then $(s+B)_{s \in S}$ is a partition of T. By 2) and 3),

$$s + B \in \mathfrak{R}, \quad \mu(s + B) = \mu(B)$$

for every $s \in S$. Thus, by 1),

$$\mu(B) \operatorname{Card} D = \sum_{s \in D} \mu(s + B) = \mu\Big(\bigcup_{s \in D} (s + B)\Big) \leq \sup_{A \in \mathfrak{R}} \mu(A)$$

for every finite subset D of S. Hence

$$\mu(s + B) = \mu(B) = 0$$

for every $s \in S$. ■

Lemma 1.3.2.4 *Let T be an infinite additive group and $\mu : \mathfrak{P}(T) \to \mathbb{R}_+$ a map such that*

$$\mu(A \cup B) = \mu(A) + \mu(B)$$

for disjoint $A, B \in \mathfrak{P}(T)$ and

$$\mu(t + A) = \mu(A)$$

for every $A \in \mathfrak{P}(T)$ *and* $t \in T$ *. Then there is a partition* $(A_n)_{n\in\mathbb{N}}$ *of* T *, such that*

$$\mu(A_n) = 0$$

for every $n \in \mathbb{N}$ *.*

Let S be a countable infinite subgroup of T. The assertion follows from Lemma 1.3.2.3 by setting

$$\mathfrak{R} := \mathfrak{P}(T), \quad C := T. \qquad \blacksquare$$

Example 1.3.2.5 *Let* T *be a compact additive group and* λ *be its Haar measure normed by* $\lambda(T) = 1$ *. Given* $x \in \ell^\infty(T)$ *and* $t \in T$ *, define*

$$u_t x : T \longrightarrow \mathbb{K}, \quad s \longmapsto x(s + t),$$

$$ux : T \longrightarrow \mathbb{K}, \quad s \longmapsto x(-s).$$

Then there is an $x' \in \ell^\infty(T)'$ *with the following properties:*

1) $\|x'\| = 1$ *,*

2) $x' \circ u_t = x' \circ u = x'$ *for every* $t \in T$ *,*

3) $x'(x) \in \mathbb{R}_+$ *for every positive real function in* $\ell^\infty(T)$ *,*

4) $x'(x) = \int x\, d\lambda$ *for every* $x \in \mathcal{L}^1(\lambda) \cap \ell^\infty(T)$ *.*

If T *is infinite, then for every such* x' *there is a partition* $(A_n)_{n\in\mathbb{N}}$ *of* T *such that*

$$x'(e_{A_n}) = 0$$

for every $n \in \mathbb{N}$ *. (Hence the Lebesgue Convergence Theorem does not hold for such an* x' *.)*

First consider the case $\mathbb{K} = \mathbb{R}$. Let p be the function defined in Lemma 1.3.2.1. By Lemma 1.3.2.1 f),

$$\int x\, d\lambda \leq p(x)$$

for every $x \in \mathcal{L}^1(\lambda) \cap \ell^\infty(T)$ and the existence of x' with the desired properties follows from Example 1.3.2.2.

If $\mathbb{K} = \mathbb{C}$, then

$$\ell^\infty(T) \longrightarrow \mathbb{C}\,, \quad x \longmapsto x'(\operatorname{re} x) + ix'(\operatorname{im} x)$$

has the required properties.

The final assertion follows from Lemma 1.3.2.4. ■

Example 1.3.2.6 *Let T be a topological additive group with the Baire property. Given $x \in \ell^\infty(T)$ and $t \in T$, define*

$$u_t x : T \longrightarrow \mathbb{K}\,, \quad s \longmapsto x(s+t)\,,$$

$$ux : T \longrightarrow \mathbb{K}\,, \quad s \longmapsto x(-s)\,.$$

Then there is an $x' \in \ell^\infty(T)'$ with the following properties:

1) $\|x'\| = 1$,

2) $x' \circ u_t = x' \circ u = x'$ *for every* $t \in T$,

3) $x'(x) \in \mathbb{R}_+$ *for every positive real function* x *in* $\ell^\infty(T)$,

4) $x'(e_A) = 1$ *for every dense* G_δ*–set* A *of* T.

If T is infinite, then for every such x' there is a partition $(A_n)_{n\in\mathbb{N}}$ of T such that

$$x'(e_{A_n}) = 0$$

for every $n \in \mathbb{N}$. (Hence the Lebesgue Convergence Theorem does not hold for such an x'.)

First consider the case $\mathbb{K} = \mathbb{R}$. Let p be the function defined in Lemma 1.3.2.1, $\mathfrak{A}$ the set of dense G_δ–sets of T, $\mathcal{F}$ the vector subspace of $\ell^\infty(T)$ generated by $\{e_A \mid A \in \mathfrak{A}\}$, and

$$y' : \mathcal{F} \longrightarrow \mathbb{R}\,, \quad \sum_{A\in\mathfrak{A}} \alpha_A e_A \longmapsto \sum_{A\in\mathfrak{A}} \alpha_A\,.$$

(In the above sums,

$$\{A \in \mathfrak{A} \mid \alpha_A \neq 0\}$$

is obviously finite and

$$\bigcap_{\substack{A\in\mathfrak{A}\\ \alpha_A\neq 0}} A \neq \emptyset$$

so the value of y' does not depend on the representation.) By the last assertion of Lemma 1.3.2.1, $y' \leq p \mid \mathcal{F}$. The existence of x' with the desired properties now follows from Example 1.3.2.2.

If $\mathbb{K} = \mathbb{C}$, then

$$\ell^\infty(T) \longrightarrow \mathbb{C}, \quad x \longmapsto x'(\operatorname{re} x) + ix'(\operatorname{im} x)$$

has the required properties.

The final assertion follows from Lemma 1.3.2.4. ∎

Example 1.3.2.7 *Let $n \in \mathbb{N}$, λ the Lebesgue measure on $\mathbb{R}^n$, and*

$$\mathcal{F} := \{x \in \ell^\infty(\mathbb{R}^n) \mid \{x \neq 0\} \quad \textit{is bounded}\}.$$

Define

$$u_t x : \mathbb{R}^n \longrightarrow \mathbb{K}, \quad s \longmapsto x(s+t),$$

$$ux : \mathbb{R}^n \longrightarrow \mathbb{K}, \quad s \longmapsto x(-s)$$

for $x \in \mathcal{F}$ and $t \in \mathbb{R}^n$. Then $\mathcal{F}$ is a vector subspace of $\ell^\infty(\mathbb{R}^n)$, $u_t x$, $ux \in \mathcal{F}$ for every $x \in \mathcal{F}$ and $t \in \mathbb{R}^n$, and there is a linear form x' on $\mathcal{F}$ with the following properties:

1) $x' \circ u_t = x' \circ u = x'$ *for every* $t \in T$,

2) $x'(x) \in \mathbb{R}_+$ *for every positive real function* x *in* $\mathcal{F}$,

3) $x'(x) = \int x \, d\lambda$ *for every* $x \in \mathcal{F} \cap \mathcal{L}^1(\lambda)$.

We may replace 3) *by*

3′) *There is a* λ*-null set* A *with* $x'(e_A) = 1$.

For every such x' there is a dosjoint sequence $(A_n)_{n\in\mathbb{N}}$ of subsets of $\mathbb{R}^n$ such that $\bigcup_{n\in\mathbb{N}} A_n$ is bounded,

$$x'\left(e_{\bigcup_{n\in\mathbb{N}} A_n}\right) > 0$$

and

$$x'(e_{A_n}) = 0$$

for every $n \in \mathbb{N}$. *(Hence the Lebesgue Convergence Theorem does not hold for such an* x'.*)*

Let T be the compact additive group $\mathbb{R}^n/\mathbb{Z}^n$, $\varphi : \mathbb{R}^n \to T$ the quotient map, and x' the linear form from Example 1.3.2.5 (resp. Example 1.3.2.6). For every $p = (p_k)_{k\in\mathbb{N}_n} \in \mathbb{Z}^n$ the map

$$\prod_{k=1}^{n} [p_k, 1 + p_k[\longrightarrow T, \quad t \longmapsto \varphi(t)$$

is bijective. Let φ_p denote its inverse. The map

$$\mathcal{F} \longrightarrow \mathbb{K}, \quad x \longmapsto \sum_{p\in\mathbb{Z}^n} x'(x \circ \varphi_p)$$

has the required properties (for $3'$, it is sufficient to take as A a G_δ–set of $\mathbb{R}^n$ which is dense in $[0, 1[^n$ and which is a λ–null set).

In order to prove the last assertion, put

$$S := \mathbb{Q}^n, \quad C = [0, 1[^n,$$

$$\mathfrak{R} := \{A \subset \mathbb{R}^n \mid A \text{ is bounded}\},$$

$$\mu : \mathfrak{R} \longrightarrow \mathbb{R}_+, \quad A \longmapsto x'(e_A).$$

By Lemma 1.3.2.3, there is a $B \subset C$ such that $(s + B)_{s\in S}$ is a partition of T and

$$\mu(s + B) = 0$$

for every $s \in S$. Given $s \in S$, we put

$$A_s := (s + B) \cap [0, 1[^n.$$

Then

$$x'(A_s) = 0$$

for every $s \in S$ and

$$x'\left(e_{\bigcup_{n\in\mathbb{N}} A_n}\right) = x'\left(e_{[0,1[^n}\right) > 0. \qquad \blacksquare$$

1.3.3 The Hahn–Banach Theorem

Theorem 1.3.3.1 (0) (Hahn 1927, Banach 1929) *Let F be a subspace of the normed space E. Take $y' \in F'$. Then there is a continuous linear extention x' of y' to E with $\|x'\| = \|y'\|$.*

Case 1 $\mathbb{K} = \mathbb{R}$

Define

$$p : E \longrightarrow \mathbb{R}, \quad x \longmapsto \|y'\| \, \|x\| .$$

Then

$$p(x+y) = \|y'\| \, \|x+y\| \leq \|y'\|(\|x\| + \|y\|) \leq p(x) + p(y) ,$$

$$p(\alpha x) = \|y'\| \, \|\alpha x\| = \|y'\| \alpha \|x\| = \alpha p(x)$$

for every $x, y \in E$ and $\alpha \in \mathbb{R}_+$. Moreover

$$y'(x) \leq \|y'\| \, \|x\| = p(x)$$

for every $x \in F$. (Proposition 1.2.1.4 a)). By the Banach Theorem, there is a continuous linear extension x' of y' to E with

$$x'(x) \leq p(x)$$

for every $x \in E$. Then

$$x'(x) \leq \|y'\| \, \|x\|$$

and

$$-x'(x) = x'(-x) \leq \|y'\| \, \| - x\| = \|y'\| \, \|x\|$$

and so

$$|x'(x)| \leq \|y'\| \, \|x\|$$

for every $x \in E$. Hence x' is continuous and $\|x'\| \leq \|y'\|$. By Proposition 1.2.1.4 c), $\|x'\| = \|y'\|$.

Case 2 $\mathbb{K} = \mathbb{C}$

$\operatorname{re} y'$ is a continuous real linear form on F. By the above proof, there is a continuous $\mathbb{R}$–linear form z' on E extending $\operatorname{re} y'$ with $\|z'\| = \|\operatorname{re} \circ y'\|$. Define

$$x' : E \longrightarrow \mathbb{C}, \quad x \longmapsto z'(x) - iz'(ix).$$

Then x' is linear (Lemma 1.3.1.5) and continuous. By Corollary 1.2.1.6,

$$\|x'\| = \|\operatorname{re} \circ x'\| = \|z'\| = \|\operatorname{re} \circ y'\| = \|y'\|.$$

Then

$$\operatorname{re} \circ x'(x) = z'(x) = \operatorname{re} \circ y'(x),$$

$$\operatorname{im} \circ x'(x) = -z'(ix) = -\operatorname{re} \circ y'(ix) = \operatorname{re} \circ (-iy')(x') = \operatorname{im} \circ y'(x),$$

$$x'(x) = \operatorname{re} \circ x'(x) + i \operatorname{im} \circ x'(x) = \operatorname{re} \circ y'(x) + i \operatorname{im} \circ y'(x) = y'(x)$$

for every $x \in F$, i.e. x' is an extention of y'. ■

Corollary 1.3.3.2 (0) *Let F be a finite–dimensional vector subspace of the normed space E. Let G be a closed vector subspace of E with $F \cap G = \{0\}$. Let $(x_\iota)_{\iota \in I}$ be an algebraic basis of F.*

a) *There is a family $(x'_\iota)_{\iota \in I}$ in E' such that x'_ι vanishes on G for every $\iota \in I$ and*

$$x'_\iota(x_\lambda) = \delta_{\iota\lambda}$$

for every $\iota, \lambda \in I$.

b) *The map*

$$p : E \longrightarrow E, \quad x \longmapsto \sum_{\iota \in I} x'_\iota(x) x_\iota$$

is a projection of E onto F, vanishing on G.

c) *F is a complemented subspace of E.*

d) *If the map*

$$F \times G \longrightarrow E, \quad (x, y) \longmapsto x + y$$

is surjective, then E is the direct sum of F and G and $1 - p$ is the projection of E onto G, vanishing on F.

a) Put

$$H := F + G\,.$$

Take $\iota \in I$ and

$$x' : F \longrightarrow \mathbb{K}\,, \quad \sum_{\lambda\in I} \alpha_\lambda x_\lambda \longmapsto \alpha_\iota\,.$$

By Lemma 1.3.1.1, there is a linear form y' on H such that

$$y'|F = x'\,, \quad y'|G = 0\,.$$

$\operatorname{Ker} y'$ is the vector subspace of H generated by $G \cup \{x_\lambda \mid \lambda \in I\backslash\{\iota\}\}$. It is thus a closed vector subspace of H (Corollary 1.2.4.3). Hence y' is continuous (Corollary 1.2.4.12 b $\Rightarrow$ a). By the Hahn–Banach Theorem, there is a continuous linear extention x'_ι of y' to E. x'_ι vanishes on G and

$$x'_\iota(x_\lambda) = \delta_{\iota\lambda}$$

for every $\lambda \in I$.

b) p is linear and continuous. It vanishes on G and $\operatorname{Im} p \subset F$. Moreover, given $\iota \in I$,

$$px_\iota = \sum_{\lambda\in I} x'_\lambda(x_\iota)x_\lambda = x_\iota\,.$$

Hence $\operatorname{Im} p = F$. Thus

$$p \circ p(x) = p\Big(\sum_{\iota\in I} x'_\iota(x)x_\iota\Big) = \sum_{\iota\in I} x'_\iota(x)px_\iota = \sum_{\iota\in I} x'_\iota(x)x_\iota = px$$

for every $x \in E$, so that

$$p \circ p = p\,.$$

Hence p is a projection of E onto F.

c) follows from b) and Murray's Theorem (Theorem 1.2.5.8).

d) Since

$$\operatorname{Ker} p = G\,,$$

E is the direct sum of F and G, and $1 - p$ is the projection of E onto G, vanishing on F (Murray's Theorem). ■

Definition 1.3.3.3 $(\,0\,)$ *Let E, F be normed spaces. Set*

$$\langle\cdot, x'\rangle y : E \longrightarrow F\,, \quad x \longmapsto \langle x, x'\rangle y$$

for every $(x', y) \in E' \times F$ and

$$\sum_{\iota\in I}\langle\cdot, x'_\iota\rangle y_\iota : E \longrightarrow F\,, \quad x \longmapsto \sum_{\iota\in I}\langle x, x'_\iota\rangle y_\iota$$

for every finite family $((x'_\iota, y_\iota))_{\iota\in I}$ in $E' \times F$.

Corollary 1.3.3.4 $(\,0\,)$ *Let E, F be normed spaces and take $u \in \mathcal{L}(E, F)$. Then u belongs to $\mathcal{L}_f(E, F)$ iff there is a finite family $((x'_\iota, y_\iota))_{\iota\in I}$ in $E' \times F$ such that*

$$u = \sum_{\iota\in I}\langle\cdot, x'_\iota\rangle y_\iota\,.$$

Assume $u \in \mathcal{L}_f(E, F)$. By Lemma 1.2.4.6, $\operatorname{Ker} u$ has finite codimension in E. Thus by Corollary 1.2.5.6, there is a finite–dimensional vector subspace G of E such that

$$E = G \oplus \operatorname{Ker} u\,.$$

Let $(x_\iota)_{\iota\in I}$ be an algebraic basis of G. By Corollary 1.3.3.2, there is a family $(x'_\iota)_{\iota\in I}$ in E', such that for each $\iota \in I$, x'_ι vanishes on $\operatorname{Ker} u$ and

$$x'_\iota(x_\lambda) = \delta_{\iota\lambda}$$

for every $\iota, \lambda \in I$, the map

$$p : E \longrightarrow E\,, \quad x \longmapsto \sum_{\iota\in I}\langle x, x'_\iota\rangle x_\iota$$

is a projection of E onto G vanishing on $\operatorname{Ker} u$, and $1 - p$ is a projection of E onto $\operatorname{Ker} u$ vanishing on G. Given $\iota \in I$, put

$$y_\iota := u x_\iota\,.$$

Then

$$ux = upx + u(1-p)x = u\left(\sum_{\iota\in I}\langle x, x'_\iota\rangle x_\iota\right) = \sum_{\iota\in I}\langle x, x_\iota\rangle y_\iota$$

for every $x \in E$, i.e.

$$u = \sum_{\iota\in I}\langle\cdot, x'_\iota\rangle y_\iota\,.$$

The reverse implication is trivial. ■

Corollary 1.3.3.5 (0) *Let F be a vector subspace of the normed space E. Take $x \in E$ with $d_F(x) > 0$. Then there is an $x' \in E'$ such that*

$$\|x'\| = 1\,, \quad x' \mid F = 0\,, \quad x'(x) = d_F(x)\,.$$

Put

$$G := \{\alpha x + y \mid \alpha \in \mathbb{K}\,,\, y \in F\}$$

and let y' denote the linear form on G which vanishes on F and takes the value $d_F(x)$ at x (Lemma 1.3.3.1). Take $(\alpha, y) \in \mathbb{K} \times F$ with

$$\|\alpha x + y\| \leq 1\,, \quad \alpha \neq 0\,.$$

Then

$$1 \geq \|\alpha x + y\| = |\alpha|\, \|x - \frac{1}{-\alpha} y\| \geq |\alpha| d_F(x) = |\alpha d_F(x)| = |y'(\alpha x + y)|\,.$$

Hence y' is continuous (Proposition 1.2.1.1 d $\Rightarrow$ a) and $\|y'\| \leq 1$ (Proposition 1.2.1.4 b)). Given $y \in F$,

$$d_F(x) = y'(x - y) \leq \|y'\|\, \|x - y\|$$

(Proposition 1.2.1.4 a)) and so

$$d_F(x) \leq \|y'\| d_F(x)\,, \quad 1 \leq \|y'\|\,, \quad \|y'\| = 1\,.$$

By the Hahn–Banach Theorem, there is a continuous linear extention x' of y' to E with $\|x'\| = \|y'\|$. x' has the desired properties. ■

Corollary 1.3.3.6 (0) *Let F be a vector subspace of the normed space E. If F is not dense, then there is an $x' \in E' \backslash \{0\}$ which vanishes on F.*

Let $x \in E \backslash \overline{F}$. Then $d_F(x) > 0$. By Corollary 1.3.3.5, there is an $x' \in E'$, vanishing on F, such that

$$x'(x) = d_F(x)\,.$$

Then $x' \neq 0$. ■

Corollary 1.3.3.7 (0) *The normed space E is separable whenever its dual E' is.*

Let $(x'_n)_{n\in\mathbb{N}}$ be a dense sequence in E'. For each $n \in \mathbb{N}$, there is an $x_n \in E$ with

$$\|x_n\| \le 1, \quad |x'_n(x_n)| \ge \frac{1}{2}\|x'_n\|$$

(Proposition 1.2.1.4 b)). Let F be the closed vector subspace of E generated by $(x_n)_{n\in\mathbb{N}}$ and let x' be a continuous linear form on E vanishing on F. $(x'_n)_{n\in\mathbb{N}}$ contains a subsequence $(x'_{k_n})_{n\in\mathbb{N}}$ which converges to x'. Then

$$\frac{1}{2}\|x'_{k_n}\| \le |x'_{k_n}(x_{k_n})| = |\langle x_{k_n}, x'_{k_n}\rangle - \langle x_{k_n}, x'\rangle| = |\langle x_{k_n}, x'_{k_n} - x'\rangle| \le \|x'_{k_n} - x'\|,$$

$$\|x'\| \le \|x' - x'_{k_n}\| + \|x'_{k_n}\| \le 3\|x'_{k_n} - x'\|$$

for every $n \in \mathbb{N}$. Thus

$$\|x'\| \le \lim_{n\to\infty} 3\|x'_{k_n} - x'\| = 0.$$

Thus $x' = 0$ and $F = E$ (Corollary 1.3.3.6). Hence E is separable (Corollary 1.1.5.5). ■

Remark. The separability of E does not imply that of E' as is shown by taking $E := \ell^1$ (Examples 1.1.2.5, 1.2.2.3 d), and 1.1.2.2).

Corollary 1.3.3.8 (0) *Let E be a normed space and take $x \in E\setminus\{0\}$.*

a) *There is an $x' \in E'$ with $\|x'\| = 1$, $x'(x) = \|x\|$.*

b) $\|x\| = \sup_{x'\in E'^{\#}} |x'(x)| = \sup_{\substack{x'\in E' \\ \|x'\|=1}} |x'(x)|$.

a) follows from Corollary 1.3.3.5.

b) Given $x' \in E'^{\#}$,

$$|x'(x)| \le \|x'\|\,\|x\| \le \|x\|$$

(Proposition 1.2.1.4 a)) and so

$$\sup_{x'\in E'^{\#}} |x'(x)| \le \|x\|.$$

By a), there is an $y' \in E'$ with

$$\|y'\| = 1, \quad y'(x) = \|x\|.$$

Hence

$$\|x\| \le \sup_{\substack{x'\in E' \\ \|x'\|=1}} |x'(x)| \le \sup_{x'\in E'^{\#}} |x'(x)| \le \|x\|.$$ ■

Corollary 1.3.3.9 (0) *Let E be a normed space and take distinct x and y in E. Then there is an $x' \in E'$ with*

$$x'(x) \neq x'(y).$$

Since $x - y \neq 0$, the assertion follows from Corollary 1.3.3.8 a). ■

Corollary 1.3.3.10 (0) *Let E be a Banach space and $\sum_{n=0}^{\infty} t^n x_n$ a power series in E. If there is an $r > 0$ smaller than the radius of convergence of the above power series such that*

$$\sum_{n=0}^{\infty} \alpha^n x_n = 0$$

for every $\alpha \in U_r^{\mathbb{K}}(0)$, then

$$x_n = 0$$

for every $n = \mathbb{N} \cup \{0\}$.

Take $n \in \mathbb{N} \cup \{0\}$ and $x' \in E'$. Then

$$0 = x'\left(\sum_{n=0}^{\infty} \alpha^n x_n\right) = \sum_{n=0}^{\infty} \alpha^n x'(x_n)$$

for every $\alpha \in U_r^{\mathbb{K}}(0)$ (Proposition 1.2.1.16), so that

$$x'(x_n) = 0$$

by a classical result of function theory. Since x' is arbitrary,

$$x_n = 0$$

(Corollary 1.3.3.8 b)). ■

Corollary 1.3.3.11 *Let E, F be Banach spaces. Take $u \in \mathcal{L}(E, F)$. Let U be a domain of $\mathbb{K}$ and take $f : U \to E$.*

a) *If f is differentiable at $\alpha \in U$ then $u \circ f$ is differentiable at α and*

$$(u \circ f)'(\alpha) = uf'(\alpha).$$

b) *f is constant iff f is differentiable and $f' = 0$.*

a) is trivial.

b) The necessitiy is trivial. Take $\alpha, \beta \in U$ and assume that $f(\alpha) \neq f(\beta)$. By Corollary 1.3.3.9, there is an $x' \in F'$ with

$$x'(f(\alpha)) \neq x'(f(\beta)).$$

By a), $x' \circ f$ is differentiable, and by the above, its derivative does not vanish identically. Hence, by a), the derivative of f does not vanish identically.

■

Corollary 1.3.3.12 *Let E be a normed space and p a norm on E' equivalent to the canonical norm on E'. Put*

$$q(x) := \sup \{|x'(x)| \mid x' \in E', p(x') \leq 1\}.$$

Then

$$E \longrightarrow \mathbb{R}_+ , \ x \longmapsto q(x)$$

is a norm of E equivalent to the initial norm of E. Let E_q (resp. E'_p) denote the vector space E (resp. E') endowed with the norm q (resp. p). If p is lower semicontinuous on E'_E, then E'_p is the dual of E_q.

Since p is equivalent to the canonical norm on E', there is an $\alpha > 0$ such that

$$\frac{1}{\alpha} p \leq \|\cdot\| \leq \alpha p.$$

Hence

$$\frac{1}{\alpha} q \leq \|\cdot\| \leq \alpha q.$$

(Corollary 1.3.3.8 b)). In particular, q is finite and

$$x \in E, \ q(x) = 0 \Longrightarrow x = 0.$$

Take $x, y \in E$. Then

$$|x'(x+y)| = |x'(x) + x'(y)| \leq |x'(x)| + |x'(y)| \leq q(x) + q(y)$$

for every $x' \in E'$, $p(x') \leq 1$, so that

$$q(x+y) \leq q(x) + q(y).$$

Moreover,

$$q(\beta x) = \sup\{|x'(\beta x)| \mid x' \in E', p(x') \leq 1\} =$$

$$= |\beta| \sup\{|x'(x)| \mid x' \in E', p(x') \leq 1\} = |\beta| q(x)$$

for every $\beta \in \mathbb{K}$. Hence q is a norm on E equivalent to the initial norm of E.

Now assume that p is lower semicontinuous on E'_E. Put

$$p' : E' \longrightarrow \mathbb{R}_+, \quad x' \longmapsto \sup\left\{|x'(x)| \mid x \in E, \quad q(x) \leq 1\right\},$$

$$A' := \{x' \in E' \mid p(x') \leq 1\}.$$

Is is easy to see that $p' \leq p$. Take $x' \in E' \backslash A'$. Since A' is a closed absolutely convex set of E'_E (Proposition 1.2.7.2), there is an $x \in E$ with

$$\sup_{y' \in A'} \operatorname{re} y'(x) < \operatorname{re} x'(x)$$

(Proposition 1.3.1.8). Hence

$$q(x) = \sup_{y' \in A'} |y'(x)| = \sup_{y' \in A'} \operatorname{re} y'(x) < \operatorname{re} x'(x) \leq p'(x') q(x),$$

$$p'(x') > 1.$$

Take $y' \in E' \backslash \{0\}$ and let $\alpha \in]0, p(y')[$. Then $\frac{1}{\alpha} y' \in E' \backslash A'$ so

$$\frac{1}{\alpha} p'(y') = p'\left(\frac{1}{\alpha} y'\right) > 1$$

by the above considerations. It follows successively that

$$p'(y') > \alpha, \quad p'(y') \geq p(y'), \quad p = p',$$

and E'_p is the dual of E_q. ■

Remark. The final assertion does not hold if p is not lower semicontinuous (see Example 1.3.1.16 c)).

Corollary 1.3.3.13 (E. Helly) *Let E be a normed space, $(x'_\iota)_{\iota \in I}$ a finite linearly independent family in E', and $\alpha > 0$. Given $a \in \mathbb{K}^I$, the following are equivalent:*

a) *For every* $\varepsilon > 0$, *there is an* $x \in E$ *such that* $\|x\| \le \alpha + \varepsilon$ *and*

$$x'_\iota(x) = a_\iota$$

for every $\iota \in I$.

b) $|\sum_{\iota\in I} \alpha_\iota a_\iota| \le \alpha \|\sum_{\iota\in I} \alpha_\iota x'_\iota\|$ *for every* $(\alpha_\iota)_{\iota\in I} \in \mathbb{K}^I$.

a $\Rightarrow$ b. Take $\varepsilon > 0$ and x satisfying a). Then

$$|\sum_{\iota\in I} \alpha_\iota a_\iota| = \left|\sum_{\iota\in I} \alpha_\iota x'_\iota(x)\right| = \left|\langle x, \sum_{\iota\in I} \alpha_\iota x'_\iota\rangle\right| \le$$

$$\le \|x\| \left\|\sum_{\iota\in I} \alpha_\iota x'_\iota\right\| \le (\alpha+\varepsilon)\left\|\sum_{\iota\in I} \alpha_\iota x'_\iota\right\|.$$

Since ε is arbitrary,

$$\left|\sum_{\iota\in I} \alpha_\iota a_\iota\right| \le \alpha \left\|\sum_{\iota\in I} \alpha_\iota x'_\iota\right\|.$$

b $\Rightarrow$ a. Put

$$A := \left\{ \left(x'_\iota(x)\right)_{\iota\in I} \,\middle|\, x \in \alpha E^\# \right\},$$

$$p : \mathbb{K}^I \longrightarrow \mathbb{R}_+, \quad b \longmapsto \inf\left\{\beta \in \mathbb{R}_+ \mid b \in \beta A\right\}.$$

Since A is absolutely convex, p is a norm. Endow $\mathbb{K}^I$ with this norm. By Corollary 1.3.3.8 a), there is an $x' \in (\mathbb{K}^I)'$ such that

$$\|x'\| = 1, \quad x'(a) = p(a).$$

Take $(\alpha_\iota)_{\iota\in I} \in \mathbb{K}^I$ such that

$$x'(b) = \sum_{\iota\in I} \alpha_\iota b_\iota$$

for every $b \in \mathbb{K}^I$. By b) (and Proposition 1.2.1.4 b)),

$$p(a) = x'(a) = \sum_{\iota\in I} \alpha_\iota a_\iota \le \alpha \|\sum_{\iota\in I} \alpha_\iota x'_\iota\| = \sup\left\{\left|\sum_{\iota\in I} \alpha_\iota x'_\iota(x)\right| \,\middle|\, x \in \alpha E^\#\right\} =$$

$$= \sup_{b\in A} \left|\sum_{\iota\in I} \alpha_\iota b_\iota\right| = \|x'\| = 1.$$

Hence $a \in (1+\varepsilon)A$ for every $\varepsilon > 0$, and a) now follows. ∎

Proposition 1.3.3.14 *Let E, F be normed spaces.*

a) $(x', y) \in E' \times F \Rightarrow \|\langle \cdot, x' \rangle y\| = \|x'\| \, \|y\|$.

b) *If $E \neq \{0\}$ then F is complete iff $\mathcal{L}(E, F)$ is complete.*

a) follows from

$$\|\langle \cdot, x' \rangle y\| = \sup_{x \in E^{\#}} \|\langle x, x' \rangle y\| = \sup_{x \in E^{\#}} |\langle x, x' \rangle| \, \|y\| = \|x'\| \, \|y\| .$$

b) Assume $\mathcal{L}(E, F)$ complete and let $(y_n)_{n \in \mathbb{N}}$ be a Cauchy sequence in F. Take $x \in E \backslash \{0\}$. By Corollary 1.3.3.5, there is an $x' \in E'$ such that

$$\|x'\| = 1 , \quad \langle x, x' \rangle = 1 .$$

By a), $(\langle \cdot, x' \rangle y_n)_{n \in \mathbb{N}}$ is a Cauchy sequence in $\mathcal{L}(E, F)$. Denote by u its limit. Then

$$ux = \lim_{n \to \infty} \langle x, x' \rangle y_n = \lim y_n .$$

Hence $(y_n)_{n \in \mathbb{N}}$ converges and F is complete.

The converse implication follows from Theorem 1.2.1.9 b). ■

1.3.4 The Transpose of an Operator

Definition 1.3.4.1 (0) (Banach 1929) *Let* E, F *be normed spaces and take* $u \in \mathcal{L}(E, F)$. *The map*

$$u' : F' \longrightarrow E' , \quad y' \longmapsto y' \circ u$$

is called ***the transpose (operator) of*** u.

Theorem 1.3.4.2 (0) *Let* E, F *be normed spaces and take* $u \in \mathcal{L}(E, F)$.

a) u' *is the unique map* $F' \to E'$, *such that*

$$\langle x, u'y' \rangle = \langle ux, y' \rangle$$

for every $(x, y') \in E \times F'$.

b) $u' \in \mathcal{L}(F', E')$, $\|u'\| = \|u\|$.

c) u *is determined by* u'.

a) Given $(x, y') \in E \times F'$,

$$\langle x, u'y' \rangle = \langle x, y' \circ u \rangle = y' \circ u(x) = y'(ux) = \langle ux, y' \rangle .$$

Take $f : F' \to E'$, such that

$$\langle x, f(y') \rangle = \langle ux, y' \rangle$$

for every $(x, y') \in E \times F'$. Then

$$\langle x, f(y') \rangle = \langle x, u'y' \rangle$$

for every $(x, y') \in E \times F'$, so that $f(y') = u'y'$ for all $y' \in F'$. Thus $f = u'$.

b) Take $x', y' \in F'$ and $\alpha, \beta \in \mathbb{K}$. By a),

$$\langle x, u'(\alpha x' + \beta y') \rangle = \langle ux, \alpha x' + \beta y' \rangle = \alpha \langle ux, x' \rangle + \beta \langle ux, y' \rangle =$$

$$= \alpha \langle x, u'x' \rangle + \beta \langle x, u'y' \rangle = \langle x, \alpha u'x' + \beta u'y' \rangle$$

for every $x \in E$. In other words,

$$u'(\alpha x' + \beta y') = \alpha u'x' + \beta u'y' .$$

Hence u' is linear.

Given $y' \in F'$,

$$\|u'y'\| = \|y' \circ u\| \le \|y'\|\,\|u\|$$

(Corollary 1.2.1.5), and so u' is continuous (Proposition 1.2.1.1 d $\Rightarrow$ a) and $\|u'\| \le \|u\|$.

Take $x \in E$ with $ux \neq 0$. By Corollary 1.3.3.8 a), there is a $y' \in F'$ such that

$$\|y'\| = 1\,, \quad y'(ux) = \|ux\|\,.$$

Thus

$$\|ux\| = y'(ux) = y' \circ u(x) = \langle x, u'y' \rangle \le \|x\|\,\|u'y'\| \le \|x\|\,\|u'\|$$

(Proposition 1.2.1.4 b)), so

$$\|u\| \le \|u'\|\,, \qquad \|u\| = \|u'\|\,.$$

c) follows from a) and Corollary 1.3.3.9. ■

Corollary 1.3.4.3 (0) *If E, F are normed spaces, then the map*

$$\mathcal{L}(E,F) \longrightarrow \mathcal{L}(F',E')\,, \quad u \longmapsto u'$$

(Theorem 1.3.4.2 b)) is linear and continuous.

Take $u, v \in \mathcal{L}(E,F)$ and $\alpha, \beta \in \mathbb{K}$. Then

$$\langle (\alpha u + \beta v)x, y' \rangle = \langle \alpha ux + \beta vx, y' \rangle = \alpha \langle ux, y' \rangle + \beta \langle vx, y' \rangle =$$

$$= \alpha \langle x, u'y' \rangle + \beta \langle x, v'y' \rangle = \langle x, \alpha u'y' + \beta v'y' \rangle = \langle x, (\alpha u' + \beta v')y' \rangle$$

for every $(x, y') \in E \times F'$ (Theorem 1.3.4.2 a)). Thus

$$\alpha u' + \beta v' = (\alpha u + \beta v)'$$

(Theorem 1.3.4.2 a)). Hence the map

$$\mathcal{L}(E,F) \longrightarrow \mathcal{L}(F',E')\,, \quad u \longmapsto u'$$

is linear. By Theorem 1.3.4.2 b), it is continuous. ■

Corollary 1.3.4.4 (0) *If E is a normed space, then*

$$(1_E)' = 1_{E'}\,.$$ ■

Corollary 1.3.4.5 (0) *Let E, F, G be normed spaces and take*

$$u \in \mathcal{L}(E, F), \quad v \in \mathcal{L}(F, G).$$

Then

$$(v \circ u)' = u' \circ v'.$$

Given $(x, z') \in E \times G'$,

$$\langle v \circ u(x), z' \rangle = \langle vux, z' \rangle = \langle ux, v'z' \rangle = \langle x, u'v'z' \rangle = \langle x, u' \circ v'(z') \rangle$$

(Theorem 1.3.4.2 a)). Thus

$$u' \circ v' = (v \circ u)'$$

(Theorem 1.3.4.2 a)). ■

Corollary 1.3.4.6 *The transpose of a projection is a projection.*

This is an immediate consequence of Corollary 1.3.4.5. ■

Corollary 1.3.4.7 (0) *Let E, F be normed spaces and take $u \in \mathcal{L}(E, F)$. If u is an isomorphism (isometry), then u' is an isomorphism (isometry).*

Let u be an isomorphism and let $v := u^{-1}$. Then

$$u \circ v = 1_F, \quad v \circ u = 1_E.$$

Thus

$$v' \circ u' = 1_{F'}, \quad u' \circ v' = 1_{E'}$$

(Corollary 1.3.4.5, Corollary 1.3.4.4), i.e. u' is an isomorphism.

If u is an isometry, then for $y' \in F'$,

$$\|u'y'\| = \sup_{x \in E^\#} |\langle x, u'y' \rangle| = \sup_{x \in E^\#} |\langle ux, y' \rangle| = \sup_{y \in E^\#} |\langle y, y' \rangle| = \|y'\|$$

(Proposition 1.2.1.4 b), Theorem 1.3.4.2 a)), i.e. u' is an isometry. ■

Remark. The reverse implication,

$$u' \text{ isomorphism (isometry)} \Longrightarrow u \text{ isomorphism (isometry)},$$

holds whenever E and F are complete (Corollary 1.4.2.5).

Corollary 1.3.4.8 (2) (3) *Let E, F, G, H be normed spaces. Take $u \in \mathcal{L}(E,F)$, $v \in \mathcal{L}(G,H)$ and let $((y'_\iota, z_\iota))_{\iota\in I}$ be a finite family in $F' \times G$. Then*

$$\Big(\sum_{\iota\in I}\langle\cdot, y'_\iota\rangle z_\iota\Big)\circ u = \sum_{\iota\in I}\langle\cdot, u'y'_\iota\rangle z_\iota\,, \qquad v\circ\Big(\sum_{\iota\in I}\langle\cdot, y'_\iota\rangle z_\iota\Big) = \sum_{\iota\in I}\langle\cdot, y'_\iota\rangle v z_\iota\,.$$

Given $x \in E$ and $y \in F$,

$$\Big(\sum_{\iota\in I}\langle\cdot, y'_\iota\rangle z_\iota\Big)ux = \sum_{\iota\in I}\langle ux, y'_\iota\rangle z_\iota = \sum_{\iota\in I}\langle x, u'y'_\iota\rangle z_\iota =$$

$$= \Big(\sum_{\iota\in I}\langle\cdot, u'y'_\iota\rangle z_\iota\Big)x$$

(Theorem 1.3.4.2 a)), and

$$v\Big(\sum_{\iota\in I}\langle\cdot, y'_\iota\rangle z_\iota\Big)y = v\Big(\sum_{\iota\in I}\langle y, y'_\iota\rangle z_\iota\Big) = \sum_{\iota\in I}\langle y, y'_\iota\rangle v z_\iota = \Big(\sum_{\iota\in I}\langle\cdot, y'_\iota\rangle v z_\iota\Big)y$$

which proves the assertion. ■

Corollary 1.3.4.9 (0) *Let E and F be normed spaces. Take*

$$u \in \mathcal{L}(F', E')\,.$$

Then the following are equivalent:

a) *The map $u : F'_F \to E'_E$ is continuous.*

b) *There is a $v \in \mathcal{L}(E,F)$ such that $u = v'$.*

The operator u of b) *is unique and is called* ***pretranspose of*** u.

a $\Rightarrow$ b. Take $x \in E$. By a), the linear map

$$F'_F \longrightarrow \mathbb{K}\,, \quad y' \longmapsto \langle x, uy'\rangle$$

is continuous. By Corollary 1.2.6.5, there is a $vx \in F$ such that

$$\langle x, uy'\rangle = \langle vx, y'\rangle$$

for every $y' \in F'$. It is obvious that $v : E \to F$ is linear. If $x \in E^\#$, then

$$|\langle vx, y'\rangle| = |\langle x, uy'\rangle| \le \|uy'\| \le \|u\|\,\|y'\|$$

for every $y' \in F'$. By Corollary 1.3.3.8 b),

$$\|vx\| \le \|u\| \, .$$

Hence v is continuous. By Theorem 1.3.4.2 a), $v' = u$.

b $\Rightarrow$ a. By Theorem 1.3.4.2 a),

$$\langle x, uy' \rangle = \langle vx, y' \rangle$$

for every $(x, y') \in E \times F'$. a) now follows.

The uniqueness of v follows from Theorem 1.3.4.2 c). ∎

Theorem 1.3.4.10 (Banach–Stone) *Let S, T be compact spaces. Given $u \in \mathcal{L}(\mathcal{C}(S), \mathcal{C}(T))$, the following are equivalent:*

a) *u is an isometry.*

b) *There is a homeomorphism $f : T \to S$ and a $y \in \mathcal{C}(T)$ such that*

$$|y(t)| = 1$$

for every $t \in T$ and

$$ux = y(x \circ f)$$

for every $x \in \mathcal{C}(S)$.

a $\Rightarrow$ b. We identify $\mathcal{C}(S)'$, $\mathcal{C}(T)'$ with the Banach space of Radon measures on S and T, respectively. By Corollary 1.3.4.7, u' is an isometry. Given $s \in S$ (resp. $t \in T$), let δ_s (resp. δ_t) denote the Dirac measure on S (resp. on T) at s (resp. t). Take $t \in T$. Then δ_t is an extreme point of $\mathcal{C}(T)'^{\#}$ (Example 1.2.7.14) and so $u'\delta_t$ is an extreme point of $\mathcal{C}(S)'^{\#}$. By Example 1.2.7.14, there are $f(t) \in S$ and $y(t) \in \mathbb{K}$ such that

$$|y(t)| = 1 \, , \quad u'\delta_t = y(t)\delta_{f(t)} \, .$$

f is injective, for otherwise u' would not be injective, f is also surjective, since u' is surjective. Take $x \in \mathcal{C}(S)$. Then

$$y(t)x(f(t)) = \langle x, y(t)\delta_{f(t)} \rangle = \langle x, u'\delta_t \rangle = \langle ux, \delta_t \rangle = ux(t) \, .$$

We see from this relation that y and f are continuous. Since f is bijective, it is a homeomorphism. Moreover,

$$ux = y(x \circ f) \, .$$

b $\Rightarrow$ a is easy to see. ∎

Example 1.3.4.11 *Take $p \in [1, \infty[$, and let q be the conjugate exponent of p. Let u be the right (left) shift in ℓ^p. Then u' is the left (right) shift in ℓ^q, where ℓ^q is identified with $(\ell^p)'$ (Example 1.2.2.3 d)).*

Let v be the left (resp. right) shift in ℓ^q. Then

$$\langle ux, y \rangle = \sum_{n=2}^{\infty} x_{n-1} y_n = \sum_{n=1}^{\infty} x_n y_{n+1} = \langle x, vy \rangle$$

$$(\text{resp. } \langle ux, y \rangle = \sum_{n=1}^{\infty} x_{n+1} y_n = \sum_{n=2}^{\infty} x_n y_{n-1} = \langle x, vy \rangle)$$

for every $(x, y) \in \ell^p \times \ell^q$ and the assertion follows from Theorem 1.3.4.2 a). ■

Example 1.3.4.12 *Let S, T be sets. Take $p, q \in [1, \infty] \cup \{0\}$ and let p' and q' be the conjugate exponents of p and q, respectively. Take*

$$k \in \ell^{p,q'}(S, T)$$

and put

$$\overset{\cap}{k} : \ell^q(T) \longrightarrow \ell^p(S), \quad x \longmapsto \overset{\cap}{k} x$$

(Proposition 1.2.3.4 b)). If either S or q is finite, then

$$(\overset{\cap}{k})' x' = \overset{\cup}{k} x'$$

for every $x' \in \ell^{p'}(S)$, where $\ell^{p'}(S)$ has been identified with a subset of $\ell^p(S)'$ (Example 1.2.2.3 b)).

Given $x \in \ell^q(T)$,

$$\langle x, (\overset{\cap}{k})' x' \rangle = \langle \overset{\cap}{k} x, x' \rangle = \langle \sum_{t \in T} k(\cdot, t) x(t), x' \rangle = \sum_{s \in S} \Big(\sum_{t \in T} k(s,t) x(t) \Big) x'(s) =$$

$$= \sum_{t \in T} x(t) \Big(\sum_{s \in S} k(s,t) x'(s) \Big) = \langle x, \overset{\cup}{k} x' \rangle$$

(Theorem 1.3.4.2 a)). ■

Remark. If S and T are finite, then k is the matrix associated to $\overset{\cap}{k}$ and the transpose of k is the matrix associated to $(\overset{\cap}{k})'$.

Example 1.3.4.13 *Take $n \in \mathbb{N}$, and let $u : \mathbb{K}^n \to \mathbb{K}^2$ be a linear map, with associated matrix $[a_{ij}]_{i\in\mathbb{N}_2, j\in\mathbb{N}_n}$. If we endow $\mathbb{K}^n$ and $\mathbb{K}^2$ with the Euclidean norms, then*

$$\|u\|^2 = \frac{1}{2}\left(\sum_{j=1}^{n} |a_{1j}|^2 + \sum_{j=1}^{n} |a_{2j}|^2 + \right.$$

$$\left. + \sqrt{\begin{array}{c}\left(\sum_{j=1}^{n} |a_{1j}|^2 + \sum_{j=1}^{n} |a_{2j}|^2\right)^2 - \\ -4\left(\left(\sum_{j=1}^{n} |a_{1j}|^2\right)\left(\sum_{j=1}^{n} |a_{2j}|^2\right)^2 - |\sum_{j=1}^{n} \alpha_{1j}\overline{\alpha}_{2j}|^2\right)\end{array}}\ \right) =$$

$$= \frac{1}{2}\left(\sum_{j=1}^{n}\left(|\alpha_{1j}|^2 + |\alpha_{2j}|^2\right) + \right.$$

$$\left. + \sqrt{\left(\sum_{j=1}^{n}\left(|\alpha_{1j}|^2 + |\alpha_{2j}|^2\right)\right)^2 - 2\sum_{i,j=1}^{n} |\alpha_{1i}\alpha_{2j} - \alpha_{1j}\alpha_{2j}|^2}\ \right).$$

By Example 1.3.4.12, the matrix associated to $u' : \mathbb{K}'^2 \to \mathbb{K}^n$ is the transpose of the matrix $[\alpha_{ij}]_{i\in\mathbb{N}_2, j\in\mathbb{N}_n}$, and by Theorem 1.3.4.2 b),

$$\|u'\| = \|u\|.$$

The assertion follows now from Example 1.2.2.7 d). ■

Example 1.3.4.14 (1) *Let S, T be sets, p, p' weakly conjugate exponents, and q, q' be conjugate exponents. Take $k \in \ell^{p',q}(S,T)$ and put*

$$\overset{\cup}{k} : \ell^p(S) \longrightarrow \ell^q(T), \quad x \longmapsto \overset{\cup}{k}x$$

(Proposition 1.2.3.8 b)). If T or p is finite, then

$$(\overset{\cup}{k})'x' = \overset{\cap}{k}x'$$

for every $x' \in \ell^{q'}(T)$, where $\ell^{q'}(T)$ has been identified with a subset of $\ell^q(T)'$ (Example 1.2.2.3 b)).

Given $x \in \ell^p(S)$,

$$\langle x, (\overset{\cup}{k})' x' \rangle = \langle \overset{\cup}{k} x, x' \rangle = \langle \sum_{s \in S} k(s, \cdot) x(s), x' \rangle = \sum_{t \in T} \Big(\sum_{s \in S} k(s,t) x(s) \Big) x'(t) =$$

$$= \sum_{s \in S} x(s) \Big(\sum_{t \in T} k(s,t) x'(t) \Big) = \langle x, \overset{\cap}{k} x' \rangle$$

(Theorem 1.3.4.2 a)). ∎

Example 1.3.4.15 *Let S, T be locally compact spaces, $f : S \rightarrow T$ a proper continuous map and put*

$$u : \mathcal{C}_0(T) \longrightarrow \mathcal{C}_0(S), \quad x \longmapsto x \circ f .$$

Then for each $\mu \in \mathcal{M}_b(S)$, $u'\mu$ is the image $f(\mu)$ of μ (Example 1.2.2.10).

For the proof see, for example, N. Bourbaki, Intégration (1956), Ch. V, §6.4. ∎

1.3.5 Polar Sets

Definition 1.3.5.1 (0) *Let E be a normed space, F a vector subspace of E, and G a vector subspace of E'. Put*

$$F^\circ := \{x' \in E' \mid x'|F = 0\} \quad (\textbf{\textit{polar of}}\ \ F),$$

$$^\circ G := \bigcap_{x' \in G} \operatorname{Ker} x' \quad (\textbf{\textit{prepolar of}}\ G).$$

F° is a closed vector subspace of E'_E and $^\circ G$ is a closed vector subspace of E. We have

$$\{0\}^\circ = E', \quad ^\circ\{0\} = E,$$

$$E^\circ = \{0\}, \quad ^\circ E' = \{0\},$$

(the last equality follows e.g. from Corollary 1.3.3.8 a)).

Proposition 1.3.5.2 (0) *Let E be a normed space, F a closed vector subspace of E, and $q : E \to E/F$ the quotient map. Then*

$$\operatorname{Im} q' = F^\circ$$

and the map

$$(E/F)' \longrightarrow F^\circ, \quad y' \longmapsto q'y'$$

is an isometry. In particular, F° is a dual space.

The assertion follows from Proposition 1.2.4.7. ∎

Remark. $(E/F)'$ and F° are frequently identified using the above isometry.

Proposition 1.3.5.3 (0) *Let E be a normed space, F a vector subspace of E', and $(x_\iota)_{\iota\in I}$ a finite family in E such that no nontrivial linear combination of $(x_\iota)_{\iota\in I}$ belongs to $^\circ F$. Then there is a family $(x'_\iota)_{\iota\in I}$ in F, with*

$$x'_\iota(x_\lambda) = \delta_{\iota\lambda}$$

for every $\iota, \lambda \in I$.

We prove the assertion by induction on the cardinality of I. Take $\lambda \in I$ and put $J := I \backslash \{\lambda\}$. By the inductive hypothesis, there is a family $(y'_\iota)_{\iota \in J}$ with

$$y'_\iota(x_\mu) = \delta_{\iota\mu}$$

for every $\iota, \mu \in J$. Given $\iota \in I$, put

$$\widetilde{x}_\iota : F \longrightarrow \mathbb{K}, \quad x' \longmapsto x'(x_\iota).$$

Assume that

$$\bigcap_{\iota \in J} \operatorname{Ker} \widetilde{x}_\iota \subset \operatorname{Ker} \widetilde{x}_\lambda .$$

(If $J = \emptyset$, replace the intersection by F.) Then, by Lemma 1.2.6.3, there is a family $(\alpha_\iota)_{\iota \in J}$ in $\mathbb{K}$ with

$$\widetilde{x}_\lambda = \sum_{\iota \in J} \alpha_\iota \widetilde{x}_\iota .$$

Thus

$$x'\Big(x_\lambda - \sum_{\iota \in J} \alpha_\iota x_\iota\Big) = \widetilde{x}_\lambda(x') - \sum_{\iota \in J} \alpha_\iota \widetilde{x}_\iota(x') = 0$$

for every $x' \in F$ and so

$$x_\lambda - \sum_{\iota \in J} \alpha_\iota x_\iota \in {}^\circ F .$$

This contradicts the hypothesis of the proposition . Hence, we can find an

$$x' \in \Big(\bigcap_{\iota \in J} \operatorname{Ker} \widetilde{x}_\iota\Big) \backslash \operatorname{Ker} \widetilde{x}_\lambda .$$

Now put

$$x'_\lambda := \frac{1}{x'(x_\lambda)} x'$$

and

$$x'_\iota := y'_\iota - y'_\iota(x_\lambda) x'_\lambda$$

for $\iota \in J$. The family $(x'_\iota)_{\iota \in I}$ has the required properties. ■

Proposition 1.3.5.4 (0) *Let E be a normed space and F a vector subspace of E'. Then $({}^\circ F)^\circ$ is the closure of F in E'_E.*

Let $\overline{F}$ be the closure of F in E'_E. It follows from $F \subset ({}^\circ F)^\circ$ that $\overline{F} \subset ({}^\circ F)^\circ$. Take $x' \in E' \backslash \overline{F}$. We prove that $x' \notin ({}^\circ F)^\circ$. Assume the contrary. Let A be a finite subset of E with

$$\bigcap_{x \in A} \left\{ y' \in E \,\middle|\, |y'(x) - x'(x)| < 1 \right\} \subset E' \backslash \overline{F}$$

(Proposition 1.2.6.2). Let G be the (finite–dimensional) vector subspace of E generated by A. Let $(x_\iota)_{\iota \in I}$ be an algebraic basis of G such that $\{x_\iota \mid \iota \in I, \, x_\iota \in {}^\circ F\}$ is an algebraic base of $G \cap {}^\circ F$. Put

$$J := \{\iota \in I \mid x_\iota \notin {}^\circ F\}.$$

By Proposition 1.3.5.3, there is a family $(x'_\iota)_{\iota \in J}$ in F with

$$x'_\iota(x_\lambda) = \delta_{\iota\lambda}$$

for $\iota, \lambda \in J$. Put

$$y' := \sum_{\iota \in J} x'(x_\iota) x'_\iota \in F.$$

Then

$$y'(x_\iota) = x'(x_\iota)$$

for every $\iota \in J$ and

$$y'(x_\iota) = 0 = x'(x_\iota)$$

for every $\iota \in I \backslash J$. Hence $y' - x' = 0$ on G, so that $y' \notin \overline{F}$, which is the contradiction we sought. ■

Corollary 1.3.5.5 (0) *Let E be a normed space, F a subspace of E' closed in E'_E, and $q : E \to E/{}^\circ F$ the quotient map. Then the map*

$$(E/{}^\circ F)' \longrightarrow F, \quad x' \longmapsto q'x'$$

is an isometry.

The assertion follows immediately from Proposition 1.3.5.2 and Proposition 1.3.5.4. ■

Corollary 1.3.5.6 (0) *Let E be a normed space. Then a subspace of E' is a dual space whenever it is closed in E'_E.*

The assertion follows immediately from Corollary 1.3.5.5. ■

Proposition 1.3.5.7 (0) *Let F be a vector subspace of the normed space E. Then*

$$^\circ(F^\circ) = \overline{F}.$$

The inclusion

$$F \subset {}^\circ(F^\circ)$$

is trivial, so

$$\overline{F} \subset {}^\circ(F^\circ).$$

Assume that

$$\overline{F} \neq {}^\circ(F^\circ).$$

Then there is a $y' \in E'$ with

$$y' \mid \overline{F} = 0, \quad y' \mid {}^\circ(F^\circ) \neq 0$$

(Corollary 1.3.3.5). But then $y' \in (\overline{F})^\circ = F^\circ = ({}^\circ(F^\circ))^\circ$ (Proposition 1.3.5.4), which is a contradiction. Hence

$$\overline{F} = {}^\circ(F^\circ).$$ ■

Proposition 1.3.5.8 (0) *Let E, F be normed spaces and take $u \in \mathcal{L}(E, F)$. Then*

$$\operatorname{Ker} u' = (\operatorname{Im} u)^\circ, \quad \operatorname{Ker} u = {}^\circ(\operatorname{Im} u'), \quad \overline{\operatorname{Im} u} = {}^\circ(\operatorname{Ker} u').$$

Take $(x, y') \in E \times F'$. Then

$$\langle ux, y' \rangle = \langle x, u'y' \rangle$$

(Theorem 1.3.4.2 a)). It follows immediately from this, that

$$y' \in \operatorname{Ker} u' \Longleftrightarrow y' \in (\operatorname{Im} u)^\circ,$$

so that

$$\operatorname{Ker} u' = (\operatorname{Im} u)^\circ$$

and

$$\overline{\operatorname{Im} u} = {}^\circ((\operatorname{Im} u)^\circ) = {}^\circ(\operatorname{Ker} u')$$

(Proposition 1.3.5.7).

By the above equality and Corollary 1.3.3.8 a), it further follows that

$$x \in \operatorname{Ker} u \Longleftrightarrow x \in {}^\circ(\operatorname{Im} u')\,.$$

Hence

$$\operatorname{Ker} u = {}^\circ(\operatorname{Im} u')\,. \qquad \blacksquare$$

Corollary 1.3.5.9 (0) *Let E, F be normed spaces and take $u \in \mathcal{L}(E,F)$. Then u' is injective iff $\operatorname{Im} u$ is dense.*

$$u' \text{ is injective } \Longleftrightarrow \operatorname{Ker} u' = \{0\} \Longleftrightarrow {}^\circ(\operatorname{Ker} u') = F \Longleftrightarrow \overline{\operatorname{Im} u} = F$$

(Corollary 1.3.3.8 a), Proposition 1.3.5.8). ■

Remark. The injectivity of u does not imply that u' is surjective, as the inclusion map $\ell^1 \to c_0$ shows.

Corollary 1.3.5.10 (0) *Let E, F be normed spaces. Take $u \in \mathcal{L}(E,F)$ such that $\operatorname{Im} u$ is closed. Let $q : F \to F/\operatorname{Im} u$ be the quotient map. Then*

$$\operatorname{Im} q' = \operatorname{Ker} u'$$

and the map

$$(F/\operatorname{Im} u)' \longrightarrow \operatorname{Ker} u'\,, \quad x' \longmapsto q'x'$$

is an isometry.

We have

$$\operatorname{Im} q' = (\operatorname{Im} u)^\circ = \operatorname{Ker} u'$$

(Proposition 1.3.5.2, Proposition 1.3.5.8), and the assertion follows from Proposition 1.3.5.2. ■

Corollary 1.3.5.11 *Let E be a normed space and p a projection in E. Then p' is a projection in E' with*

$$\operatorname{Ker} p' = (\operatorname{Im} p)^\circ, \quad \operatorname{Im} p' = (\operatorname{Ker} p)^\circ .$$

By Corollary 1.3.4.6, p' is a projection, and by Proposition 1.3.5.8,

$$\operatorname{Ker} p' = (\operatorname{Im} p)^\circ, \quad \operatorname{Im} p' \subset (^\circ(\operatorname{Im} p'))^\circ = (\operatorname{Ker} p)^\circ .$$

Take $x' \in (\operatorname{Ker} p)^\circ$. Then

$$x - px \in \operatorname{Ker} p$$

so that

$$\langle x, p'x' \rangle = \langle px, x' \rangle = \langle x, x' \rangle$$

for every $x \in E$ (Theorem 1.3.4.2 a)) and

$$x' = p'x' \in \operatorname{Im} p', \quad (\operatorname{Ker} p)^\circ \subset \operatorname{Im} p', \quad \operatorname{Im} p' = (\operatorname{Ker} p)^\circ . \qquad \blacksquare$$

Theorem 1.3.5.12 (0) *Let F be a subspace of the normed space E and $u : F \to E$ the inclusion map.*

a) $\operatorname{Im} u' = F'$.

b) $\operatorname{Ker} u' = F^\circ$.

c) *The factorization $E'/F^\circ \to F'$ of u' through E'/F° is an isometry.*

d) *If G is a closed vector subspace of E'_E and $v : {}^\circ G \to E$ is the inclusion map, then the factorization $E'/G \to (^\circ G)'$ of v' through E'/G is an isometry. In particular, E'/G is a dual space.*

a) follows from the Hahn–Banach Theorem.
b) follows from Proposition 1.3.5.8.
c) Let v be the factorization of u' through E'/F°. Then

$$\|v\| = \|u'\| = \|u\| \le 1$$

(b), Proposition 1.2.4.7, Theorem 1.3.4.2 b)). Take $X' \in E'/F^\circ$. Then

$$\|vX\| \le \|v\| \, \|X'\| \le \|X'\|$$

(Proposition 1.2.1.4 a)). Take $x' \in X'$. Since

$$x'|F = x' \circ u = u'x' = vX',$$

it follows that

$$\langle x, x' \rangle = \langle x, vX' \rangle \leq \|vX'\| \, \|x\|$$

for every $x \in F$ (Proposition 1.2.1.4 a)). Hence

$$\|x'|F\| \leq \|vX'\|.$$

By the Hahn–Banach theorem, there is a $y' \in E'$ with

$$y'|F = x'|F, \quad \|y'\| = \|x'|F\|.$$

Then $y' - x' \in F^\circ$, i.e. $y' \in X'$. Hence

$$\|X'\| \leq \|y'\| = \|x'|F\| \leq \|vX'\|, \quad \|vX'\| = \|X'\|.$$

By a), v is surjective. Hence v is an isometry.

d) follows from c) and Proposition 1.3.5.4. ■

Proposition 1.3.5.13 (0) *Let T be a compact space and let $\mathcal{C}(T)'$ be identified with the Banach space of Radon measures on T. Let $\mathcal{F}$ be a vector subspace of $\mathcal{C}(T)$ such that $x\overline{x}$, $xy \in \mathcal{F}$ for every $x, y \in \mathcal{F}$ and let μ be an extreme point of $\mathcal{F}^\circ \cap \mathcal{C}(T)'^\#$. Then the functions of $\mathcal{F}$ are constant on the support of μ.*

We may assume that $\mu \neq 0$. Let x be a positive real function in $\mathcal{F}$. Assume x is not constant on $\operatorname{Supp}\mu$. Put

$$\alpha := \sup\left\{|x(t)| \,\middle|\, t \in \operatorname{Supp}\mu\right\},$$

$$y := \frac{1}{\alpha}x, \quad \nu := y \cdot \mu.$$

Then

$$\nu \neq 0, \quad \mu - \nu \neq 0,$$

and

$$\|\nu\| + \|\mu - \nu\| = \int y d|\mu| + \int (1-y) d|\mu| = \int d|\mu| = \|\mu\| = 1$$

(Proposition 1.2.7.11). Then $zy, z(1-y) \in \mathcal{F}$, so that

$$\langle z, \nu \rangle = \int zy \, d\mu = 0 \,,$$

$$\langle z, \mu - \nu \rangle = \int z(1-y) d\mu = 0$$

for every $z \in \mathcal{F}$. Hence

$$\nu, \mu - \nu \in \mathcal{F}^\circ \,,$$

$$\frac{1}{\|\nu\|}\nu \,, \; \frac{1}{\|\mu - \nu\|}(\mu - \nu) \in \mathcal{F}^\circ \cap \mathcal{C}(T)'^{\#} \,.$$

Since

$$\mu = \|\nu\| \frac{1}{\|\nu\|} \nu + \|\mu - \nu\| \frac{1}{\|\mu - \nu\|} (\mu - \nu)$$

and since μ is an extreme point of $\mathcal{F}^\circ \cap \mathcal{C}(T)'^{\#}$,

$$\frac{1}{\|\nu\|}\nu = \mu \,, \quad y = \|\nu\| \text{ on } \operatorname{Supp}\mu \,, \quad x = \alpha\|\nu\| \text{ on } \operatorname{Supp}\mu \,.$$

Hence x is constant on $\operatorname{Supp}\mu$.

Now let x be an arbitrary function in $\mathcal{F}$ and take $s, t \in \operatorname{Supp}\mu$ with

$$x(s) \neq x(t) \,.$$

Since $x\overline{x}$ is a positive real function in $\mathcal{F}$, it follows from the above that $x\overline{x}$ takes the same values at s and t. Put

$$y := \begin{cases} x\overline{x} & \text{if } x(s)x(t) = 0 \\ |x\overline{x} - \overline{x(s)}x|^2 & \text{if } x(s)x(t) \neq 0 \,. \end{cases}$$

Then y is a positive real function in $\mathcal{F}$ taking different values at s and t, which contradicts the above result. Hence x is constant on $\operatorname{Supp}\mu$. ∎

Theorem 1.3.5.14 (0) *Let T be a compact space and $\mathcal{F}$ a vector subspace of $\mathcal{C}(T)$ such that $x\overline{x}, xy \in \mathcal{F}$ for every $x, y \in \mathcal{F}$. Put*

$$S := \bigcap_{x \in \mathcal{F}} \overset{-1}{x}(0)$$

and

$$s \sim t :\Longleftrightarrow \Big(x \in \mathcal{F} \Longrightarrow x(s) = x(t) \Big)$$

for $s, t \in T \backslash S$. Let $\mathcal{G}$ denote the set of $x \in \mathcal{C}(T)$ which vanish on S and for which

$$x(s) = x(t)$$

whenever $s, t \in T \backslash S$ satisfy $s \sim t$. Then $\overline{\mathcal{F}} = \mathcal{G}$.

Take $x \in \mathcal{G}$. Identify $\mathcal{C}(T)'$ with the Banach space of Radon measures on T. Let μ be an extreme point of $\mathcal{F}^\circ \cap \mathcal{C}(T)'^{\#}$. By Proposition 1.3.5.13, the functions in $\mathcal{F}$ are constant on $\operatorname{Supp}\mu$. Hence there is a $y \in \mathcal{F}$ with $x = y$ on $\operatorname{Supp}\mu$ and so

$$\langle x, \mu \rangle = \langle y, \mu \rangle = 0 \,.$$

$\mathcal{F}^\circ$ is a closed vector subspace of $\mathcal{C}(T)'_{\mathcal{C}(T)}$. Hence by the Alaoglu–Bourbaki Theorem, $\mathcal{F}^\circ \cap \mathcal{C}(T)'^{\#}$ is a compact set of $\mathcal{C}(T)'_{\mathcal{C}(T)}$. By the Krein–Milman Theorem, $\mathcal{F}^\circ \cap \mathcal{C}(T)'^{\#}$ is the smallest convex closed set of $\mathcal{C}(T)'_{\mathcal{C}(T)}$ containing the extreme points of $\mathcal{F}^\circ \cap \mathcal{C}(T)'^{\#}$. By the above,

$$\langle x, \mu \rangle = 0$$

for every $\mu \in \mathcal{F}^\circ \cap \mathcal{C}(T)'^{\#}$. Hence

$$x \in {}^\circ(\mathcal{F}^\circ) = \overline{\mathcal{F}}$$

(Proposition 1.3.5.7) and

$$\mathcal{G} \subset \overline{\mathcal{F}} \,.$$

The reverse inclusion is easy to see. ■

Remark. The idea of using extreme points for such denseness problems is due to de Branges (1959).

Corollary 1.3.5.15 (0) *Let T be a locally compact space and $\mathcal{F}$ be a vector subspace of $\mathcal{C}_0(T)$ such that:*

1) *If $x, y \in \mathcal{F}$, then $x\overline{x}, xy \in \mathcal{F}$;*

2) *Given distinct $s, t \in T$ there are $x, y \in \mathcal{F}$ such that $x(s)y(t) \neq x(t)y(s)$.*

Then $\mathcal{F}$ is a dense set of $\mathcal{C}_0(T)$.

Let T^* be the Alexandroff compactification of T and extend each function in $\mathcal{C}_0(T)$ by setting it equal to 0 at the Alexandroff point of T. By 2), S of Theorem 1.3.5.14 contains only the Alexandroff point of T, and the equivalence classes of $\sim$ are one point sets. ■

Corollary 1.3.5.16 (0) (Weierstrass–Stone Theorem, 1885, 1937). *Let T be a compact space and $\mathcal{F}$ a vector subspace of $\mathcal{C}(T)$ such that:*

1) $x, y \in \mathcal{F} \Rightarrow x\overline{x}\,,\ xy \in \mathcal{F}$.

2) *Given distinct* $s, t \in T$, *there are* $x, y \in \mathcal{F}$ *with* $x(s)y(t) \neq x(t)y(s)$.

Then $\mathcal{F}$ *is a dense set of* $\mathcal{C}(T)$. ■

Corollary 1.3.5.17 (0) *Let* T *be a set and* $\mathcal{F}$ *a closed vector subspace of* $\ell^\infty(T)$ *such that* $x\overline{x}\,,\ xy \in \mathcal{F}$ *for every* $x, y \in \mathcal{F}$. *Take* $x \in \mathcal{F}$ *and* $f \in \mathcal{C}(\overline{x(T)})$ *such that*

$$0 \in \overline{x(T)}\,,\ e_T \notin \mathcal{F} \Longrightarrow f(0) = 0\,.$$

Then $f \circ x \in \mathcal{F}$.

Let $\mathcal{G}$ be the smallest vector subspace of $\mathcal{C}(\overline{x(T)})$ such that:

1) the function

$$\overline{x(T)} \longrightarrow \mathbb{K}\,, \quad \alpha \longmapsto \alpha$$

is in $\mathcal{G}$,

2) $e_{\overline{x(T)}} \in \mathcal{G}$ whenever $e_T \in \mathcal{F}$,

3) $g\overline{g}\,,\ gh \in \mathcal{G}$ for every $g, h \in \mathcal{G}$.

Then $f \circ x \in \mathcal{F}$ for every $f \in \mathcal{G}$.By Theorem 1.3.5.14, f belongs to the closure of $\mathcal{G}$ in $\mathcal{C}(\overline{x(T)})$. Hence there is a sequence $(f_n)_{n\in\mathbb{N}}$ in $\mathcal{G}$ converging uniformly to f. Then $(f_n \circ x)_{n\in\mathbb{N}}$ converges uniformly to $f \circ x$. Since $\mathcal{F}$ is closed and $f_n \circ x \in \mathcal{F}$ for every $n \in \mathbb{N}$, it follows that $f \circ x \in \mathcal{F}$. ■

Corollary 1.3.5.18 *Let* T *be a compact space and* $\mathcal{F}$ *a vector subspace of* $\mathcal{C}(T)$ *such that* $xy \in \mathcal{F}$ *for every* $(x, y) \in \mathcal{C}(T) \times \mathcal{F}$. *Put*

$$S := \bigcap_{x\in\mathcal{F}} \overset{-1}{x}(0)\,.$$

Then

$$\overline{\mathcal{F}} = \{x \in \mathcal{C}(T) \mid x|S = 0\}\,.$$

Given distinct $s, t \in T\backslash S$, there is an $x \in \mathcal{F}$ with

$$x(s) \neq x(t)\,.$$

The assertion thus follows from Theorem 1.3.5.14. ■

Proposition 1.3.5.19 *Let T be a locally compact space and $\mathcal{F}$ a vector subspace of $\mathcal{C}(T)$ such that:*

1) $\overline{x}x$, $xy \in \mathcal{F}$ *for every* $x, y \in \mathcal{F}$.

2) *For distinct* $s, t \in T$, *there are* $x, y \in \mathcal{F}$ *with*

$$x(s)y(t) - x(t)y(s) \neq 0 .$$

3) $e_T \in \mathcal{F}$.

Let $\mathfrak{T}$ be the coarsest topology on $\mathcal{C}(T)$ for which the functions

$$\mathcal{C}(T) \longrightarrow \mathbb{K}, \quad x \longmapsto \int x d\mu$$

are continuous for every bounded Radon measure μ on T. Then $\mathcal{F}$ is dense in $\mathcal{C}(T)$ with respect to $\mathfrak{T}$.

Let βT be the Stone–Čech compactification of T. We consider the functions in $\mathcal{F}$ to be extended continuously on βT and put

$$\varphi : \beta T \longrightarrow \mathbb{K}^{\mathcal{F}}, \quad t \longmapsto (x(t))_{x \in \mathcal{F}} .$$

Then φ is continuous and, by 2), it is injective on T.

Let $x \in \mathcal{C}(T)$ have compact support. Let $(\mu_\iota)_{\iota \in I}$ be a finite family in $\mathcal{M}_b(T)$ and take $\varepsilon > 0$. T has a compact set K such that

$$\operatorname{Supp} x \subset K$$

and

$$|\mu_\iota|(T \backslash K) < \frac{\varepsilon}{2(1 + \|x\|)}$$

for every $\iota \in I$. Then the map

$$\psi : K \longrightarrow \varphi(K), \quad t \longmapsto \varphi(t)$$

is a homeomorphism. Hence $x|K = y \circ \psi$ for some $y \in \mathcal{C}(\varphi(K))$. By Tietze's Theorem, y can be extended to a continuous function on $\varphi(\beta T)$ such that

$$\|y\| = \|x\| .$$

Then

$$|\langle x - y \circ \varphi, \mu_\iota \rangle| \leq \int |x - y \circ \varphi| d|\mu_\iota| =$$

$$= \int_{T \backslash K} |y \circ \varphi| d|\mu_\iota| \leq \|y\| \, |\mu_\iota|(T \backslash K) \leq \|x\| \frac{\varepsilon}{2(1 + \|x\|)} < \frac{\varepsilon}{2}$$

for every $\iota \in I$. Take $s, t \in \beta T$ such that $z(s) = z(t)$ for every $z \in \mathcal{F}$. Then $\varphi(s) = \varphi(t)$ so that

$$y \circ \varphi(s) = y \circ \varphi(t) .$$

By 3) and Theorem 1.3.5.14, there is a $z \in \mathcal{F}$ such that

$$\|y \circ \varphi - z\| < \frac{\varepsilon}{2\left(1 + \sum_{\iota \in I} \|\mu_\iota\|\right)} .$$

Then

$$|\langle x - z, \mu_\iota \rangle| \leq |\langle x - y \circ \varphi, \mu_\iota \rangle| + |\langle y \circ \varphi - z, \mu_\iota \rangle| <$$

$$< \frac{\varepsilon}{2} + \frac{\varepsilon}{2\left(1 + \sum_{\iota \in I} \|\mu_\iota\|\right)} \|\mu_\iota\| < \varepsilon$$

for every $\iota \in I$. Hence x is in the closure of $\mathcal{F}$ with respect to $\mathfrak{T}$. Since the set of functions of $\mathcal{C}(T)$ with compact support is dense in $\mathcal{C}(T)$ with respect to $\mathfrak{T}$, it follows that $\mathcal{F}$ is dense in $\mathcal{C}(T)$ with respect to $\mathfrak{T}$. ■

1.3.6 The Bidual

Definition 1.3.6.1 (0) *Let E be a normed space. The dual of E' is called* ***the bidual of*** E *and is denoted by E''* (Hahn, 1927). *The dual of E'' is called* ***the tridual of*** E *and is denoted by E'''.*

Example 1.3.6.2 (3) *If T is a set, then the bidual of $c_0(T)$ is isometric to $\ell^\infty(T)$. If we endow T with the discrete topology, then the tridual of $c_0(T)$ is isometric to the Banach space of Radon measures on the Stone-Čech compactification of T.*

The first assertion follows from Example 1.2.2.3 d), e). The second assertion follows from the first one and Example 1.2.2.11. ■

Theorem 1.3.6.3 (0) (Hahn, 1927) *Let E be a normed space.*

a) $\langle x, \cdot \rangle \in E''$ *and* $\|\langle x, \cdot \rangle\| = \|x\|$ *for every* $x \in E$.

b) *The map*

$$E \longrightarrow E'', \quad x \longmapsto \langle x, \cdot \rangle$$

is injective, linear, and continuous with norm 1 *if* $E \neq \{0\}$. *It is called* ***the evaluation of*** E *and is denoted by* j_E.

a) $\langle x, \cdot \rangle$ is linear and

$$\|x\| = \sup_{x' \in E^\#} |\langle x, x' \rangle| = \|\langle x, \cdot \rangle\|$$

(Corollary 1.3.3.8 b), Proposition 1.2.1.4 b)).

b) The map is obviously linear. The other assertions follow from a) and Proposition 1.2.1.1 d $\Rightarrow$ a. ■

Corollary 1.3.6.4 (0) *A normed space is finite–dimensional iff its dual is finite–dimensional.*

The necessity was proved in Corollary 1.2.4.10. If the dual of a normed space E is finite–dimensional, then its bidual is finite–dimensional too. Thus, by Theorem 1.3.6.3 b), E is finite–dimensional. ■

Corollary 1.3.6.5 (0) *If E is a normed space, then* $\operatorname{Im} j_E$ *is a dense set of* $E''_{E'}$ *and the map*

$$E \longrightarrow \operatorname{Im} j_E, \quad x \longmapsto j_E x$$

is an isometry. E is complete iff $\operatorname{Im} j_E$ *is closed in* E''.

We have

$$^\circ(\operatorname{Im} j_E) = E^\circ = \{0\},$$

so that

$$\left(^\circ(\operatorname{Im} j_E)\right)^\circ = E''.$$

By Proposition 1.3.5.4, $\operatorname{Im} j_E$ is dense in $E''_{E'}$.

By Theorem 1.3.6.3, the map

$$E \longrightarrow \operatorname{Im} j_E, \quad x \longmapsto j_E x$$

is an isometry. Hence E is complete iff $\operatorname{Im} j_E$ is complete, and this is equivalent to $\operatorname{Im} j_E$ being closed in E'' (Corollary 1.2.1.10). ■

Remark. a) E is frequently identified with $\operatorname{Im} j_E$ via the above isometry.

b) The first assertion of the corollary will be strengthened in Corollary 1.3.6.8.

Corollary 1.3.6.6 (3) *If E, F are normed spaces, then*

$$\left(\sum_{\iota\in I}\langle\cdot, x'_\iota\rangle y_\iota\right)' = \sum_{\iota\in I}\langle\cdot, j_E y_\iota\rangle x'_\iota$$

for every finite family $((x'_\iota, y_\iota))_{\iota\in I}$ in $E' \times F$.

We have

$$\left\langle\left(\sum_{\iota\in I}\langle\cdot, x'_\iota\rangle y_\iota\right)x, y'\right\rangle = \left\langle\sum_{\iota\in I}\langle x, x'_\iota\rangle y_\iota, y'\right\rangle =$$

$$= \sum_{\iota\in I}\langle x, x'_\iota\rangle\langle y_\iota, y'\rangle = \sum_{\iota\in I}\langle x, x'_\iota\rangle\langle y', j_F y_\iota\rangle =$$

$$= \left\langle x, \sum_{\iota\in I}\langle y', j_F y_\iota\rangle x'_\iota\right\rangle = \left\langle x, \left(\sum_{\iota\in I}\langle\cdot, j_F y_\iota\rangle x'_\iota\right)y'\right\rangle$$

for every $(x, y') \in E \times F'$ and the assertion now follows from Theorem 1.3.4.2 a). ■

Proposition 1.3.6.7 *Let E be a normed space. Take $x'' \in E''$. Let $(x'_\iota)_{\iota\in I}$ be a finite family in E' and take $\varepsilon > 0$. Then there is an $x \in E$ such that*

$$\|x\| < \|x''\| + \varepsilon$$

and

$$\langle x, x'_\iota \rangle = \langle x'', x'_\iota \rangle$$

for every $\iota \in I$.

Given $(\alpha_\iota)_{\iota\in I} \in \mathbb{K}^I$,

$$\left|\sum_{\iota\in I} \alpha_\iota \langle x'', x'_\iota \rangle\right| = \left|\langle x'', \sum_{\iota\in I} \alpha_\iota x'_\iota \rangle\right| \le \|x''\| \left\|\sum_{\iota\in I} \alpha_\iota x'_\iota\right\|$$

and the assertion follws from Corollary 1.3.3.13 b $\Rightarrow$ a. ■

Corollary 1.3.6.8 (Goldstine, 1938) *If E is a normed space, then $j_E(E^\#)$ is dense in $E''^\#_{E'}$.* ■

Remark. The above result also follows from Proposition 1.3.1.8.

Definition 1.3.6.9 $\big(\,0\,\big)$ *Let E be a normed space, A a subset of E, and A' a subset of E'. We identify A with $j_E(A)$ (Theorem 1.3.6.3) and set*

$$A_{A'} := (j_E(A))_{A'}$$

(Definition 1.2.6.1). $A_{A'}$ is the set A endowed with the topology of pointwise convergence in A'. If $A' = E'$, then the corresponding topology is called ***the weak topology****. The words* ***weak*** *and* ***weakly*** *used in conjunction with a topological term will signify that this term is considered with respect to the weak topology.*

By Theorem 1.3.6.3, the weak topology is coarser than the norm topology. By Corollary 1.3.3.5, the weak topology is completely regular.

Proposition 1.3.6.10 *Let A be a convex set of the normed space E. Then A is closed iff it is weakly closed.*

Let A be closed. We may assume that $A \neq \emptyset$. Take $x \in E\backslash A$. By Corollary 1.3.1.7, there is an $x' \in E'$ such that

$$\sup_{y\in A} \operatorname{re} x'(y) < \operatorname{re} x'(x).$$

Hence x does not belong to the weak closure of A and so A is weakly closed.

The converse implication is trivial. ■

Theorem 1.3.6.11 (I. Schur, 1920) *Let T be a set. Every weak Cauchy sequence in $\ell^1(T)$ is norm convergent.*

Let $(x_n)_{n\in\mathbb{N}}$ be a weak Cauchy sequence in $\ell^1(T)$, i.e.

$$\lim_{n\to\infty} \langle x_n, x' \rangle$$

exists for every $x' \in \ell^1(T)'$. By Example 1.2.2.3 d),

$$\lim_{n\to\infty} \sum_{t\in T} x_n(t)y(t)$$

exists for every $y \in \ell^\infty(T)$ and the assertion now follows from Corollary 1.2.3.13. ■

Proposition 1.3.6.12 (0) *If E is a normed space, then the continuous and the weakly continuous linear forms on E coincide.* ■

Proposition 1.3.6.13 *Let E be a normed space and G a vector subspace of E'. Then ${}^\circ G$ is a weakly closed vector subspace of E.*

Take $x' \in E'$. Then x' is weakly continuous and so $\operatorname{Ker} x'$ is weakly closed. It follows from

$${}^\circ G = \bigcap_{x'\in G} \operatorname{Ker} x'$$

that ${}^\circ G$ is weakly closed. ■

Proposition 1.3.6.14 (0) *Let E be a normed space and F a vector subspace of E. Then*

$$F^\circ = {}^\circ(j_E(F))$$

and $F^{\circ\circ}$ is the closure of $j_E(F)$ in $E''_{E'}$.

Given $(x, x') \in E \times E'$,

$$\langle x, x' \rangle = \langle j_E x, x' \rangle$$

and so

$$F^\circ = {}^\circ(j_E(F)).$$

Hence

$$F^{\circ\circ} = \left({}^\circ(j_E(F))\right)^\circ$$

and $F^{\circ\circ}$ is the closure of $j_E(F)$ in $E''_{E'}$ (Proposition 1.3.5.4). ■

Definition 1.3.6.15 (0) *Let E, F be normed spaces and take $u \in \mathcal{L}(E,F)$. The transpose of u' is called* **the bitranspose of** *u and is denoted by u''.*

Proposition 1.3.6.16 (0) *Let E, F be normed spaces, and take $u \in \mathcal{L}(E,F)$. Then*

$$u'' \circ j_E = j_F \circ u\,.$$

For $x \in E$,

$$\langle u''(j_E x), y' \rangle = \langle j_E x, u'y' \rangle = \langle x, u'y' \rangle = \langle ux, y' \rangle = \langle j_F(ux), y' \rangle$$

whenever $y' \in F'$ (Theorem 1.3.4.2 a)). Thus

$$u'' \circ j_E(x) = u''(j_E x) = j_F(ux) = j_F \circ u(x)$$

and

$$u'' \circ j_E = j_F \circ u\,. \qquad \blacksquare$$

Proposition 1.3.6.17 *Let E be a normed space, F a subspace of E, and $j : F \to E$ the inclusion map. Then*

$$\operatorname{Im} j'' = F^{\circ\circ}$$

and the map

$$F'' \longrightarrow F^{\circ\circ}\,, \quad y'' \longmapsto j''y''$$

is an isometry.

Let $q : E' \to E'/F^\circ$ be the quotient map and $u : E'/F^\circ \to F'$ the factorization of j' through E'/F° (Theorem 1.3.5.12 a)). Then

$$j' = u \circ q\,,$$

so that

$$j'' = q' \circ u'$$

(Corollary 1.3.4.5). By Theorem 1.3.5.12 c) and Corollary 1.3.4.7, u' is an isometry. By Proposition 1.3.5.2,

$$\operatorname{Im} q' = F^{\circ\circ}$$

and the map

$$(E'/F^\circ)' \longrightarrow F^{\circ\circ}, \quad x'' \longmapsto q'x''$$

is an isometry. Hence

$$\operatorname{Im} j'' = F^{\circ\circ},$$

and the map

$$F'' \longrightarrow F^{\circ\circ}, \quad y'' \longmapsto j''y''$$

is an isometry. ■

Proposition 1.3.6.18 *Let E be a normed space, F a closed subspace of E, and $q : E \to E/F$ the quotient map. Then*

$$\operatorname{Ker} q'' = F^{\circ\circ}$$

and the factorization of q'' through $E''/F^{\circ\circ}$ is an isometry.

By Proposition 1.3.5.2,

$$\operatorname{Im} q' = F^\circ$$

and the map

$$u : (E/F)' \longrightarrow F^\circ, \quad y' \longmapsto q'y'$$

is an isometry. Let $j : F^\circ \to E'$ be the inclusion map. Then

$$q' = j \circ u$$

so that

$$q'' = u' \circ j'$$

(Corollary 1.3.4.5). Now

$$\operatorname{Ker} j' = F^{\circ\circ}$$

and the factorization v of j' through $E''/F^{\circ\circ}$ is an isometry (Theorem 1.3.5.12 b), c)). Since u' is an isometry (Corollary 1.3.4.7),

$$\operatorname{Ker} q'' = \operatorname{Ker} j' = F^{\circ\circ}.$$

The factorization of q'' through $E''/F^{\circ\circ}$ is $u' \circ v$ and therefore an isometry. ■

Proposition 1.3.6.19 (0) *Let E be a normed space and let*

$$u := j_{E'} \circ j'_E .$$

a) $j'_E \circ j_{E'} = 1_{E'}$.

b) *u is the projection of E''' onto $\operatorname{Im} j_{E'}$ and $\|u\| \leq 1$; u is called* ***the canonical projection of*** *E''' (better:* ***of the tridual of*** *E).*

c) $\operatorname{Ker} u = (\operatorname{Im} j_E)^\circ$.

d) $E''' = (\operatorname{Im} j_{E'}) \oplus (\operatorname{Im} j_E)^\circ$.

e) *If E is complete and*

$$q : E'' \longrightarrow E''/\operatorname{Im} j_E , \quad r : E''' \longrightarrow E'''/\operatorname{Im} j_{E'}$$

are the quotient maps (Corollary 1.3.6.5), then $r \circ q'$ is an isometry.

a) Given $(x, x') \in E \times E'$,

$$\langle x, j'_E \circ j_{E'}(x') \rangle = \langle j_E x, j_{E'} x' \rangle = \langle j_E x, x' \rangle = \langle x, x' \rangle$$

(Theorem 1.3.4.2 a)), and so

$$j'_E \circ j_{E'} = 1_{E'} .$$

b) By a),

$$u \circ j_{E'} = j_{E'} \circ j'_E \circ j_{E'} = j_{E'} ,$$

$$u \circ u = u \circ j_{E'} \circ j'_E = j_{E'} \circ j'_E = u .$$

Hence u is a projection in E''' and it follows from

$$\operatorname{Im} j_{E'} = \operatorname{Im} (u \circ j_{E'}) \subset \operatorname{Im} u = \operatorname{Im} (j_{E'} \circ j'_E) \subset \operatorname{Im} j_{E'} ,$$

$$\operatorname{Im} u = \operatorname{Im} j_{E'}$$

that u is a projection of E''' onto $\operatorname{Im} j_{E'}$. By Theorem 1.3.6.3,

$$\|j_E\| \leq 1 , \quad \|j_{E'}\| \leq 1 ,$$

so that

$$\|u\| = \|j_{E'} \circ j'_E\| \le \|j_{E'}\| \, \|j'_E\| \le 1$$

(Corollary 1.2.1.5, Theorem 1.3.4.2 b)).

c) By a),

$$j'_E \circ u = j'_E \circ j_{E'} \circ j'_E = j'_E .$$

Hence

$$\operatorname{Ker} j'_E \supset \operatorname{Ker} u = \operatorname{Ker}(j_{E'} \circ j'_E) \supset \operatorname{Ker} j'_E , \quad \operatorname{Ker} u = \operatorname{Ker} j'_E = (\operatorname{Im} j_E)^\circ$$

(Proposition 1.3.5.8).

d) follows from b), c) and Murray's Theorem.

e) The maps

$$(E''/\operatorname{Im} j_E)' \longrightarrow (\operatorname{Im} j_E)^\circ , \quad x' \longmapsto q'x' ,$$

$$(\operatorname{Im} j_E)^\circ \longrightarrow E'''/\operatorname{Im} j_{E'} , \quad x''' \longmapsto rx'''$$

are isometries (d), Proposition 1.3.5.2, Proposition 1.2.5.2 a $\Rightarrow$ c), and the assertion now follows. ■

Corollary 1.3.6.20 (3) *Let βT be the Stone–Čech compactification of the discrete space T. Put*

$$\Delta := \beta T \backslash T ,$$

and take each $x \in \ell^\infty(T)$ to be extended continuously to βT. Identify $c_0(T)'$, $c_0(T)''$, and $c_0(T)'''$ with $\ell^1(T)$, $\ell^\infty(T)$, and the Banach space $\mathcal{M}(\beta T)$ of Radon measures on βT, respectively (Example 1.3.6.2). Let i and j denote the evaluation of $c_0(T)$ and $\ell^1(T)$, respectively, and put

$$u := j \circ i' .$$

Then i and j are the inclusion maps, u is the projection of $\mathcal{M}(\beta T)$ onto $\ell^1(T)$, $\operatorname{Ker} u$ is the Banach space $\mathcal{M}(\Delta)$ of Radon measures on Δ,

$$\mathcal{M}(\beta T) = \ell^1(T) \oplus \mathcal{M}(\Delta) ,$$

and $\ell^\infty(T)/c_0(T)$ and $(\ell^\infty(T)/c_0(T))'$ are canonically isometric to $\mathcal{C}(\Delta)$ and $\mathcal{M}(\Delta)$, respectively. ■

Proposition 1.3.6.21 (0) *Let E be a Banach space and F a closed vector subspace of E'. Put $G := (^\circ F)^\circ$ and let $j : F \to G$ be the inclusion map. Put $H := E/^\circ F$, and let $q : E \to H$ be the quotient map. Finally let u be the isometry*

$$H' \longrightarrow G\,, \quad \xi' \longmapsto q'\xi'$$

(Corollary 1.3.5.5).

a) $\langle x, \eta \rangle = \langle (j_H qx) \circ u^{-1}, \eta \rangle$ *for every* $(x, \eta) \in E \times G$.

b) $(j_H \xi) \circ u^{-1}$ *is continuous on* G_E *whenever* $\xi \in H$.

c) *If* $F^\#$ *is dense in* $G_E^\#$, *then*

$$\|(j_H \xi) \circ u^{-1} \mid F\| = \|\xi\|$$

for every $\xi \in H$.

d) *If* $F^\#$ *is dense in* $G_E^\#$ *and if every continuous linear form on* F *is continuous on* F_E, *then the map*

$$v : H \longrightarrow F'\,, \quad \xi \longmapsto (j_H \xi) \circ u^{-1} \mid F$$

is an isometry and

$$u \circ v' \circ j_F = j\,.$$

e) *Under the hypothesies of d), if we identified* F'' *with* G *via the isometry* $u \circ v'$ *then* j *becomes the evaluation of* F.

a) We have

$$\langle (j_H qx) \circ u^{-1}, \eta \rangle = \langle j_H qx, u^{-1}\eta \rangle = \langle qx, u^{-1}\eta \rangle = \langle x, q'u^{-1}\eta \rangle = \langle x, \eta \rangle\,.$$

b) follows from a).

c) Since $F^\#$ is dense in $G_E^\#$, it follows from b) that

$$\|(j_H \xi) \circ u^{-1} \mid F\| = \|(j_H \xi) \circ u^{-1}\| = \|j_H \xi\| = \|\xi\|$$

(Theorem 1.3.6.3 a)).

d) Let $\xi' \in F'$. By the hypothesis of *d)* and Corollary 1.2.6.5, there is an $x \in E$ such that

$$\xi'(\xi) = \langle x, \xi \rangle$$

for every $\xi \in F$. By a),

$$(j_H q x) \circ u^{-1} \mid F = \xi' .$$

Hence v is surjective. By c), it is an isometry.

Take $\xi \in F$ and $x \in E$. By a),

$$\langle x, uv' j_F \xi \rangle = \langle x, q'v' j_F \xi \rangle = \langle vqx, j_F \xi \rangle = \langle vqx, \xi \rangle = \langle (j_H q x) \circ u^{-1}, \xi \rangle = \langle x, \xi \rangle .$$

We deduce that

$$uv' j_F \xi = \xi = j\xi ,$$

$$u \circ v' \circ j_F = j .$$

e) follows from d) (and Corollary 1.3.4.7). ■

Proposition 1.3.6.22 (1) *Let T be a set and βS the Stone–Čech compactification of the infinite discrete space S. Put*

$$\varDelta := \beta S \backslash S .$$

Take $k \in \ell^{\infty,1}(S,T)$, and let $\widetilde{k}$ be the continuous extention of the map

$$S \longrightarrow \ell^1(T) , \quad s \longmapsto k(s, \cdot)$$

to

$$\beta S \longrightarrow \ell^1(T)''_{\ell^1(T)'} ,$$

where $\ell^1(T)$ is identified with a subspace of $\ell^1(T)''$ via the evaluation map.

a) *$k \in \ell^{0,1}(S,T)$ iff $\widetilde{k} \mid \varDelta = 0$.*

b) *$k \in \ell_0^{\infty,1}(S,T)$ iff $\widetilde{k}(\varDelta) \subset c_0(T)^0$, where $c_0(T)$ is identified with a subspace of $\ell^1(T)'$ via the evaluation map (and Example 1.2.2.3 e)).*

The continuous extention $\widetilde{k}$ exists by the Alaoglu–Bourbaki Theorem.

a) Let $\mathfrak{F}$ be the filter on S consisting of all cofinite subsets of S, i.e.

$$\mathfrak{F} := \{ A \in \mathfrak{P}(S) \mid S \backslash A \text{ finite} \} .$$

$\widetilde{k} | \varDelta = 0$ is equivalent to

$$x' \in \ell^1(T)' \Longrightarrow \lim_{s,\mathfrak{F}} \langle k(s,\cdot), x' \rangle = 0\,.$$

By Corollary 1.2.3.13, this is equivalent to

$$\lim_{s,\mathfrak{F}} \|k(s,\cdot)\|_1 = 0\,,$$

i.e. to $k \in \ell^{0,1}(S,T)$.

b) Take $k \in \ell_0^{\infty,1}(S,T)$ and $s \in \Delta$. Then

$$\langle e_t^T, \widetilde{k}(s) \rangle = \lim_{S \ni r \to s} \langle e_t^T, k(r,\cdot) \rangle = \lim_{S \ni r \to s} k(r,t) = 0$$

for every $t \in T$, where $\ell^1(T)'$ is canonically identified with $\ell^\infty(T)$ (Example 1.2.2.3 d)), so that $\widetilde{k}(s) \in (\mathbb{K}^{(T)})^0$. Since $\mathbb{K}^{(T)}$ is dense in $c_0(T)$ (Proposition 1.1.2.6 c)), $\widetilde{k}(s) \in c_0(T)^0$. Hence $\widetilde{k}(\Delta) \subset c_0(T)^0$.

Now suppose that $\widetilde{k}(\Delta) \subset c_0(T)^0$. Then

$$\lim_{S \ni r \to s} k(r,t) = \lim_{S \ni r \to s} \langle e_t^T, k(r,\cdot) \rangle = \langle e_t^T, \widetilde{k}(s) \rangle = 0$$

for every $s \in \Delta$ and $t \in T$. Hence $k(\cdot,t) \in c_0(S)$ for every $t \in T$. Thus $k \in \ell_0^{\infty,1}(S,T)$. ∎

Example 1.3.6.23 (1) *Let S,T be sets and βS the Stone–Čech compactification of S with respect to the discrete topology on S. Put*

$$\Delta := \beta S \backslash S$$

and for $k \in \ell_0^{\infty,1}(S,T)$ let $\widetilde{k}$ denote the continuous extension of the map

$$S \longrightarrow \ell^1(T)\,, \quad s \longmapsto k(s,\cdot)$$

to

$$\Delta \longrightarrow c_0(T)^0_{\ell^1(T)'}\,,$$

with the identifications in Proposition 1.3.6.22. Put

$$\mathcal{M} := \{\widetilde{k} \mid k \in \ell_0^{\infty,1}(S,T)\}\,,$$

$$u : \ell_0^{\infty,1}(S,T) \longrightarrow \mathcal{M}\,, \quad k \longmapsto \widetilde{k}\,,$$

and endow $\mathcal{M}$ with the norm

$$\mathcal{M} \longrightarrow \mathbb{R}_+\,, \quad \widetilde{k} \longmapsto \sup_{s \in \Delta} \|\widetilde{k}(s)\|\,.$$

Then

$$\operatorname{Ker} u = \ell^{0,1}(S,T)$$

and the factorization

$$\ell_0^{\infty,1}(S,T)/\ell^{0,1}(S,T) \longrightarrow \mathcal{M}$$

of u *through* $\ell_0^{\infty,1}(S,T)/\ell^{0,1}(S,T)$ *is an isometry.*

By Proposition 1.3.6.22 a),

$$\operatorname{Ker} u = \ell^{0,1}(S,T).$$

By Lemma 1.2.4.6, the factorization

$$v : \ell_0^{\infty,1}(S,T)/\ell^{0,1}(S,T) \longrightarrow \mathcal{M}$$

of u through $\ell_0^{\infty,1}(S,T)/\ell^{0,1}(S,T)$ is bijective. Take $K \in \ell_0^{\infty,1}(S,T)/\ell^{0,1}(S,T)$. Then

$$\|vK\| = \|\widetilde{k}\| = \sup_{s\in\Delta} \|\widetilde{k}(s)\| \leq \sup_{s\in S} \|k(s,\cdot)\|_1 = \|k\|$$

for every $k \in K$, so that

$$\|vK\| \leq \|K\|.$$

Take $\alpha \in \mathbb{R}$ with $\alpha < \|vK\|$, and $k \in K$. There is an $s \in \Delta$, such that

$$\|\widetilde{k}(s)\| > \alpha.$$

Furthermore, there is an $x' \in (\ell^1(T)')^\#$ with

$$|\langle x', \widetilde{k}(s)\rangle| > \alpha.$$

Then

$$\lim_{S\ni r\to s} |\langle x', k(r,\cdot)\rangle| = |\langle x', \widetilde{k}(s)\rangle| > \alpha.$$

Hence there is an $r \in S$ such that

$$\alpha < |\langle x', k(r,\cdot)\rangle| \leq \|k(r,\cdot)\|_1 \leq \|k\|.$$

Thus

$$\alpha \leq \|K\|$$

and

$$\|vK\| \leq \|K\|,$$

since k and α are arbitrary. Hence

$$\|vK\| = \|K\|.$$

■

Proposition 1.3.6.24 (4) *Let T be a compact space and μ a positive Radon measure on T with support equal to T. Denote by u and v the evaluation map of $\mathcal{C}(T)$ and $L^\infty(\mu)$, respectively, and by w the inclusion map $\mathcal{C}(T) \to L^\infty(\mu)$. Then*

$$(\operatorname{Im} v) \cap (\operatorname{Im} w'') = \operatorname{Im}(w'' \circ u) = \operatorname{Im}(v \circ w).$$

Take $a \in (\operatorname{Im} v) \cap (\operatorname{Im} w'')$. There are $x \in L^\infty(\mu)$ and $x'' \in \mathcal{C}(T)''$ such that

$$a = vx = w''x''.$$

By the Vitali–Lusin Theorem, there is a disjoint sequence $(K_n)_{n\in\mathbb{N}}$ of compact sets of T such that for every $n \in \mathbb{N}$, $x|K_n$ is continuous and K_n is the support of $e_{K_n}\cdot\mu$ and such that

$$\mu(T \setminus \bigcup_{n\in\mathbb{N}} K_n) = 0.$$

Let $t_0 \in T$ and suppose $x|\bigcup_{n\in\mathbb{N}} K_n$ has two distinct limits α_1 and α_2 in t_0. Put

$$\varepsilon := \frac{|\alpha_1 - \alpha_2|}{3}.$$

Take $k \in \{1,2\}$. Put

$$A_k := \left\{ t \in \bigcup_{n\in\mathbb{N}} K_n \,\middle|\, |x(t) - \alpha_k| < \varepsilon \right\},$$

$$\mathfrak{U}_k := \{U \cap A_k \mid U \text{ open neighbourhood of } t_0\},$$

denote by $\mathfrak{F}_k$ an ultrafilter on $\mathfrak{U}_k$ finer than the lower section filter of $\mathfrak{U}_k$ (ordered by inclusion), and define

$$x'_k : L^\infty(\mu) \longrightarrow \mathbb{K}, \quad y \longmapsto \lim_{A,\mathfrak{F}_k} \frac{1}{\mu(A)} \int_A y\,d\mu.$$

Then $x'_k \in L^\infty(\mu)'$ and

$$\langle y, w'x'_k \rangle = \langle wy, x'_k \rangle = y(t_0)$$

for every $y \in \mathcal{C}(T)$. Hence

$$w'x'_1 = w'x'_2.$$

Since

$$\langle x, x' \rangle = \langle vx, x' \rangle = \langle w''x'', x' \rangle = \langle x'', w'x' \rangle$$

for every $x' \in L^\infty(\mu)'$, we get

$$\langle x, x'_1 \rangle = \langle x, x'_2 \rangle .$$

But

$$|\langle x, x'_k \rangle - \alpha_k| = \left| \lim_{A,\mathfrak{F}_k} \frac{1}{\mu(A)} \int_A (x - \alpha_k) d\mu \right| \leq \varepsilon$$

for every $k \in \{1, 2\}$ and we get the contradictory relation

$$3\varepsilon = |\alpha_1 - \alpha_2| \leq |\alpha_1 - \langle x, x'_1 \rangle| + |\langle x, x'_2 \rangle - \alpha_2| \leq 2\varepsilon .$$

Hence $x|\bigcup_{n\in\mathbb{N}} K_n$ has a unique limit at t_0.

Since t_0 is arbitrary, we may extend $x|\bigcup_{n\in\mathbb{N}} K_n$ continuously on T. Thus $x \in \operatorname{Im} w$, $a \in \operatorname{Im}(v \circ w)$, and

$$(\operatorname{Im} v) \cap (\operatorname{Im} w'') \subset \operatorname{Im}(v \circ w) .$$

By Proposition 1.3.6.16, $w'' \circ u = v \circ w$, so that

$$\operatorname{Im}(v \circ w) = \operatorname{Im}(w'' \circ u) ,$$

and this implies

$$\operatorname{Im}(v \circ w) \subset (\operatorname{Im} v) \cap (\operatorname{Im} w'') .$$

■

Proposition 1.3.6.25 (7) *Let E, F be normed spaces and $u \in \mathcal{L}(E, F)$. Then*

$$(\operatorname{Im} j_{E'}) \cap (\operatorname{Im} u''') = \operatorname{Im}(j_{E'} \circ u') = \operatorname{Im}(u''' \circ j_{F'}) .$$

Take $x''' \in \operatorname{Im}(j_{E'}) \cap (\operatorname{Im} u''')$. There are $x' \in E'$ and $y''' \in F'''$ such that

$$x''' = j_{E'}x' = u'''y''' .$$

By Proposition 1.3.6.19 a) and Proposition 1.3.6.16 (and Corollary 1.3.4.5),

$$x' = j'_E j_{E'} x' = j'_E u''' y''' = (u'' \circ j_E)' y''' = (j_F \circ u)' y''' = u' j'_F y''' ,$$

$$x''' = j_{E'} x' = j_{E'} u' j'_F y''' \in j_{E'} \circ u'(F') = u''' \circ j_{F'}(F') .$$

Hence

$$(\operatorname{Im} j_{E'}) \cap (\operatorname{Im} u''') \subset \operatorname{Im}(j_{E'} \circ u') = \operatorname{Im}(u''' \circ j_{F'}) .$$

The reverse inclusion is trivial. ■

Proposition 1.3.6.26 *Let E be a complex Banach space and F' a vector subspace of E' such that $E_{F'}$ is Hausdorff. We denote by $\widetilde{E}$ the underlying real Banach space of E and by G' the set of continuous linear forms on $\widetilde{E}_{F'}$ and put*

$$\widetilde{y'} : E \longrightarrow \mathbb{C}\,, \quad x \longmapsto y'(x) - iy'(ix)$$

for every $y' \in G'$.

a) *G' is a vector subspace of $\widetilde{E}'$, $\widetilde{y'} \in F'$ for every $y' \in G'$, the map*

$$G' \longrightarrow F'\,, \quad y' \longmapsto \widetilde{y'}$$

is an isometry of real normed spaces (with respect to the induced norms), and

$$F' \longrightarrow G'\,, \quad x' \longmapsto \operatorname{re} x'$$

is its inverse.

b) *If $F' = E'$ then $G' = \widetilde{E}'$.*

c) *If the map*

$$E \longrightarrow F''\,, \quad x \longmapsto \langle x, \cdot \rangle | F'$$

is an isometry of complex Banach spaces then the map

$$\widetilde{E} \longrightarrow G''\,, \quad x \longmapsto \langle x, \cdot \rangle | G'$$

is an isometry of real Banach spaces.

a) It is obvious that G' is a vector subspace of $\widetilde{E}'$. By Lemma 1.3.1.5, $\widetilde{y'}$ is a linear form on E and by Lemma 1.2.6.4, $\widetilde{y'} \in F'$. Moreover,

$$\operatorname{re} \widetilde{y'}(x) = \operatorname{re}\,(y'(x) - iy'(ix)) = y'(x)\,,$$

for every $x \in E$, so that

$$\operatorname{re} \widetilde{y'} = y'\,.$$

For $x' \in F'$ and $x \in E$,

$$\widetilde{\operatorname{re} x'}(x) = \langle x, \operatorname{re} x' \rangle - i\langle ix, \operatorname{re} x' \rangle =$$

$$= \mathrm{re}\,\langle x, x'\rangle - i\,\mathrm{re}\,i\langle x, x'\rangle = \mathrm{re}\langle x, x'\rangle + i\,\mathrm{im}\langle x, x'\rangle = \langle x, x'\rangle\,,$$

so that

$$\widetilde{\mathrm{re}\,x'} = x'\,.$$

Hence the two given maps are bijective and everyone of them is the inverse of the other map. Since they are obviously $\mathbb{R}$–linear, it folllows from Corollary 1.2.1.6, that they are isometries of real normed spaces.

b) follows from a) and Proposition 1.3.6.12.

c) Let $y'' \in G''$. Put

$$x'' : F' \longrightarrow \mathbb{C}\,, \quad x' \longmapsto y''(\mathrm{re}\,x') - iy''(i\,\mathrm{re}\,x')\,.$$

By a), Lemma 1.3.1.5, and Corollary 1.2.1.6, $x'' \in F''$ and $\|x''\| = \|y''\|$. By the hypothesis of c), there is an $x \in E$ such that

$$\langle x, \cdot\rangle | F' = x''\,, \quad \|x\| = \|x''\| = \|y''\|\,.$$

Take $y' \in G'$. By a), $\widetilde{y'} \in F'$ and

$$\langle x, \widetilde{y'}\rangle = \langle x'', \widetilde{y'}\rangle = y''(\mathrm{re}\,\widetilde{y'}) - iy''(i\,\mathrm{re}\,\widetilde{y'})\,.$$

Since

$$\langle x, \widetilde{y'}\rangle = \langle x, y'\rangle - iy'(ix)$$

it follows

$$\langle x, y'\rangle = y''(y')\,.$$

Since y' is arbitrary

$$\langle x, \cdot\rangle | G' = y''\,.$$

Hence the map

$$\widetilde{E} \longrightarrow G''\,, \quad x \longmapsto \langle x, \cdot\rangle | G'$$

is an isometry of real Banach spaces. ■

Proposition 1.3.6.27 (0) *Let E be a Banach space, F a subspace of E', and*

$$u : E \longrightarrow F'\,, \quad x \longmapsto \langle x, \cdot\rangle | F\,.$$

If $E_F^{\#}$ is compact, then u is an isometry of Banach spaces.

The map

$$E_F \longrightarrow F'_F\,, \quad x \longmapsto ux$$

being continuous, $u(E^{\#})$ is a compact and therefore a closed set of F'_F. For every $y \in F$,

$$\|y\| = \sup_{x \in E^{\#}} |\langle y, x \rangle| = \sup_{x \in E^{\#}} |\langle y, ux \rangle|\,.$$

By Corollary 1.3.1.9, $u(E^{\#}) = F'^{\#}$. Since $E_F^{\#}$ is compact, E_F is Hausdorff and so u is injective. Since $E_F^{\#}$ is compact, E_F is Hausdorff and so u is injective. It follows that u is an isometry of Banach spaces. ■

1.3.7 The Krein–Šmulian Theorem

Proposition 1.3.7.1 (0) *Let E be a normed space and K a weakly compact convex set of E. Then*

$$\{\alpha x \mid (\alpha, x) \in \mathbb{K}^{\#} \times K\}$$

is a weakly compact, absolute convex set of E containing K.

Put

$$u : \mathbb{K} \times E_{E'} \longrightarrow E_{E'}, \quad (\alpha, x) \longmapsto \alpha x .$$

Then u is continuous, so that

$$\{\alpha x \mid (\alpha, x) \in \mathbb{K}^{\#} \times K\} = u(\mathbb{K}^{\#} \times K)$$

is weakly compact. This set is obviously absolutely convex and contains K. ∎

Theorem 1.3.7.2 (0) (Mackey) *Let E be a normed space, $\mathfrak{K}$ the set of convex weakly compact sets of E, and A' a convex set of E' which is closed with respect to the topology on E' of uniform convergence on $\mathfrak{K}$. Then A' is a closed set of E'_E.*

Take $x' \in E' \backslash A'$. There is a $K \in \mathfrak{K}$, such that

$$\{y' \in E' \mid x \in K \Longrightarrow |\langle x, x' - y' \rangle| \leq 1\} \subset E' \backslash A' .$$

Put

$$K' := \{y' \in E' \mid x \in K \Longrightarrow |x'(x)| \leq 1\} .$$

Then K' is an absolutely convex set so that

$$x' \in E' \backslash (A' + K') .$$

Every $y' \in E'$ is bounded on K. Thus there is an $\alpha \in \mathbb{R}_+$ with $y' \in \alpha K'$. By Proposition 1.3.1.6, there is a linear form x'' on E' which is bounded on K' such that

$$\sup_{y' \in A'} \operatorname{re} x''(y') < \operatorname{re} x''(x') .$$

We show that x'' belongs to $\operatorname{Im} j_E$. Put

$$\alpha := \sup_{y' \in K'} |x''(y')| .$$

We may assume that $x'' \neq 0$, so that $\alpha \neq 0$. Put

$$L := \{\beta x \mid (\beta, x) \in \mathbb{K}^\# \times K\} .$$

By Proposition 1.3.7.1, L is absolutely convex, weakly compact and contains K. Assume that $\frac{1}{\alpha}x'' \notin j_E(L)$. $j_E(L)$ is an absolutely convex, compact set of $E''_{E'}$. By Proposition 1.3.1.8, there is a $y' \in E'$ such that

$$\beta := \sup_{x \in L} \operatorname{re} \langle j_E x, y' \rangle < \operatorname{re} \frac{1}{\alpha} x''(y') .$$

Since $j_E(L)$ is absolutely convex,

$$\beta = \sup_{x \in L} |\langle j_E x, y' \rangle| = \sup_{x \in L} |\langle x, y' \rangle| .$$

In particular, $\frac{1}{\beta}y' \in K'$ and

$$\alpha \geq |x''(\frac{1}{\beta}y')| \geq \frac{1}{\beta} \operatorname{re} x''(y') > \frac{1}{\beta}\alpha\beta = \alpha ,$$

which is a contradiction. Hence $\frac{1}{\alpha}x'' \in j_E(L)$ and there is an $x \in E$ with $x'' = j_E(x)$. Then

$$\sup_{y' \in A'} \operatorname{re} y'(x) = \sup_{y' \in A'} \operatorname{re} x''(y') < \operatorname{re} x''(x') = \operatorname{re} x'(x) .$$

Therefore x' is not in the closure in E'_E of A' and so A' is a closed set of E'_E. ■

Theorem 1.3.7.3 (0) (Krein–Šmulian, 1940) *Let E be a Banach space and A' a convex set of E'. If $A' \cap nE'^\#$ is a closed set of E'_E for every $n \in \mathbb{N}$, then A' is a closed set of E'_E.*

Let $\mathfrak{T}'$ be the finest topology on E' inducing the topology of pointwise convergence on the equicontinuous sets of E' (Theorem 1.2.8.2). Since every equicontinuous set of E' is contained in a set of the form $nE'^\#$ $(n \in \mathbb{N})$, $E' \setminus A' \in \mathfrak{T}'$, i.e. A' is closed with respect to $\mathfrak{T}'$. By Corollary 1.2.8.3, A' is closed with respect to the topology on E' of uniform convergence on the convex compact sets of E and the assertion follows from Theorem 1.3.7.2 since every compact set is weakly compact. ■

Remark. The theorem no longer holds if E is not complete.

Definition 1.3.7.4 (0) *Let E be a vector space. A* **cone** *of E is a nonempty subset A of E for which $\alpha A \subset A$ whenever $\alpha \in \mathbb{R}_+$. The cone A is called* ***sharp*** *if*

$$A \cap (-A) = \{0\}.$$

0 belongs to every cone. The cone A is convex iff

$$A + A \subset A.$$

Corollary 1.3.7.5 (0) *Let E be a Banach space and A' a convex cone of E'. Then A' is a closed set of E'_E iff $A' \cap E'^{\#}$ is a closed set of E'_E.*

First assume that $A' \cap E'^{\#}$ is a closed set of E'_E. Take $n \in \mathbb{N}$ and put

$$u : E'_E \longrightarrow E'_E, \quad x' \longmapsto \frac{1}{n} x'.$$

Since u is continuous and

$$A' \cap nE'^{\#} = \overset{-1}{u}(A' \cap E'^{\#}),$$

$A' \cap nE'^{\#}$ is a closed set of E'_E. By the Krein–Šmulian Theorem, A' is a closed set of E'_E.

The reverse implication follows from Proposition 1.2.6.6. ■

Proposition 1.3.7.6 (0) *Let E be a Banach space and $(u_\iota)_{\iota \in I}$ a finite family of projections in E' such that $\operatorname{Im} u_\iota$ is a closed set of E'_E and that*

$$u_\iota \circ u_\lambda = 0$$

for distinct $\iota, \lambda \in I$. Then $\sum_{\iota \in I} \operatorname{Im} u_\iota$ is a closed set of E'_E.

First observe that $\sum_{\iota \in I} u_\iota$ is a projection in E' and that

$$\operatorname{Im} \sum_{\iota \in I} u_\iota = \sum_{\iota \in I} \operatorname{Im} u_\iota.$$

Let x' be point of adherence of $E'^{\#} \cap \sum_{\iota \in I} \operatorname{Im} u_\iota$ in E'_E. There is an ultrafilter $\mathfrak{F}$ on E' converging to x' in E'_E with

$$E'^{\#} \cap \sum_{\iota \in I} \operatorname{Im} u_\iota \in \mathfrak{F}.$$

By the Alaoglu–Bourbaki Theorem, $x' \in E'^{\#}$ and $u_\iota(\mathfrak{F})$ converges in E'_E for every $\iota \in I$. Moreover,

$$\lim u_\iota(\mathfrak{F}) \in \operatorname{Im} u_\iota ,$$

since $\operatorname{Im} u_\iota$ is a closed set of E'_E. Then

$$\lim \Big(\sum_{\iota \in I} u_\iota(\mathfrak{F}) \Big) = \sum_{\iota \in I} \lim u_\iota(\mathfrak{F}) \in \sum_{\iota \in I} \operatorname{Im} u_\iota .$$

Since $\Big(\sum_{\iota \in I} u_\iota \Big)(\mathfrak{F})$ is finer than $\sum_{\iota \in I} u_\iota(\mathfrak{F})$,

$$x' = \lim \mathfrak{F} = \lim \Big(\sum_{\iota \in I} u_\iota \Big)(\mathfrak{F}) = \lim \Big(\sum_{\iota \in I} u_\iota(\mathfrak{F}) \Big) \in \sum_{\iota \in I} \operatorname{Im} u_\iota .$$

Hence $E'^{\#} \cap \sum_{\iota \in I} \operatorname{Im} u_\iota$ is a closed set of E'_E. By Corollary1.3.7.5, $\sum_{\iota \in I} \operatorname{Im} u_\iota$ is a closed set of E'_E. ■

Proposition 1.3.7.7 (0) *Let E be a normed space. Given a linear form x'' on E', the following are equivalent:*

a) $x'' \in \operatorname{Im} j_E$.

b) $\operatorname{Ker} x''$ *is a closed set of* E'_E.

c) x'' *is continuous on* E'_E.

a $\Rightarrow$ b and a $\Rightarrow$ c are trivial.

b $\Rightarrow$ a. We may assume that $x'' \neq 0$. Take $x' \in E' \backslash \operatorname{Ker} x''$. By Proposition 1.2.6.2, there is a finite subset A of E such that

$$\{ y' \in E' \mid x \in A \Longrightarrow |\langle x, x' - y' \rangle| \leq 1 \} \subset E' \backslash \operatorname{Ker} x'' .$$

Then x'' is bounded on

$$\{ y' \in E' \mid x \in A \Longrightarrow |\langle x, y' \rangle| \leq 1 \}$$

as can be seen by factorizing x'' through $E'/\operatorname{Ker} x''$. Hence, by Lemma 1.2.6.4, there is an $x \in E$ with

$$x'' := \langle x, \cdot \rangle$$

and $x'' \in \operatorname{Im} j_E$.

c $\Rightarrow$ a. Put

$$V' := \{x' \in E' \mid |x''(x')| < 1\}.$$

By Proposition 1.2.6.2, there is a finite subset A of E such that

$$\{x' \in E' \mid x \in A \Longrightarrow |x'(x)| < 1\} \subset V'.$$

By Lemma 1.2.6.4, x'' is a linear combination of the $(\langle x, \cdot \rangle)_{x \in A}$ and so it belongs to $\mathrm{Im} j_E$. ■

Corollary 1.3.7.8 *Let E, F be normed spaces. Given $u \in \mathcal{L}(E', F')$, the following are equivalent:*

a) *There is a $v \in \mathcal{L}(F, E)$ with $u = v'$.*

b) *The map*

$$E'_E \longrightarrow F'_F, \quad x' \longmapsto ux'$$

is continuous.

c) $u'(\mathrm{Im}\, j_F) \subset \mathrm{Im}\, j_E$.

a ⇒ b. Given $(x', y) \in E' \times F$,

$$\langle y, ux' \rangle = \langle y, v'x' \rangle = \langle vy, x' \rangle$$

(Theorem 1.3.4.2 a)) and the assertion now follows.

b ⇒ c. Take $y \in F$. Then

$$\langle u' j_F y, x' \rangle = \langle j_F y, ux' \rangle = \langle y, ux' \rangle$$

for every $x' \in E'$ (Theorem 1.3.4.2 a)). Thus the map

$$E'_E \longrightarrow \mathbb{K}, \quad x' \longmapsto \langle u' j_F y, x' \rangle$$

is continuous. By Proposition 1.3.7.7 c ⇒ a, $u' j_F y$ belongs to $\mathrm{Im}\, j_E$. Hence

$$u'(\mathrm{Im}\, j_F) \subset \mathrm{Im}\, j_E.$$

c ⇒ a. Put

$$v : F \longrightarrow E, \quad y \longmapsto j_E^{-1} u' j_F y$$

(Corollary 1.3.6.5). Then $v \in \mathcal{L}(F, E)$ and

$$\langle y, v'x' \rangle = \langle vy, x' \rangle = \langle u' j_F y, x' \rangle = \langle j_F y, ux' \rangle = \langle y, ux' \rangle$$

for every $(x', y) \subset E' \times F$ (Theorem 1.3.4.2 a)) and so $u = v'$. ■

Proposition 1.3.7.9 (0) *Let E be a Banach space. Given a linear form x'' on E', the following are equivalent:*

a) *The restriction of x'' to $E'^{\#}$ is continuous at 0 with respect to the topology on $E'^{\#}$ of uniform convergence on the weakly compact convex sets of E.*

b) *$E'^{\#} \cap \operatorname{Ker} x''$ is a closed subset of E'_E.*

c) *$\operatorname{Ker} x''$ is a closed set of E'_E.*

d) *$x'' \in \operatorname{Im} j_E$.*

a $\Rightarrow$ b. Let $\mathfrak{T}'$ be the topology on E' of uniform convergence on the weakly compact convex sets of E and $\mathfrak{S}'$ the topology induced on $E'^{\#}$ by $\mathfrak{T}'$. Take $x' \in E'^{\#}$ and $\varepsilon > 0$. There is a weakly compact convex set K of E such that

$$|x''(y')| < \frac{\varepsilon}{2}$$

for every $y' \in E'^{\#} \cap K'$, where

$$K' := \{z' \in E' \mid x \in K \Longrightarrow |z'(x)| \leq 1\}.$$

$E'^{\#} \cap (x' + 2K')$ is a neighbourhood of x' in $E'^{\#}$ with respect to $\mathfrak{S}'$. Take $y' \in E'^{\#} \cap (x' + 2K')$. Then

$$\frac{1}{2}(y' - x') \in E'^{\#} \cap K',$$

so

$$|x''(y') - x''(x')| = 2\left|x''\left(\frac{1}{2}(y' - x')\right)\right| < \varepsilon.$$

Hence the restriction of x'' to $E'^{\#}$ is continuous at x' with respect to $\mathfrak{S}'$. Since x' is arbitrary, $x''|E'^{\#}$ is continuous with respect to $\mathfrak{S}'$. Hence $E'^{\#} \cap \operatorname{Ker} x''$ is closed with respect to $\mathfrak{S}'$. Since $E'^{\#}$ is closed with respect to $\mathfrak{T}'$ (Proposition 1.2.6.6), $E'^{\#} \cap \operatorname{Ker} x''$ is also closed with respect to $\mathfrak{T}'$. Being a convex set, $E'^{\#} \cap \operatorname{Ker} x''$ is a closed set of E'_E (Theorem 1.3.7.2).

b $\Rightarrow$ c follows from Corollary 1.3.7.5.

a $\Rightarrow$ d follows from Proposition 1.3.7.7 b $\Rightarrow$ a.

d $\Rightarrow$ a is trivial. ∎

Definition 1.3.7.10 (0) *Let E, F be vector spaces. The map $u : E \to F$ is called **conjugate–linear** if*

$$u(\alpha x + \beta y) = \overline{\alpha} ux + \overline{\beta} uy$$

for every $x, y \in E$ and $\alpha, \beta \in \mathbb{K}$.

Proposition 1.3.7.11 (0) *Let E, F be Banach spaces. Given a linear (resp. conjugate–linear) map $u : E' \to F'$, the following are equivalent:*

a) *The map*

$$E'_E \longrightarrow F'_F, \quad x' \longmapsto ux'$$

is continuous.

b) *The map*

$$E^{\#}_{E'} \longrightarrow F'_F, \quad x' \longmapsto ux'$$

is continuous at 0.

a $\Rightarrow$ b is trivial.

b $\Rightarrow$ a. Take $y \in F$. By b), the map

$$E^{\#}_{E'} \longrightarrow \mathbb{K}, \quad x' \longmapsto \langle y, ux' \rangle \quad (\text{resp. } \overline{\langle y, ux' \rangle})$$

is continuous at 0. By Proposition 1.3.7.9, a $\Rightarrow$ d, there is an $x \in E$ such that

$$\langle y, ux' \rangle = \langle x, x' \rangle \quad (\text{resp. } \overline{\langle y, ux' \rangle} = \langle x, x' \rangle)$$

for every $x' \in E'$. a) now follows. ■

Corollary 1.3.7.12 (0) *Let E be a Banach space. Given a projection u in E', the following are equivalent:*

a) *The map*

$$E'_E \longrightarrow E'_E, \quad x' \longmapsto ux'$$

is continuous.

b) $\operatorname{Im} u$ *and* $\operatorname{Ker} u$ *are closed sets of* E'_E.

Put

$$v := 1_E - u\,.$$

Then

$$\operatorname{Im} u = \operatorname{Ker} v\,.$$

a $\Rightarrow$ b. $\operatorname{Ker} u$ is obviously a closed set of E'_E. The map

$$E'_E \longrightarrow E'_E\,, \quad x' \longmapsto vx'$$

is continuous, so that $\operatorname{Ker} v$ is a closed set of E'_E. Hence $\operatorname{Im} u$ is a closed set of $E_{E'}$.

b $\Rightarrow$ a. Let $\mathfrak{F}$ be an ultrafilter on $E^{\#}_{E'}$ converging to 0. Let x' denote the limit of $u(\mathfrak{F})$ in $E^{\#}_{E'}$ (Alaoglu–Bourbaki Theorem). Then $v(\mathfrak{F})$ converges to $-x'$ in $E^{\#}_{E'}$. By b),

$$x' \in \operatorname{Im} u\,, \quad -x' \in \operatorname{Im} v = \operatorname{Ker} u$$

and so

$$x' = ux' = 0\,.$$

Hence the map

$$E^{\#}_{E'} \longrightarrow E'_E\,, \quad x' \longmapsto ux'$$

is continuous at 0. By Proposition 1.3.7.11 b $\Rightarrow$ a, the map

$$E'_E \longrightarrow E'_E\,, \quad x' \longmapsto ux'$$

is continuous. ∎

Lemma 1.3.7.13 *Let T be a compact space. Take $x \in \mathcal{C}(T)$, and let $(x_n)_{n\in\mathbb{N}}$ be a sequence in $\mathcal{C}(T)$ for which x is a point of adherence in the topology of pointwise convergence. If every subsequence of $(x_n)_{n\in\mathbb{N}}$ has a point of adherence in $\mathcal{C}(T)$ with respect to the topology of pointwise convergence, then there is a subsequence of $(x_n)_{n\in\mathbb{N}}$ converging to x in the topology of pointwise convergence.*

First assume that T is separable. Then, by the diagonal procedure, we may construct a subsequence $(y_n)_{n\in\mathbb{N}}$ of $(x_n)_{n\in\mathbb{N}}$ converging to x on a dense set of T. Take $t \in T$. Assume that $(y_n(t))_{n\in\mathbb{N}}$ does not converge to $x(t)$. Then there is an $\varepsilon > 0$ together with a subsequence $(z_n)_{n\in\mathbb{N}}$ of $(y_n)_{n\in\mathbb{N}}$, such that

$$|z_n(t) - x(t)| \geq \varepsilon$$

for every $n \in \mathbb{N}$. By assumption, $(z_n)_{n\in\mathbb{N}}$ has a point of adherence z in $\mathcal{C}(T)$ with respect to the topology of pointwise convergence. Then

$$|z(t) - x(t)| \geq \varepsilon,$$

and this is a contradiction, since z and x coincide on an dense set of T. Hence $(y_n(t))_{n\in\mathbb{N}}$ convergens to $x(t)$. Since t is arbitrary, $(y_n)_{n\in\mathbb{N}}$ converges to x in the topology of pointwise convergence.

Now let T be arbitrary and let

$$S := \prod_{n\in\mathbb{N}} x_n(T),$$

$$\varphi : T \longrightarrow S, \quad t \longmapsto (x_n(t))_{n\in\mathbb{N}},$$

$$\psi : T \longrightarrow \varphi(T), \quad t \longmapsto \varphi(t),$$

and

$$\pi_p : \varphi(T) \longrightarrow \mathbb{K}, \quad (s_n)_{n\in\mathbb{N}} \longmapsto s_p$$

for every $p \in \mathbb{N}$. Then φ is continuous and $\varphi(T)$ is compact. Let y be a point of adherence of $(x_n)_{n\in\mathbb{N}}$ in $\mathcal{C}(T)$ with respect to the topology of pointwise convergence, and take $t', t'' \in T$ such that

$$\psi(t') = \psi(t'').$$

Then

$$x_n(t') = x_n(t'')$$

for every $n \in \mathbb{N}$, so that

$$y(t') = y(t'').$$

Hence there is a unique map $\tilde{y}: S \to \mathbb{K}$ with

$$y = \tilde{y} \circ \psi.$$

Since $\varphi(T)$ is the quotient space of T with respect to ψ, $\widetilde{y}$ is continuous. $\widetilde{x}$ is a point of adherence of $(\pi_n)_{n\in\mathbb{N}}$ and every subsequence of $(\pi_n)_{n\in\mathbb{N}}$ has a point of adherence in $\mathcal{C}(\varphi(T))$ with respect to the topology of pointwise convergence. Since $\varphi(T)$ is separable, the first part of the proof implies the existence of a subsequence $(\pi_{k_n})_{n\in\mathbb{N}}$ of $(\pi_n)_{n\in\mathbb{N}}$ converging to $\widetilde{x}$ in the topology of pointwise convergence. It follows that the subsequence $(x_{k_n})_{n\in\mathbb{N}}$ of $(x_n)_{n\in\mathbb{N}}$ converges to x in the topology of pointwise convergence. ∎

Lemma 1.3.7.14 *Let T be a compact space and let $\mathcal{C}(T)_T$ denote the set $\mathcal{C}(T)$ endowed with the topology of pointwise convergence. Given $\mathcal{F} \subset \mathcal{C}(T)$, the following are equivalent:*

a) *Every sequence in $\mathcal{F}$ has a point of adherence in $\mathcal{C}(T)_T$.*

b) *Every sequence in $\mathcal{F}$ contains a sequence which converges in $\mathcal{C}(T)_T$.*

c) *$\mathcal{F}$ is a relatively compact set of $\mathcal{C}(T)_T$.*

b $\Rightarrow$ a and c $\Rightarrow$ a are trivial.

a $\Rightarrow$ b follows from Lemma 1.3.7.13.

a $\Rightarrow$ c. Let $\mathfrak{F}$ be an ultrafilter on $\mathcal{C}(T)$ containing $\mathcal{F}$. By a), $\{x(t)|x \in \mathcal{F}\}$ is bounded for every $t \in T$, so that the map

$$x : T \longrightarrow \mathbb{K}, \quad t \longmapsto \lim_{y,\mathfrak{F}} y(t)$$

is well-defined. We show that x is continuous. Take $t \in T$ and $\varepsilon > 0$. Assume that every neighbourhood of t contains a point s such that

$$|x(s) - x(t)| \geq \varepsilon .$$

We construct inductively a sequence $(t_n)_{n\in\mathbb{N}}$ in T starting with $t_1 := t$ and a sequence $(x_n)_{n\in\mathbb{N}}$ in $\mathcal{F}$ such that the following hold for every $n \in \mathbb{N}$:

1) $n \neq 1 \Longrightarrow |x(t_n) - x(t)| \geq \varepsilon$.

2) $|x(t_k) - x_n(t_k)| < \frac{1}{n}$ for every $k \in \mathbb{N}_{n-1}$.

3) $|x_k(t_n) - x_k(t)| < \frac{1}{n}$ for every $k \in \mathbb{N}_n$.

Choose x_1 arbitrarily. Take $n \in \mathbb{N}$, $n > 1$, and assume that the sequences have been constructed up to $n-1$. By the definition of x, there is an $x_n \in \mathcal{F}$ such that 2) is fulfilled. Since the functions in $\mathcal{F}$ are continuous,

$$\bigcap_{k=1}^{n}\{s \in T \mid |x_k(s) - x_k(t)| < \frac{1}{n}\}$$

is a neighbourhood of t. By hypothesis, there is a t_n in this neighbourhood of T satisfying 1). This finishes the inductive construction.

Let s be a point of adherence of $(t_n)_{n\in\mathbb{N}}$ and y a point of adherence of $(x_n)_{n\in\mathbb{N}}$ in $\mathcal{C}(T)_T$. Then

$$y(t_k) = x(t_k)$$

for every $k \in \mathbb{N}$, by 2). Thus

$$|y(s) - x(t)| \geq \varepsilon$$

by 1). By 3),

$$x_k(s) = x_k(t)$$

for every $k \in \mathbb{N}$, so that

$$y(s) = y(t) = y(t_1) = x(t_1) = x(t),$$

which is a contradiction.

Hence there is a neighbourhood V of t such that

$$|x(s) - x(t)| < \varepsilon$$

for every $s \in V$, and so x is continuous at t. It follows that $x \in \mathcal{C}(T)$, $\mathfrak{F}$ converges to x in $\mathcal{C}(T)_T$, and $\mathcal{F}$ is relatively compact. ■

Theorem 1.3.7.15 *Let A be a subset of the Banach space E. Then the following are equivalent:*

a) *Every sequence in A has a point of weak adherence in E.*

b) *Every sequence in A has a weakly convergent subsequence in E.*

c) *A is weakly relatively compact.*

By the Alaoglu–Bourbaki Theorem, $E^{\#}_{E'}$ is compact. Given $x \in E$, define

$$\tilde{x} : E^{\#}_{E'} \longrightarrow \mathbb{K}, \quad x' \longmapsto \langle x, x' \rangle$$

and

$$\varphi : E \longrightarrow \mathcal{C}(E^{\#}_{E'}), \quad x \longmapsto \tilde{x}.$$

a $\Rightarrow$ b. Let $(x_n)_{n\in\mathbb{N}}$ be a sequence in A. By a), every subsequence of $(\varphi(x_n))_{n\in\mathbb{N}}$ has a point of adherence in $\mathcal{C}(E^{\#}_{E'})$ with respect to the topology of pointwise convergence. By Lemma 1.3.7.13, there is a strictly increasing sequence $(k_n)_{n\in\mathbb{N}}$ in $\mathbb{N}$ such that $(\varphi(x_{k_n}))_{n\in\mathbb{N}}$ converges to some $y \in \mathcal{C}(E^{\#}_{E'})$ in the topology of pointwise convergence. Define

$$x'' : E' \longrightarrow \mathbb{K}, \quad x' \longmapsto \lim_{n\to\infty} \langle x_{k_n}, x' \rangle .$$

Then x'' is linear and

$$x''|E'^{\#} = y .$$

Hence $E'^{\#} \cap \operatorname{Ker} x''$ is a closed set of E'_E. By Proposition 1.3.7.9 b $\Rightarrow$ d, there is an $x \in E$, such that

$$j_E x = x'' .$$

Then $(x_{k_n})_{n\in\mathbb{N}}$ converges weakly to x.

a $\Rightarrow$ c. Let $\mathfrak{F}$ be an ultrafilter on E containing A. By a), $\{x'(x) \mid x \in A\}$ is a bounded set of $\mathbb{K}$ for every $x' \in E'$. Thus the map

$$x'' : E' \longrightarrow \mathbb{K}, \quad x' \longmapsto \lim_{x,\mathfrak{F}} x'(x)$$

is well–defined. It is obviously linear. By a), every sequence in $\varphi(A)$ has a point of adherence in $\mathcal{C}(E^{\#}_{E'})$ with respect to the topology of pointwise convergence. By Lemma 1.3.7.14 a $\Rightarrow$ c, $\varphi(\mathfrak{F})$ converges to some $y \in \mathcal{C}(E^{\#}_{E'})$ with respect to the topology of pointwise convergence. We have

$$x''|E'^{\#} = y ,$$

so $E'^{\#} \cap \operatorname{Ker} x''$ is a closed set of E'_E. By Proposition 1.3.7.9 b $\Rightarrow$ d, there is an $x \in E$ with

$$j_E x = x'' .$$

Then $\mathfrak{F}$ converges weakly to x, and A is weakly relatively compact.

b $\Rightarrow$ a and c $\Rightarrow$ a are trivial. ∎

Remark. a) The implication a $\Rightarrow$ b was proved by Šmulian (1940) and the implication a $\Rightarrow$ c was proved by Eberlein (1947).

b) It is possible to prove a stronger form of Lemma 1.3.7.14 (for T σ–compact instead of compact) so that the above theorem can be proved without the use of Proposition 1.3.7.9.

1.3.8 Reflexive Spaces

Definition 1.3.8.1 (0) (H. Hahn, 1927) *A normed space is called* ***reflexive*** *if its evaluation map is surjective (in which case it is an isometry (Corollary 1.3.6.5)).*

It may happen that a Banach space is isometric to its bidual without being reflexive (R.C. James, A non–reflexive Banach space isometric with its second conjugate space, Proc. Nat. Acad. Sci. USA 37 (1958) 174-177).

Proposition 1.3.8.2 (0) *Every finite–dimensional normed space is reflexive.*

This follows immediately from the fact that the dual and the algebraic dual of a finite dimensional normed space coincide (Corollary 1.2.4.10). ■

Proposition 1.3.8.3 (0) *Every reflexive space is complete and its bounded sets are weakly relatively compact.*

Let E be a reflexive space. Then, using the evaluation map, we may identify E with E'' and $E^{\#}_{E'}$ with $E''^{\#}_{E'}$. By Corollary 1.2.1.10, E is complete and, by the Alaoglu–Bourbaki Theorem, $E^{\#}_{E'}$ is compact. Hence every bounded set of E is weakly relatively compact. ■

Proposition 1.3.8.4 (0) (P.J. Pettis, 1938) *A Banach space E is reflexive iff E' is reflexive.*

If E is reflexive, then E' is obviously reflexive. Assume that E is not reflexive. Identify E with a subspace of E'' via the evaluation map. Since E is complete, it is a closed subspace of E''. By Corollary 1.3.3.6, there is an $x''' \in E'''\backslash\{0\}$ vanishing on E. Then x''' does not belong to $\operatorname{Im} j_{E'}$ and so E' is not reflexive. ■

Proposition 1.3.8.5 *Let E be a normed space and F a subspace of E.*

a) *F is reflexive iff $j_E(F) = F^{\circ\circ}$.*

b) *If E is reflexive and F is closed, then F is reflexive* (P.J. Pettis, 1938).

a) Let $j : F \to E$ be the inclusion map. Then

$$\operatorname{Im} j'' = F^{\circ\circ}$$

(Proposition 1.3.6.17).

If F is reflexive, then

$$F^{\circ\circ} = \operatorname{Im} j'' = \operatorname{Im}(j'' \circ j_F) = \operatorname{Im}(j_E \circ j) = j_E(F)$$

(Proposition 1.3.6.16).

Now suppose that $j_E(F) = F^{\circ\circ}$. Take $y'' \in F''$. Then $j''y'' \in F^{\circ\circ}$, and so there is an $x \in F$ with

$$j_E x = j''y'' .$$

Then

$$j'' j_F x = j_E j x = j_E j x = j''y''$$

(Proposition 1.3.6.16) and

$$y'' \in j_F x$$

(Proposition 1.3.6.17). Hence j_F is surjective and F is reflexive.

b) Take $x'' \in F^{\circ\circ}$. Since E is reflexive, there is an $x \in E$ with

$$x'' = j_E x .$$

Given $x' \in F^\circ$,

$$\langle x, x' \rangle = \langle j_E x, x' \rangle = \langle x'', x' \rangle = 0$$

so that

$$x \in {}^\circ(F^\circ) = \overline{F} = F$$

(Proposition 1.3.5.7). Hence

$$F^{\circ\circ} \subset j_E(F) .$$

The reverse inclusion is trivial. By a), F is reflexive. ∎

Proposition 1.3.8.6 *Let E be a normed space, F a closed vector subspace of E, and $q : E'' \to E''/F^{\circ\circ}$ the quotient map.*

a) *E/F is reflexive iff $q \circ j_E$ is surjective.*

b) *If E is reflexive, then so is E/F.*

a) Let $r : E \to E/F$ be the quotient map. Then

$$j_{E/F} \circ r = r'' \circ j_E$$

(Proposition 1.3.6.16), so that $j_{E/F}$ is surjective iff $r'' \circ j_E$ is surjective. By Proposition 1.3.6.18, the factorization of r'' through $E''/F^{\circ\circ}$ is an isometry. Thus $r'' \circ j_E$ is surjective iff $q \circ j_E$ is surjective.

b) If E is reflexive, then j_E is surjective. Thus $q \circ j_E$ is surjective. By a), E/F is reflexive. ■

Corollary 1.3.8.7 *Let F be a closed subspace of the normed space E. Then E is reflexive iff F and E/F are reflexive.*

The necessity follows from Proposition 1.3.8.5 b) and Proposition 1.3.8.6 b). For the converse, assume that F and E/F are both reflexive and let $q : E'' \to E''/F^{\circ\circ}$ be the quotient map. Take $x'' \in E''$. By Proposition 1.3.8.6 a), there is an $x \in E$ such that

$$qj_E x = qx'' .$$

Then

$$q(x'' - j_E x) = 0 ,$$

$$x'' - j_E x \in F^{\circ\circ} .$$

By Proposition 1.3.8.5 a), there is a y with

$$j_E y = x'' - j_E x .$$

Hence

$$x'' = j_E(x + y) ,$$

i.e. j_E is surjective and so E is reflexive. ■

Proposition 1.3.8.8 (0) *A Banach space, which is isomorphic to a reflexive Banach space, is itself reflexive.*

Let $u : E \to F$ be an isomorphism of Banach spaces and assume E reflexive. Then u'' is surjective (Corollary 1.3.4.7). Since

$$j_F \circ u = u'' \circ j_E$$

(Proposition 1.3.6.16), it follows that $j_F \circ u$ is surjective. Hence j_F is surjective and F is reflexive. ■

Example 1.3.8.9 (1) (7) *$\ell^p(T)$ is reflexive for every $p \in]1, \infty[$ and every set T. $c_0(T)$, $c(T)$, $\ell^1(T)$ and $\ell^\infty(T)$ are reflexive iff T is finite.*

For $\ell^p(T)$ $(p \in [1, \infty] \cup \{0\})$ this follows from Example 1.2.2.3 d),e) and Proposition 1.3.8.4. By Example 1.2.2.4 c), $c(T)$ and $c_0(T)$ are isomorphic, so that the assertion for $c(T)$ follows from that for $c_0(T)$ and from Proposition 1.3.8.8. ■

Example 1.3.8.10 *Let S, T be sets and $p, q \in]1, \infty[$ be conjugate. Then $\ell^{p,q}(S, T)$ is reflexive.*

This is an immediate consequence of Example 1.3.8.9 and Proposition 1.2.3.6 b). ■

Example 1.3.8.11 *If μ is a measure and $p \in]1, \infty[$, then $L^p(\mu)$ is reflexive.*

The assertion follows from Example 1.2.2.5 c). ■

Example 1.3.8.12 *If T is a completely regular space, then $\mathcal{C}(T)$ is reflexive iff T is finite.*

If T is finite, then $\mathcal{C}(T)$ is reflexive py Proposition 1.3.8.2. Assume that T is infinite. Replacing T by its Stone–Čech compactification, if necessary, we may assume that T is compact. There is a sequence $(t_n)_{n \in \mathbb{N}}$ in T for which

$$t_n \notin \overline{\{t_m \mid m \in \mathbb{N} \backslash \{n\}\}}$$

for every $n \in \mathbb{N}$.

First suppose that there are two distinct ultrafilters $\mathfrak{F}$, $\mathfrak{G}$ on $\mathbb{N}$ with

$$\lim_{n, \mathfrak{F}} t_n = \lim_{n, \mathfrak{G}} t_n \,.$$

Take $A \in \mathfrak{F} \backslash \mathfrak{G}$ and put

$$B := \{t_n \mid n \in A\}\,.$$

Then e_B is a Borel function on T and

$$x'' : \mathcal{M}_b(T) \longrightarrow \mathbb{K}, \quad \mu \longmapsto \int e_B \, d\mu$$

is a continuous linear form on $\mathcal{M}_b$. Take $x \in \mathcal{C}(T)$. Then

$$\lim_{n,\mathfrak{F}} x(t_n) = \lim_{n,\mathfrak{G}} x(t_n),$$

and so there is an $n \in \mathbb{N}$ with

$$x(t_n) \neq e_B(t_n).$$

Hence

$$j_{\mathcal{C}(T)} x \neq x''.$$

Thus $j_{\mathcal{C}(T)}$ is not surjective (Example 1.2.2.10), and $\mathcal{C}(T)$ is not reflexive.

Now suppose that for any two distinct ultrafilters $\mathfrak{F}$, $\mathfrak{G}$ on $\mathbb{N}$

$$\lim_{n,\mathfrak{F}} t_n \neq \lim_{n,\mathfrak{G}} t_n.$$

Then $\overline{\{t_m \mid n \in \mathbb{N}\}}$ is homeomorphic to the Stone–Čech compactification of $\mathbb{N}$. Define

$$\mathcal{F} := \{x \in \mathcal{C}(T) \mid n \in \mathbb{N} \Longrightarrow x(t_n) = 0\},$$

$$u : \mathcal{C}(T) \longrightarrow \ell^\infty, \quad x \longmapsto \big(x(t_n)\big)_{n \in \mathbb{N}}.$$

Then

$$\operatorname{Ker} u = \mathcal{F}$$

and the factorization of u through $\mathcal{C}(T)/\mathcal{F}$ is an isometry (Tietze's Theorem). By Example 1.3.8.9, ℓ^∞ is not reflexive, so that $\mathcal{C}(T)/\mathcal{F}$ is not reflexive. By Proposition 1.3.8.6 b), $\mathcal{C}(T)$ is not reflexive. ∎

Proposition 1.3.8.13 *Every bounded sequence in a reflexive Banach space has a weakly convergent subsequence.*

By Proposition 1.3.8.3, every bounded set of a reflexive space is weakly relatively compact and the assertion now follows from Theorem 1.3.7.15 c $\Rightarrow$ b. ∎

1.3.9 Completion of Normed Spaces

Definition 1.3.9.1 (0) *Let E be a normed space. A* **completion of** *E is a Banach space F such that E is a dense subspace of F.*

Theorem 1.3.9.2 (0) *Let E be a normed space. If E is identified with $\operatorname{Im} j_E$ (Corollary 1.3.6.5), then $\overline{\operatorname{Im} j_E}$ is a completion of E.* ■

Theorem 1.3.9.3 (0) *Let E be a normed space and let F, G be completions of E. Then there is a unique isometry $u : F \to G$ with*

$$ux = x$$

for every $x \in E$.

The uniqueness of u is trivial. Let $j_1 : E \to F$, $j_2 : E \to G$ be the inclusion maps. Then, by Proposition 1.2.1.13, we can extend them to operators $\overline{j_1} \in \mathcal{L}(G, F)$, $\overline{j_2} \in \mathcal{L}(F, G)$, respectively, with

$$\|\overline{j_1}\| = \|\overline{j_2}\| = 1 .$$

We have

$$\overline{j_1} \circ \overline{j_2}(x) = x , \quad \overline{j_2} \circ \overline{j_1}(x) = x$$

for every $x \in E$. We deduce

$$\overline{j_1} \circ \overline{j_2} = 1_F , \quad \overline{j_2} \circ \overline{j_1} = 1_G .$$

$u := \overline{j_2}$ now has the required properties. ■

Remark. The above theorem allows us to identify all completions of E. This justifies the use of the term *the completion of E*.

1.3.10 Analytic Functions

Definition 1.3.10.1 (0) *Let E be a Banach space and U an open set of $\mathbb{K}$. A function $f : U \to E$ is called* **analytic** *if for given $\alpha_0 \in U$ there is a power series $\sum_{n=0} t^n x_n$ in E and an $r > 0$ such that r is smaller than the radius of convergence of this power series,*

$$U_r^{\mathbb{K}}(\alpha_0) \subset U ,$$

and

$$f(\alpha) = \sum_{n=0}^{\infty} (\alpha - \alpha_0)^n x_n$$

for every $\alpha \in U_r^{\mathbb{K}}(\alpha_0)$.

By Proposition 1.1.6.11, if $f : U \to E$ and $g : U \to E$ are analytic, then $\alpha f + \beta g : U \to E$ is analytic for all $\alpha, \beta \in \mathbb{K}$.

Proposition 1.3.10.2 *Analytic functions are differentiable and their derivatives are analytic.*

The proposition follows immediately from Proposition 1.1.6.25. ■

Proposition 1.3.10.3 (0) *Let E be a Banach space, $\sum_{n=0}^{\infty} t^n x_n$ a power series in E, r its radius of convergence, $\alpha_0 \in \mathbb{K}$, and*

$$f : U_r^{\mathbb{K}}(\alpha_0) \longrightarrow E, \quad \alpha \longmapsto \sum_{n=0}^{\infty} (\alpha - \alpha_0)^n x_n .$$

Then f is analytic.

Take $\beta_0 \in U_r^{\mathbb{K}}(\alpha_0)$ and put

$$r' := r - |\alpha_0 - \beta_0| .$$

Take $\alpha \in U_{r'}(\beta_0)$ and let

$$\rho := |\alpha - \beta_0| + |\beta_0 - \alpha_0| < r' + |\alpha_0 - \beta_0| = r .$$

Then

$$\sum_{m=0}^{n} \binom{n}{m} |\alpha - \beta_0|^m |\beta_0 - \alpha_0|^{n-m} = (|\alpha - \beta_0| + |\beta_0 - \alpha_0|)^n = \rho^n$$

for $n \in \mathbb{N} \cup \{0\}$, and so

$$\left\| \binom{n}{m} |\beta_0 - \alpha_0|^{n-m} x_n \right\| |\alpha - \beta_0|^m \leq \|x_n\| \rho^n$$

for $n \in \mathbb{N} \cup \{0\}$ and $m \in \mathbb{N}_n \cup \{0\}$. Hence

$$\sum_{n=m}^{\infty} \left\| \binom{n}{m} |\beta_0 - \alpha_0|^{n-m} x_n \right\| < \infty .$$

Given $m \in \mathbb{N} \cup \{0\}$, put

$$y_m := \sum_{n=m}^{\infty} \binom{n}{m} (\beta_0 - \alpha_0)^{n-m} x_n$$

(Corollary 1.1.6.10 a $\Rightarrow$ c) and

$$\binom{n}{m} := 0$$

for $n \in \mathbb{N}$ with $m > n$. Take $p \in \mathbb{N}$. Then

$$\sum_{m=0}^{p} |\alpha - \beta_0|^m \|y_m\| \leq \sum_{m=0}^{p} \left(\sum_{n=0}^{\infty} \binom{n}{m} |\beta_0 - \alpha_0|^{n-m} \|x_n\| \right) |\alpha - \beta_0|^m =$$

$$= \sum_{n=0}^{\infty} \|x_n\| \left(\sum_{m=0}^{p} \binom{n}{m} |\beta_0 - \alpha_0|^{n-m} |\alpha - \beta_0|^m \right) \leq \sum_{n=0}^{\infty} \|x_n\| \rho^n < \infty .$$

(Corollary 1.1.6.10). Since p is arbitrary, the family $\left((\alpha - \beta_0)^m y_m \right)_{m \in \mathbb{N} \cup \{0\}}$ is absolutely summable. It follows that the radius of convergence of the power series $\sum_{m=0}^{\infty} t^m y_m$ is greater than r' (Theorem 1.1.6.23).

Take $\varepsilon > 0$. There is a $p \in \mathbb{N}$ such that

$$\left\| f(\alpha) - \sum_{n=0}^{p} (\alpha - \alpha_0)^n x_n \right\| < \frac{\varepsilon}{3}, \quad \sum_{m=p+1}^{\infty} \|x_n\| \rho^n < \frac{\varepsilon}{3} .$$

Then (Proposition 1.1.6.11, Corollary 1.1.6.10),

$$\sum_{m=0}^{p} (\alpha - \beta_0)^m y_m = \sum_{m=0}^{p} (\alpha - \beta_0)^m \left(\sum_{n=0}^{\infty} \binom{n}{m} (\beta_0 - \alpha_0)^{n-m} x_n \right) =$$

$$= \sum_{n=0}^{\infty} \left(\sum_{m=0}^{p} \binom{n}{m} (\alpha - \beta_0)^m (\beta_0 - \alpha_0)^{n-m} \right) x_n =$$

$$= \sum_{n=0}^{p} (\alpha - \alpha_0)^n x_n + \sum_{n=p+1}^{\infty} \left(\sum_{m=0}^{p} \binom{n}{m} (\alpha - \beta_0)^m (\beta_0 - \alpha_0)^{n-m} \right) x_n ,$$

$$\| \sum_{m=0}^{p} (\alpha - \beta_0)^m y_m - \sum_{n=0}^{p} (\alpha - \alpha_0)^n x_n \| \leq$$

$$\leq \sum_{n=p+1}^{\infty} \|x_n\| \left(\sum_{m=0}^{p} \binom{n}{m} |\alpha - \beta_0|^m |\beta_0 - \alpha_0|^{n-m} \right) \leq \sum_{n=p+1}^{\infty} \|x_n\| \rho^n < \frac{\varepsilon}{3} ,$$

$$\|f(\alpha) - \sum_{n=0}^{\infty} (\alpha - \beta_0)^m y_m\| \leq \|f(\alpha) - \sum_{n=0}^{p} (\alpha - \alpha_0)^n x_n\| +$$

$$+ \| \sum_{n=0}^{p} (\alpha - \alpha_0)^n x_n - \sum_{m=0}^{p} (\alpha - \beta_0)^m y_m \| +$$

$$+ \sum_{m=p+1}^{\infty} \|y_m\| \, |\alpha - \beta_0|^m < \frac{\varepsilon}{3} + \frac{\varepsilon}{3} + \frac{\varepsilon}{3} = \varepsilon .$$

Since ε is arbitrary,

$$f(\alpha) = \sum_{m=0}^{\infty} (\alpha - \alpha_0)^m y_m .$$

Since α and β_0 are arbitrary, f is analytic. ■

Proposition 1.3.10.4 (0) *Let E, F be Banach spaces. Take $u \in \mathcal{L}(E, F)$. Let U be an open set of $\mathbb{K}$ and $f : U \to E$ an analytic function. Then $u \circ f$ is analytic.*

Take $\alpha_0 \in U$. There is a power series $\sum\limits_{n=0}^{\infty} t^n x_n$ in E and an $r > 0$, such that r is smaller than the radius of convergence of the power series,

$$U_r^{\mathbb{K}}(\alpha_0) \subset U ,$$

and

$$f(\alpha) = \sum_{n=0}^{\infty} (\alpha - \alpha_0)^n x_n$$

for every $\alpha \in U_r^{\mathbb{K}}(\alpha_0)$. By Corollary 1.2.1.17, r is smaller than the radius of convergence of the power series $\sum_{n=0}^{\infty} t^n u x_n$ and

$$u \circ f(\alpha) = \sum_{n=0}^{\infty} (\alpha - \alpha_0)^n u x_n$$

for every $\alpha \in U_r^{\mathbb{K}}(\alpha_0)$. Hence $u \circ f$ is analytic. ■

Corollary 1.3.10.5 *Let E be a Banach space, U a domain in $\mathbb{K}$, and $f : U \to E$ an analytic function. If f vanishes on an open nonempty subset of U, then f vanishes identically.*

Take $x' \in E'$. By Proposition 1.3.10.4, $x' \circ f$ is analytic. Since it vanishes on an open nonempty subset of U, it vanishes identically. By Corollary 1.3.3.8 a), f vanishes identically. ■

Theorem 1.3.10.6 (0) (Liouville's Theorem) *Let E be a complex Banach space. Every bounded analytic function $\mathbb{C} \to E$ is constant.*

Let $f : \mathbb{C} \to E$ be a bounded analytic function and take $\alpha, \beta \in \mathbb{C}$. Assume that

$$f(\alpha) \neq f(\beta).$$

Then there is an $x' \in E'$ with

$$x' \circ f(\alpha) \neq x' \circ f(\beta)$$

(Corollary 1.3.3.9). By Proposition 1.3.10.4, $x' \circ f$ is analytic. Since it is bounded, it is constant by the classical form of Liouville's Theorem. Hence

$$x' \circ f(\alpha) = x' \circ f(\beta)$$

which is a contradiction. ■

Remark. The above theorem was proved by Cauchy (1844) for $E = \mathbb{C}$.

Corollary 1.3.10.7 (0) *Let E be a complex Banach space and $f : \mathbb{C} \to E$ an analytic function. If*

$$\lim_{\alpha \to \infty} f(\alpha) = 0$$

then f is identically zero.

f is bounded, so it is constant (Theorem 1.3.10.6). ■

Theorem 1.3.10.8 (6) (Laurent's Theorem, 1843) *Let E be a complex Banach space. Take $\alpha_0 \in \mathbb{C}$ and $0 < r_1 < r_2$. Put*

$$U := \{\alpha \in \mathbb{C} \mid r_1 < |\alpha - \alpha_0| < r_2\},$$

and let $f : U \to E$ be an analytic function. Then there is a unique family $(x_n)_{n\in\mathbb{Z}}$ in E such that

$$f(\alpha) = \sum_{n=-\infty}^{\infty} (\alpha - \alpha_0)^n x_n$$

for every $\alpha \in U$. The radius of convergence of the power series

$$\sum_{n=0}^{\infty} t^n x_n \quad (\text{ resp. } \sum_{n=1}^{\infty} t^n x_{-n})$$

is greater than r_2 (resp. $\frac{1}{r_1}$). The expression

$$\sum_{n=-\infty}^{\infty} (t - \alpha_0)^n x_n ;$$

*is called **the Laurent series of** f,*

$$\sum_{n=-\infty}^{-1} (t - \alpha_0)^n x_n$$

*is called its **principal part** and x_{-1} is called its **residue**.*

Take $r \in]r_1, r_2[$. Given $n \in \mathbb{N}$, put

$$x_n := \frac{1}{2\pi r^n} \int_0^{2\pi} f(\alpha_0 + re^{it})e^{-int} dt,$$

where the integral is defined (as in the classical case) with the help of the Riemann sums.

Take $n \in \mathbb{N}$. Then

$$\langle x_n, x' \rangle = \frac{1}{2\pi r^n} \int_0^{2\pi} x' \circ f(\alpha_0 + re^{it})e^{-int} dt$$

for every $x' \in E'$. By Cauchy's Theorem, $\langle x_n, x' \rangle$ does not depend on r. Hence, by Corollary 1.3.3.9, x_n does not depend on r.

Set

$$\beta := \sup_{t\in\mathbb{R}} \|f(\alpha_0 + re^{it})\| < \infty .$$

Then

$$|\langle x_n, x'\rangle| \le \frac{\|x'\|\beta}{r^n}$$

for $n \in \mathbb{Z}$ and $x' \in E'$. Hence

$$\|x_n\| \le \frac{\beta}{r^n}$$

for every $n \in \mathbb{N}$ (Corollary 1.3.3.8 b)). Thus

$$\limsup_{n\to\infty} \|x_n\|^{\frac{1}{n}} \le \frac{1}{r} ,$$

$$\limsup_{n\to\infty} \|x_{-n}\|^{\frac{1}{n}} \le r .$$

Since r is arbitrary, the radius of convergence of

$$\sum_{n=0}^{\infty} t^n x_n \quad (\text{resp. } \sum_{n=1}^{\infty} t^n x_{-n})$$

is greater than r_2 (resp. $\frac{1}{r_1}$).

Take $\alpha \in U$. By the classical Laurent's Theorem,

$$x' \circ f(\alpha) = \sum_{n=-\infty}^{\infty} \langle x_n, x'\rangle \, (\alpha - \alpha_0)^n$$

for $x' \in E'$. Hence

$$\langle f(\alpha), x'\rangle = \langle \sum_{n=-\infty}^{\infty} (\alpha - \alpha_0)^n x_n, x'\rangle$$

for every $x' \in E'$ (Corollary 1.2.1.17) and so

$$f(\alpha) = \sum_{n=-\infty}^{\infty} (\alpha - \alpha_0)^n x_n$$

(Corollary 1.3.3.9).

To prove the uniqueness, let $(x_n)_{n\in\mathbb{Z}}$ be a family in E, such that

$$f(\alpha) = \sum_{n=-\infty}^{\infty} (\alpha - \alpha_0)^n x_n$$

for every $\alpha \in U$. Choose $x' \in E'$. Then

$$x' \circ f(\alpha) = \sum_{n=-\infty}^{\infty} \langle x_n, x' \rangle (\alpha - \alpha_0)^n$$

(Corollary 1.2.1.17). Then

$$\langle x_n, x' \rangle = \frac{1}{2\pi r^n} \int_0^{2\pi} x' \circ f(\alpha_0 + re^{it}) e^{-int} dt = \langle \frac{1}{2\pi r^n} \int_0^{2\pi} f(\alpha_0 + re^{it}) e^{-int} dt, x' \rangle$$

for every $n \in \mathbb{Z}$ by the Theorem of Residues. Since x' is arbitrary,

$$x_n = \frac{1}{2\pi r^n} \int_0^{2\pi} f(\alpha_0 + re^{it}) e^{-int} dt$$

for every $n \in \mathbb{Z}$ (Corollary1.3.3.9). ■

Remark. In the proof we have used only the fact that f is continuous and $x' \circ f$ is analytic for every $x' \in E'$.

Definition 1.3.10.9 (6) *Let E be a complex Banach space, U an open set of $\mathbb{C}$, $\alpha_0 \in U$, $f : U \backslash \{\alpha_0\} \to E$ an analytic function, and $r > 0$ such that*

$$U_r^{\mathbb{C}}(\alpha_0) \subset U .$$

The Laurent series of $f|(U_r^{\mathbb{C}}(\alpha_0) \backslash \{\alpha_0\})$ is called ***the Laurent series of** f **in** α_0 and its principal part (residue)* ***the principal part (the residue) of** f **at** α_0. If there is a $p \in \mathbb{N}$, such that the principal part of f at α_0 has the form*

$$\sum_{n=-1}^{-p} (t - \alpha_0)^n x_n$$

with $x_{-p} \neq 0$, then we say that f ***has a pole at** α_0 **of order** p.*

By Laurent's Theorem, the principal part of f at α_0 and the residue of f at α_0 are fully determined by f, and the principal part of f is convergent for every $\alpha \in \mathbb{C} \backslash \{\alpha_0\}$. In particular, if α_0 is a pole of f, then its order is determined by f.

Proposition 1.3.10.10 (0) *Let E, F, G be Banach spaces, $u : E \times F \to G$ a continuous bilinear map, U an open set of $\mathbb{K}$, and f, g analytic functions on U with values in E and F, respectively. Then the map*

$$U \longrightarrow G, \quad \alpha \longmapsto u(f(\alpha), g(\alpha))$$

is also analytic.

Take $\alpha_0 \in U$. There are power series $\sum\limits_{n=0}^{\infty} t^n x_n$, $\sum\limits_{n=0}^{\infty} t^n y_n$ in E and F, respectively, and an $r > 0$ such that r is smaller than the radii of convergence of these power series,

$$U_r^{\mathbb{K}}(\alpha_0) \subset U,$$

and

$$f(\alpha) = \sum_{n=0}^{\infty} (\alpha - \alpha_0)^n x_n, \quad g(\alpha) = \sum_{n=0}^{\infty} (\alpha - \alpha_0)^n y_n$$

for every $\alpha \in U_r^{\mathbb{K}}(\alpha_0)$. Given $n \in \mathbb{N} \cup \{0\}$, set

$$z_n := \sum_{k=0}^{n} u(x_k, y_{n-k}).$$

By Proposition 1.2.9.6 and Theorem 1.1.6.23, the radius of convergence of the power series $\sum\limits_{n=0}^{\infty} t^n z_n$ is greater than r and

$$\sum_{n=0}^{\infty} (\alpha - \alpha_0)^n z_n = u\left(\sum_{n=0}^{\infty} (\alpha - \alpha_0)^n x_n, \sum_{n=0}^{\infty} (\alpha - \alpha_0)^n y_n\right) = u(f(\alpha), g(\alpha))$$

for every $\alpha \in U_r^{\mathbb{K}}(\alpha_0)$. Hence the map

$$U \longrightarrow G, \quad \alpha \longmapsto u(f(\alpha), g(\alpha))$$

is analytic. ■

Exercises

E 1.3.1 Let E be a real vector space and p a real function on E satisfying

$$p(x+y) \le p(x) + p(y)\,,$$

$$p(\alpha x) = \alpha p(x)$$

for every $x, y \in E$ and $\alpha > 0$. Show that

a) $p(0) = 0$.

b) $p(-x) \ge -p(x)$ for every $x \in E$.

E 1.3.2 Let E be a normed space. Take $\alpha > 0$ and let $(x_n)_{n\in\mathbb{N}}$ be a sequence in E. Let $(\alpha_n)_{n\in\mathbb{N}}$ be a sequence in $\mathbb{K}$. Prove that the following are equivalent:

a) There is an $x' \in E'$, such that $\|x'\| \le \alpha$ and

$$x'(x_n) = \alpha_n$$

for every $n \in \mathbb{N}$.

b)
$$\left|\sum_{k=1}^{n} \alpha_k \beta_k\right| \le \alpha \left\|\sum_{k=1}^{n} \beta_k x_k\right\|$$

for every $n \in \mathbb{N}$ and $(\beta_k)_{k\in\mathbb{N}_n} \in \mathbb{K}^n$.

E 1.3.3 (Phillips, 1940) Let T be a set, E a normed space and F a subspace of E. Take $u \in \mathcal{L}(F, \ell^\infty(T))$. Show that there is a $v \in \mathcal{L}(E, \ell^\infty(T))$, such that

$$v|F = u\,, \quad \|u\| = \|v\|\,.$$

E 1.3.4 Let G be a normed space. Show that the following are equivalent:

a) Given a subspace F of the normed space E, and $u \in \mathcal{L}(F, G)$, there is a $v \in \mathcal{L}(E, G)$ with $v|F = u$.

b) If E is a normed space, then every subspace of E which is isomorphic to G is a complemented subspace of E.

c) Given a set T, every subspace of $\ell^\infty(T)$ which is isometric to G is a complemented subspace of $\ell^\infty(T)$.

Show further that G is complete whenever the above conditions are fulfilled. (Hint: Use E 1.3.3 and E 1.2.15 to prove c $\Rightarrow$ a.)

E 1.3.5 (Kottman) Show that every infinite–dimensional normed space contains a sequence $(x_n)_{n\in\mathbb{N}}$ such that

$$\|x_n\| = 1\,, \qquad \|x_m - x_n\| > 1$$

for distinct $m, n \in \mathbb{N}$. (This is an improvement on Corollary 1.1.4.3.)

E 1.3.6 Take $p \in [1, \infty[$. Let q be the conjugate expontent of p, and u_r, u_ℓ, respectively, the right and left shift on $\ell^p(\mathbb{Z})$. Show that u'_r and u'_ℓ are the left and right shift of $\ell^q(\mathbb{Z})$, respectively.

E 1.3.7 Take $n \in \mathbb{N}$. Let K be a compact set of $\mathbb{K}^n$. Given $P \in \mathbb{K}[s_1, \ldots, s_n, t_1, \ldots, t_n]$, let $\widetilde{P}$ be the map

$$K \longrightarrow \mathbb{K}\,, \quad \alpha \longmapsto P(\alpha_1, \ldots, \alpha_n, \overline{\alpha}_1, \ldots, \overline{\alpha}_n)\,.$$

Show that $\{\widetilde{P} \mid P \in \mathbb{K}[s_1, \ldots, s_n, t_1, \ldots, t_n]\}$ is a dense set of $\mathcal{C}(K)$.

E 1.3.8 Let $(T_\iota)_{\iota\in I}$ be a finite family of compact spaces and for $(x_\iota)_{\iota\in I} \in \prod_{\iota\in I} \mathcal{C}(T_\iota)$ define

$$\bigotimes_{\iota\in I} x_\iota : \prod_{\iota\in I} T_\iota \longrightarrow \mathbb{K}\,, \quad t \longmapsto \prod_{\iota\in I} x_\iota(t_\iota)\,.$$

Show that $\{\bigotimes_{\iota\in I} x_\iota \mid (x_\iota)_{\iota\in I} \in \prod_{\iota\in I} \mathcal{C}(T_\iota)\}$ is a dense set of $\mathcal{C}(\prod_{\iota\in I} T_\iota)$.

E 1.3.9 Let E, F be normed spaces. Let G and H be vector subspaces of E and F, respectively. Take $u \in \mathcal{L}(E, F)$. Show:

a) $u(G)^\circ = \overset{-1}{u}(G^\circ)$.

b) $u'(H^\circ) \subset \overset{-1}{u}(H)^\circ$.

c) The inclusion in b) may be strict.

E 1.3.10 Let $\mathbb{K} = \mathbb{C}$, U an open set of $\mathbb{K}$, and

$$E := \{f \in \ell^\infty(U) \mid f \text{ is analytic}\}\,.$$

Prove the following:

a) E is a closed vector subspace of $\ell^\infty(U)$.

b) E is complete with respect to the normed induced form $\ell^\infty(U)$.

c) The above assertions not longer hold when $\mathbb{C}$ is replaced by $\mathbb{R}$.

1.4 Applications of Baire's Theorem

Baire's Theorem has two very important consequences in the theory of Banach spaces, namely the Principle of Uniform Boundedness — a set of operators is bounded whenever it is pointwise bounded — and the Principle of Inverse Operators — the inverse function to a bijective operator is continuous. The latter principle, which was proved by Banach, is deep and probably the most frequently quoted theorem in this book. This is because invertible operators play a central role in the theory and the Principle of Inverse Operators provides a remarkably simple criterion for determining whether an operator is invertible: an operator is invertible in the category of Banach spaces if and only if it is invertible in the category of vector spaces.

1.4.1 The Banach–Steinhaus Theorem

Proposition 1.4.1.1 (0) *Let E be a Banach space and A an absolutely convex closed set of E such that*

$$E = \bigcup_{n\in\mathbb{N}} (nA)\,.$$

Then 0 is an interior point of A.

By Baire's Theorem, there is an $n \in \mathbb{N}$ such that nA has nonempty interior. Hence there are $x \in E$ and $\varepsilon > 0$ with

$$U_\varepsilon(x) \subset A\,.$$

Since A is absolutely convex,

$$U_\varepsilon(-x) \subset A\,.$$

It follows that

$$y = \frac{1}{2}(x+y) + \frac{1}{2}(-x+y) \in A$$

for every $y \in U_\varepsilon(0)$, i.e. $U_\varepsilon(0) \subset A$ and 0 is an interior point of A. ∎

Theorem 1.4.1.2 (0) (Banach–Steinhaus) *Let E be a Banach space, F a normed space, and $\mathcal{F}$ a set of $\mathcal{L}(E,F)$, such that*

$$\sup_{u\in\mathcal{F}} \|ux\| < \infty$$

for every $x \in E$. Then $\mathcal{F}$ is bounded.

Put

$$A := \bigcap_{u \in \mathcal{F}} \overset{-1}{u}(F^{\#}) .$$

A is an absolutely convex closed set of E (Proposition 1.2.7.2, Proposition 1.2.7.7, Proposition 1.2.7.4). Take $x \in E$. Then

$$\sup_{u \in \mathcal{F}} \|ux\| \leq n ,$$

for some $n \in \mathbb{N}$. Then $\frac{1}{n} x \in A$ and $x \in nA$. Hence

$$\bigcup_{n \in \mathbb{N}} (nA) = E .$$

By Proposition 1.4.1.1, 0 is an interior point of A, i.e.

$$\overline{U_\varepsilon(0)} \subset A ,$$

for some $\varepsilon > 0$. Thus $\varepsilon x \in A$ and

$$\|ux\| = \frac{1}{\varepsilon}\|u(\varepsilon x)\| \leq \frac{1}{\varepsilon}$$

for every $u \in \mathcal{F}$ and $x \in E^{\#}$. Hence

$$\|u\| \leq \frac{1}{\varepsilon}$$

for every $u \in \mathcal{F}$ (Proposition 1.2.1.4 b)), i.e. $\mathcal{F}$ is bounded. ■

Remark. The above theorem is occasionally called **the Principle of Uniform Boundedness**. It was proved in 1923 by Banach for sequences of operators and in 1927 by Banach and Steinhaus in the above form. The condition that E be complete cannot be relaxed (see Exercise 1.4.1).

Corollary 1.4.1.3 (0) *Let E, F be Banach spaces. Take a sequence $(u_n)_{n \in \mathbb{N}}$ in $\mathcal{L}(E, F)$ such that $(u_n x)_{n \in \mathbb{N}}$ converges for every $x \in E$ and consider*

$$u : E \longrightarrow F , \quad x \longmapsto \lim_{n \to \infty} u_n x .$$

Then $u \in \mathcal{L}(E, F)$,

$$\|u\| \leq \liminf_{n \to \infty} \|u_n\| \leq \sup_{n \in \mathbb{N}} \|u_n\| < \infty ,$$

and $(u_n)_{n \in \mathbb{N}}$ converges to u uniformly on every compact set of E.

By the Banach–Steinhaus Theorem,

$$\sup_{n\in\mathbb{N}} \|u_n\| < \infty .$$

The other assertions follow from Proposition 1.2.1.7 and Proposition 1.1.2.15. ■

Corollary 1.4.1.4 (0) *Let E be a normed space and A a weakly bounded set of E, i.e.*

$$\sup_{x\in A} |x'(x)| < \infty$$

for every $x' \in E'$. Then A is bounded. In particular, the weakly compact sets of E are bounded.

Given $x' \in E'$,

$$\sup_{x\in A} |\langle j_E x, x'\rangle| = \sup_{x\in A} |\langle x, x'\rangle| < \infty .$$

By the Banach–Steinhaus Theorem, $j_E(A)$ is a bounded set of E''. Hence A is a bounded set of E (Theorem 1.3.6.3 a)). ■

Remark. This corollary was proved for sequences bv Hahn in 1922.

Corollary 1.4.1.5 (0) *Let E be a Banach space, F a normed space, and take $\mathcal{F} \subset \mathcal{L}(E,F)$. If*

$$\sup_{u\in\mathcal{F}} |\langle ux, y'\rangle| < \infty$$

for every $(x,y') \in E \times F'$, then $\mathcal{F}$ is bounded.

By Corollary 1.4.1.4, $\{ux \mid u \in \mathcal{F}\}$ is bounded whenever $x \in E$. Thus, by the Banach–Steinhaus Theorem, $\mathcal{F}$ is bounded. ■

Corollary 1.4.1.6 (0) *Let $u : E \to F$ be a linear map between normed spaces. Then the following are equivalent:*

a) *u is continuous.*

b) *$y' \circ u$ is continuous for every $y' \in F'$.*

c) *u is continuous with respect to the weak topologies on E and F.*

a $\Rightarrow$ b is trivial.

b $\Rightarrow$ c. Take $x \in E$ and let $\mathfrak{F}$ be a filter on E converging weakly to x. Finally, take $y' \in F'$. By b) and Proposition 1.3.6.12, $y' \circ u$ is weakly continuous and so $y'(u(\mathfrak{F}))$ converges to $y'(ux)$. Since y' is arbitrary, $u(\mathfrak{F})$ converges weakly to ux and u is continuous with respect to the weak topologies on E and F.

c $\Rightarrow$ a. Take $y' \in F'$. By Proposition 1.3.6.12, y' is weakly continuous. Hence, by c), $y' \circ u$ is weakly continuous too. Using Proposition 1.3.6.12, again, we see that $y' \circ u$ is continuous and so

$$\sup_{x \in E^\#} |y'(ux)| = \sup_{x \in E^\#} |y' \circ u(x)| < \infty$$

(Proposition 1.2.1.1 a $\Rightarrow$ e). By Corollary 1.4.1.4, $u(E^\#)$ is bounded, so that u is continuous (Proposition 1.2.1.1 e $\Rightarrow$ a). ■

Corollary 1.4.1.7 (0) *Let E be a vector space and p, q norms on E. Let E_p (resp. E_q) be the vector space E endowed with the norm p (resp. q), and p' (resp. q') be the canonical norm on $(E_p)'$ (resp. $(E_q)'$). Then the following are equivalent:*

a) *$p \le \alpha q$ for some $\alpha \in \mathbb{R}_+$.*

b) *$(E_p)' \subset (E_q)'$ and $q'|(E_p)' \le \beta p'$ for some $\beta \in \mathbb{R}_+$.*

c) *$(E_p)' \subset (E_q)'$.*

In particular, p and q are equivalent iff the vector spaces $(E_p)'$, $(E_q)'$ coincide.

a $\Rightarrow$ b. Let $j : E_q \to E_p$ be the identity map on E. By a) (and Proposition 1.2.1.1 d) $\Rightarrow$ a), j is continuous. $j' : (E_p)' \to (E_q)'$ is the inclusion map and the existence of β follows from Proposition 1.2.1.1 a $\Rightarrow$ d.

b $\Rightarrow$ c is trivial.

c $\Rightarrow$ a follows from Corollary 1.4.1.6 b $\Rightarrow$ a (and Proposition 1.2.1.1 a $\Rightarrow$ d).

The final assertion follows from a $\Leftrightarrow$ c. ■

Corollary 1.4.1.8 *Let E, F, G be Banach spaces and $u : E \times F \to G$ a bilinear map. If u is continuous in each variable, then u is continuous.*

Given $y \in F$,

$$\sup_{x \in E^\#} |u(x, y)| \le \sup_{x \in E^\#} \|u(\cdot, y)\| \, \|x\| = \|u(\cdot, y)\| < \infty$$

(Proposition 1.2.1.4 a)). By the Banach–Steinhaus Theorem

$$\alpha := \sup_{x \in E^{\#}} \|u(x, \cdot)\| < \infty .$$

Take $(x, y) \in E \times F$ with $x \neq 0$. Then $\frac{1}{\|x\|} x \in E^{\#}$, so that

$$\|u(x, y)\| = \|x\| \, \|u(\frac{1}{\|x\|} x, y)\| \leq \|x\| \, \|u(\frac{1}{\|x\|} x, \cdot)\| \, \|y\| \leq \alpha \|x\| \, \|y\|$$

(Proposition 1.2.1.4 b)). By Proposition 1.2.9.2 c $\Rightarrow$ a, u is continuous. ■

Proposition 1.4.1.9 (Gelfand) *Let A be a subset of the separable Banach space E. Then the following are equivalent.*

a) *A is relatively compact.*

b) *Every sequence in E' which converges pointwise to 0 converges uniformly to 0 on A.*

a $\Rightarrow$ b follows from Corollary 1.4.1.3.

b $\Rightarrow$ a. We first show that A is bounded. By Corollary 1.4.1.4, it is sufficient to show that A is weakly bounded, i.e. that

$$\sup_{x \in A} |x'(x)| < \infty$$

for every $x' \in E'$. Assume the contrary. Then there is a sequence $(x_n)_{n \in \mathbb{N}}$ in A and an $x' \in E'$ such that

$$|x'(x_n)| > n$$

for every $n \in \mathbb{N}$. Then $(\frac{1}{n} x')_{n \in \mathbb{N}}$ is a sequence in E' which converges pointwise to 0. But it does not converge to 0 uniformly on A, which contradicts b). Hence A is bounded.

Let $(x_n)_{n \in \mathbb{N}}$ be a dense sequence in E and put

$$B := \{x_n \mid n \in \mathbb{N}\} .$$

Then $E'^{\#}_B$ is a compact space (Alaoglu–Bourbaki Theorem and Proposition 1.1.2.15). Set

$$u : E \longrightarrow \mathcal{C}(E'^{\#}_B), \quad x \longmapsto j_E x \mid E'^{\#} .$$

u preserves norms (Corollary 1.3.3.8 b), Theorem 1.3.6.3 a)) and so $\operatorname{Im} u$ is a closed set of $\mathcal{C}(E'^{\#}_B)$ (since E is complete). In order to show that A is relatively compact, we must therefore prove that $u(A)$ is a relatively compact set of $\mathcal{C}(E'^{\#}_{B})$. Since $u(A)$ is a bounded set of $\mathcal{C}(E'^{\#}_{B})$, it is sufficient to show that $u(A)$ is equicontinuous (Ascoli Theorem). Take $x' \in E'^{\#}_{B}$ and $\varepsilon > 0$. We show that there is a neighbourhood U of x' in $E'^{\#}_{B}$ such that

$$|\langle x, y'\rangle - \langle x, x'\rangle| < \varepsilon$$

for every $y' \in U$ and $x \in A$. Assume the contrary and take $n \in \mathbb{N}$. Then

$$U_n := \left\{ y' \in E'^{\#} \,\middle|\, k \in \mathbb{N}_n \Longrightarrow |\langle x_k, x' - y'\rangle| < \frac{1}{n} \right\}$$

is a neighbourhood of x' in $E'^{\#}_{B}$. There are $y_n \in A$ and $x'_n \in U_n$ such that

$$|\langle y_n, x'_n - x'\rangle| \geq \varepsilon .$$

Then

$$\lim_{n\to\infty} x'_n(x_k) = x'(x_k)$$

for every $k \in \mathbb{N}$ and so $(x'_n)_{n\in\mathbb{N}}$ converges to x' in $E'^{\#}_{B}$. By Proposition 1.1.2.15, $(x'_n)_{n\in\mathbb{N}}$ converges to x' in E'_E and by b), $(x'_n)_{n\in\mathbb{N}}$ converges to x' uniformly on A. Hence

$$\varepsilon \leq \lim_{n\to\infty} |\langle y_n, x'_n - x'\rangle| = 0$$

and this is a contradicition. ■

Example 1.4.1.10 (4) *Let*

$$S_1 := \{\alpha \in \mathbb{C} \mid |\alpha| = 1\}$$

and

$$\widehat{x}_n := \frac{1}{2\pi} \int_0^{2\pi} e^{-int} x(t)dt$$

for $x \in \mathcal{C}(S_1)$ and $n \in \mathbb{Z}$. There is an $x \in \mathcal{C}(S_1)$ such that

$$\left(\sum_{n=-p}^{p} \widehat{x}_n \right)_{p\in\mathbb{N}}$$

does not converge, i.e. the Fourier series of x is not pointwise convergent.

Given $p \in \mathbb{N}$, set

$$f_p : S_1 \longrightarrow \mathbb{K}, \quad \alpha \longmapsto \sum_{n=-p}^{p} \alpha^n$$

and

$$x'_p : \mathcal{C}(S_1) \longrightarrow \mathbb{K}, \quad x \longmapsto \sum_{n=-p}^{p} \widehat{x}_n .$$

For $p \in \mathbb{N}$ and $t \in]0, 2\pi[$

$$f_p(e^{it}) = \frac{\sin(p+\frac{1}{2})t}{\sin\frac{t}{2}} .$$

x'_p is linear and

$$x'_p(x) = \frac{1}{2\pi}\int_0^{2\pi} f_p(e^{it})x(t)dt .$$

Thus

$$|x'_p(x)| \leq \|x\|_\infty \frac{1}{2\pi}\int_0^{2\pi} |f_p(e^{it})|dt$$

for every $x \in \mathcal{C}(S_1)$. Hence x'_p is continuous and

$$\|x'_p\| \leq \frac{1}{2\pi}\int_0^{2\pi} |f_p(e^{it})|dt$$

(Proposition 1.2.1.1 d $\Rightarrow$ a). Define

$$f : S_1 \longrightarrow \mathbb{K}, \quad \alpha \longmapsto \begin{cases} 1 & \text{if } f_p(\alpha) \geq 0 \\ -1 & \text{if } f_p(\alpha) < 0 . \end{cases}$$

Take $\varepsilon > 0$. There is an $x \in \mathcal{C}(S_1)$ such that $\|x\|_\infty \leq 1$ and

$$\left| \frac{1}{2\pi}\int_0^{2\pi} f_p(e^{it})\Big(x(e^{it}) - f(e^{it})\Big)dt \right| < \varepsilon .$$

Then

$$\left| x'_p(x) - \frac{1}{2\pi}\int_0^{2\pi} |f_p(e^{it})|dt \right| < \varepsilon$$

and

$$\|x'_p\| \geq |x'_p(x)| \geq \frac{1}{2\pi}\int_0^{2\pi} |f_p(e^{it})|dt - \varepsilon$$

(Proposition 1.2.1.4 b)). Since ε is arbitrary,

$$\|x'_p\| \geq \frac{1}{2\pi}\int_0^{2\pi} |f_p(e^{it})|dt > \frac{1}{\pi}\int_0^{2\pi} \frac{|\sin(p+\frac{1}{2})t|}{t}\,dt =$$

$$= \frac{1}{\pi}\int_0^{(2p+1)\pi} \frac{|\sin t|}{t}dt = \frac{1}{\pi}\sum_{k=0}^{2p}\int_{k\pi}^{(k+1)\pi} \frac{|\sin nt|}{t}\,dt \geq$$

$$\geq \frac{1}{\pi}\sum_{k=0}^{2p}\frac{1}{(k+1)\pi}\int_{k\pi}^{(k+1)\pi} |\sin nt|dt = \frac{2}{\pi^2}\sum_{k=0}^{2p}\frac{1}{k+1}.$$

Hence

$$\lim_{p\to\infty} \|x'_p\| = \infty .$$

By Corollary 1.4.1.3, there is an $x \in \mathcal{C}(S_1)$ for which $(x'_p(x))_{p\in\mathbb{N}}$ does not converge. ■

1.4.2 Open Mapping Principle

Proposition 1.4.2.1 (0) *Let E be a normed space, F a Banach space, and $u : E \to F$ a linear surjective map. Then 0 is an interior point of $\overline{u(E^{\#})}$.*

$\overline{u(E^{\#})}$ is absolutely convex (Propositions 1.2.7.2, 1.2.7.7, and 1.2.7.5), and

$$\bigcup_{n\in\mathbb{N}} n\overline{u(E^{\#})} \supset \bigcup_{n\in\mathbb{N}} nu(E^{\#}) = \bigcup_{n\in\mathbb{N}} u(nE^{\#}) = u(\bigcup_{n\in\mathbb{N}} nE^{\#}) = u(E) = F\,.$$

By Proposition 1.4.1.1, 0 is an interior point of $\overline{u(E^{\#})}$. ■

Proposition 1.4.2.2 (0) *Let E be a Banach space, F a normed space and take $u \in \mathcal{L}(E,F)$. If 0 is an interior point of $\overline{u(E^{\#})}$, then 0 is an interior point of $u(E^{\#})$.*

By hypothesis, there is an $\varepsilon > 0$ with

$$\varepsilon F^{\#} \subset \overline{u(E^{\#})}\,.$$

We prove that

$$\frac{\varepsilon}{2}F^{\#} \subset u(E^{\#})\,.$$

Take $y \in \frac{\varepsilon}{2}F^{\#}$. We construct inductively a sequence $(x_n)_{n\in\mathbb{N}}$ in $E^{\#}$ such that

$$\left\| y - u\left(\sum_{m=1}^{n-1} \frac{1}{2^m}x_m\right)\right\| < \frac{\varepsilon}{2^n}$$

for every $n \in \mathbb{N}$. Take $n \in \mathbb{N}$ and suppose that $x_1, \ldots, x_{n-1}$ have been constructed. Then

$$y - u\left(\sum_{m=1}^{n-1} \frac{1}{2^m}x_m\right) \in \frac{\varepsilon}{2^n}F^{\#} \subset \frac{1}{2^n}\overline{u(E^{\#})}\,.$$

Hence there is an $x_n \in E^{\#}$ with

$$\left\| y - u\left(\sum_{m=1}^{n-1} \frac{1}{2^m}x_m\right) - \frac{1}{2^n}ux_n\right\| < \frac{\varepsilon}{2^{n+1}}\,,$$

i.e.

$$\left\| y - u\left(\sum_{m=1}^{n} \frac{1}{2^m}x_m\right)\right\| < \frac{\varepsilon}{2^{n+1}}\,.$$

This completes the inductive construction.

$\left(\frac{1}{2^n}x_n\right)_{n\in\mathbb{N}}$ is an absolutely convergent sequence in E. Put

$$x := \sum_{n\in\mathbb{N}} \frac{1}{2^n} x_n$$

(Corollary 1.1.6.10 a $\Rightarrow$ c). Then

$$\|x\| \le \sum_{n\in\mathbb{N}} \frac{1}{2^n}\|x_n\| \le 1$$

(Corollary 1.1.6.10) and

$$y = \lim_{n\to\infty} u\Big(\sum_{m=1}^{n} \frac{1}{2^m} x_m\Big) = ux \in u(E^{\#}).$$

Hence

$$\frac{\varepsilon}{2}F^{\#} \subset u(E^{\#})$$

and 0 is an interior point of $u(E^{\#})$. ■

Theorem 1.4.2.3 (0) (Open Mapping Principle, Banach 1932) *Every surjective operator between two Banach spaces is open, i.e. maps open sets into open sets.*

Let E, F be Banach spaces and $u : E \to F$ a surjective operator. Let U be an open set of E and $y \in u(U)$. Take $x \in U$ with

$$ux = y.$$

Then

$$x + \varepsilon E^{\#} \subset U,$$

for some $\varepsilon > 0$. Hence

$$y + \varepsilon u(E^{\#}) = u(x + \varepsilon E^{\#}) \subset u(U).$$

By Proposition 1.4.2.1 and 1.4.2.2, 0 is an interior point of $u(E^{\#})$. Hence y is an interior point of $u(U)$ and $u(U)$ is thus open. ■

Corollary 1.4.2.4 (0) (Principle of Inverse Operators) *Every bijective operator between Banach spaces is an isomorphism.*

Let E, F be Banach spaces and $u : E \to F$ a bijective operator. Since u is open (Theorem 1.4.2.3), u^{-1} is continuous. Hence u is an isomorphism. ■

Corollary 1.4.2.5 *Let E, F be Banach spaces. An operator $u : E \to F$ is an isomorphism (isometry) iff u' is an isomorphism (isometry).*

The necessity was proved in Corollary 1.3.4.7. So assume that u' is an isomorphism (isometry). By Corollary 1.3.4.7, u'' is an isomorphism (isometry). Then

$$u'' \circ j_E = j_F \circ u$$

(Proposition 1.3.6.16), so that

$$j_E = u''^{-1} \circ j_F \circ u .$$

Hence

$$\|x\| = \|j_E x\| = \|u''^{-1} \circ j_F \circ u(x)\| \leq \|u''^{-1} \circ j_F\| \, \|ux\|$$

$$(\|x\| = \|j_E x\| = \|j_F ux\| = \|ux\|)$$

for every $x \in E$ (Theorem 1.3.6.3 a), Proposition 1.2.1.4 a)). Hence u is injective and $\operatorname{Im} u$ is closed (Proposition 1.2.1.18 a), c)). Since u' is injective, $\operatorname{Im} u$ is dense (Corollary 1.3.5.9). Thus u is surjective. By Corollary 1.4.2.4, u is an isomorphism (isometry). ■

Corollary 1.4.2.6 (0) *Let F, G be closed subspaces of the Banach space E such that the map*

$$F \times G \longrightarrow E, \quad (x, y) \longmapsto x + y$$

is bijective. Then

$$E = F \oplus G .$$

F, G being Banach spaces, $F \times G$, is also a Banach space (Proposition 1.1.5.1). The map

$$F \times G \longrightarrow E, \quad (x, y) \longmapsto x + y$$

is linear and continuous (Proposition 1.2.5.1) and so it is an isomorphism(Corollary 1.4.2.4). ■

Corollary 1.4.2.7 (0) *Let E, F be Banach spaces and take $u \in \mathcal{L}(E,F)$. If $F/\operatorname{Im} u$ is finite–dimensional, then there is a finite–dimensional vector subspace G of F such that*

$$F = G \oplus \operatorname{Im} u\,.$$

Let v be the factorization of u through $E/\operatorname{Ker} u$, $q : F \to F/\operatorname{Im} u$ the quotient map, and $(Y_\iota)_{\iota\in I}$ an algebraic basis of $F/\operatorname{Im} u$. Given $\iota \in I$, take $y_\iota \in Y_\iota$ and let G be the vector subspace of F generated by $(y_\iota)_{\iota\in I}$. Being finite–dimensional, G is a Banach space (Corollary 1.1.3.5), so that

$$(E/\operatorname{Ker} u) \times G$$

is also a Banach space (Theorem 1.2.4.2 e), Proposition 1.1.5.1). Define

$$w : (E/\operatorname{Ker} u) \times G \longrightarrow F\,, \quad (x,y) \longmapsto vx + y\,.$$

Being the composition of the operators

$$(E/\operatorname{Ker} u) \times G \longrightarrow (\operatorname{Im} u) \times G\,, \quad (x,y) \longmapsto (vx, y)\,,$$

$$(\operatorname{Im} u) \times G \longrightarrow F\,, \quad (z,y) \longmapsto z + y$$

(Propositions 1.2.4.7 and 1.2.5.1) w is itself an operator.

Take $y \in F$. There is a family $(\alpha_\iota)_{\iota\in I}$ in $\mathbb{K}$ such that

$$qy = \sum_{\iota\in I} \alpha_\iota Y_\iota\,.$$

Since

$$q(y - \sum_{\iota\in I} \alpha_\iota y_\iota) = qy - \sum_{\iota\in I} \alpha_\iota Y_\iota = 0$$

it follows that

$$y - \sum_{\iota\in I} \alpha_\iota y_\iota \in \operatorname{Ker} q = \operatorname{Im} u = \operatorname{Im} v\,.$$

Take $x \in E/\operatorname{Ker} u$ such that

$$y - \sum_{\iota\in I} \alpha_\iota y_\iota = vx\,.$$

Then

$$y = vx + \sum_{\iota\in I} \alpha_\iota y_\iota = w(x, \sum_{\iota\in I} \alpha_\iota y_\iota) \in \operatorname{Im} w\,.$$

Hence w is surjective.

Take $(x,y) \in \operatorname{Ker} w$. We have successively

$$vx + y = 0\,,$$

$$qy = -qvx = 0\,,$$

$$y = 0\,,$$

$$vx = 0\,,$$

$$x = 0\,.$$

Hence w is injective. By the Principle of Inverse Operators w is an isomorphism and so $\operatorname{Im} u = \operatorname{Im} v$ is closed. Since the map

$$G \longrightarrow F/\operatorname{Im} u\,, \quad y \longmapsto qy$$

is an isomorphism, it follows from Proposition 1.2.5.2 c $\Rightarrow$ a, that

$$F = G \oplus \operatorname{Im} u\,.$$

■

Corollary 1.4.2.8 (0) *Let E, F be Banach spaces and $u \in \mathcal{L}(E, F)$. The following assertions are equivalent:*

a) $\operatorname{Im} u$ *is closed and* u *is injective.*

b) u *is lower bounded, i.e. there is an* $\alpha > 0$ *such that*

$$\|ux\| \geq \alpha \|x\|$$

for every $x \in E$.

a $\Rightarrow$ b. The operator

$$v : E \longrightarrow \operatorname{Im} u\,, \quad x \longmapsto ux$$

is bijective. By the Principle of Inverse Operators, it is therefore an isomorphism. Put

$$\alpha := \frac{1}{\|v^{-1}\|}\,.$$

Then

$$\|x\| = \|v^{-1}(ux))\| \leq \|v^{-1}\|\,\|ux\|\,,$$

$$\|ux\| \geq \frac{1}{\|v^{-1}\|}\|x\| = \alpha\|x\|$$

for every $x \in E$.

b $\Rightarrow$ a follows from Proposition 1.2.1.18 a),c). ■

Proposition 1.4.2.9 (0) *Let* E, F *be Banach spaces and take* $u \in \mathcal{L}(E, F)$ *such that* $\operatorname{Im} u$ *is closed. Let* $j : \operatorname{Ker} u \to E$ *be the inclusion map.*

a) *The associated algebraic isomorphism of* u *is an isomorphism of Banach spaces.*

b) $\operatorname{Im} u' = (\operatorname{Ker} u)^\circ = \operatorname{Ker} j'$, $\quad (\operatorname{Ker} u)^{\circ\circ} = \operatorname{Ker} u''$.

c) *The factorization* $E'/\operatorname{Im} u' \to (\operatorname{Ker} u)'$ *of* j' *through* $E'/\operatorname{Im} u'$ *is a isometry.*

a) follows from the Principle of Inverse Operators (and Theorem 1.2.4.2 e)).

b) Take $x' \in (\operatorname{Ker} u)^\circ$. Let $r : E \to E/\operatorname{Ker} u$ be the quotient map and $\overline{u}$ the associated algebraic isomorphism of u. By Proposition 1.3.5.2, there is a $y' \in (E/\operatorname{Ker} u)'$ such that

$$x' = r'y' = y' \circ r \,.$$

By a), $\overline{u}^{-1}$ is continuous, so that $y' \circ \overline{u}^{-1}$ belongs to $(\operatorname{Im} u)'$. By the Hahn–Banach Theorem, there is a continuous linear form z' on F which extends $y' \circ \overline{u}^{-1}$.

Then

$$\langle x, u'z' \rangle = \langle ux, z' \rangle = \langle ux, y' \circ \overline{u}^{-1} \rangle = y'(\overline{u}^{-1}(ux)) = y'(rx) = x'(x)$$

for every $x \in E$, and so

$$x' = u'z' \in \operatorname{Im} u' \,,$$

$$(\operatorname{Ker} u)^\circ \subset \operatorname{Im} u' \,.$$

Since the reverse inclusion follows from Proposition 1.3.5.8, we have that

$$(\operatorname{Ker} u)^\circ = \operatorname{Im} u' \,,$$

$$(\operatorname{Ker} u)^{\circ\circ} = (\operatorname{Im} u')^\circ = \operatorname{Ker} u''$$

(Proposition 1.3.5.8). The fact that

$$(\operatorname{Ker} u)^\circ = \operatorname{Ker} j'$$

was proved in Theorem 1.3.5.12 b).

c) follows from b) and Theorem 1.3.5.12 c). ∎

Corollary 1.4.2.10 *Let E and F be Banach spaces. Take $u \in \mathcal{L}(E,F)$ with $\operatorname{Im} u$ closed. Let $q : E \to E/\operatorname{Ker} u$, $r : F' \to F'/\operatorname{Ker} u'$ be the quotient maps and $i : \operatorname{Im} u \to F$, $j : \operatorname{Im} u' \to E'$ the inclusion maps. Let $\overline{u} : E/\operatorname{Ker} u \to \operatorname{Im} u$ and $\overline{u'} : F'/\operatorname{Ker} u \to \operatorname{Im} u'$ be the algebraic isomorphisms associated to u and u', respectively. Define*

$$v : (E/\operatorname{Ker} u)' \longrightarrow (\operatorname{Ker} u)^\circ\,, \quad x' \longmapsto q'x'$$

(Proposition 1.3.5.2) and let

$$w : F'/(\operatorname{Im} u)^\circ \longrightarrow (\operatorname{Im} u)'$$

(Proposition 1.3.5.8) be the factorization of i' through $F'/(\operatorname{Im} u)^\circ$.

a) $\operatorname{Ker} u' = (\operatorname{Im} u)^\circ\,, \quad \operatorname{Im} u' = (\operatorname{Ker} u)^\circ\,,$

$$\operatorname{Ker} u = {}^\circ(\operatorname{Im} u')\,, \quad \operatorname{Im} u = {}^\circ(\operatorname{Ker} u')\,.$$

b) *$\overline{u}$ and $\overline{u}'$ are isomorphisms of Banach spaces.*

c) *v and w are isometries.*

d) $\overline{u'} = v \circ \overline{u}' \circ w\,.$

a) By Proposition 1.3.5.8

$$\operatorname{Ker} u' = (\operatorname{Im} u)^\circ\,, \quad \operatorname{Ker} u = {}^\circ(\operatorname{Im} u')\,, \quad \operatorname{Im} u = {}^\circ(\operatorname{Ker} u')\,,$$

and, by Proposition 1.4.2.9 b),

$$\operatorname{Im} u' = (\operatorname{Ker} u)^\circ\,.$$

b) By Proposition 1.4.2.9 a), $\overline{u}$ is an isomorphism of Banach spaces. Hence, by Corollary 1.3.4.7, the same holds for $\overline{u}'$.

c) follows from Proposition 1.3.5.2 and Theorem 1.3.5.12 c).

d) Given $x \in E$ and $y' \in F'$,

$$\langle x, jv\overline{u}'wry'\rangle = \langle x, q'\overline{u}'i'y'\rangle = \langle i\overline{u}qx, y'\rangle = \langle ux, y'\rangle = \langle x, u'y'\rangle = \langle x, j\overline{u'}ry'\rangle\,,$$

so

$$j \circ v \circ \overline{u}' \circ w \circ r = j \circ \overline{u'} \circ r\,.$$

Since j is injective and r is surjective,

$$v \circ \overline{u}' \circ w = \overline{u'}\,. \qquad \blacksquare$$

Proposition 1.4.2.11 (0) *Let E, F be Banach spaces and take $u \in \mathcal{L}(E,F)$. If $\operatorname{Im} u'$ is closed, then $\operatorname{Im} u$ is also closed.*

Define

$$G := \overline{\operatorname{Im} u},$$

$$v : E \longrightarrow G, \quad x \longmapsto ux,$$

$$w : G \longrightarrow F, \quad y \longmapsto y.$$

Then v' is injective (Corollary 1.3.5.9) and w' is surjective (Theorem 1.3.5.12 a)). Since

$$v'(G') = v'(w'(F')) = u'(F')$$

(Corollary 1.3.4.5), $v'(G')$ is closed. By Corollary 1.4.2.8 a $\Rightarrow$ b, there is an $\alpha > 0$ with

$$\|v'y'\| \geq \alpha \|y'\|$$

for every $y' \in G'$.

Take $y \in \alpha G^{\#} \backslash \overline{v(E^{\#})}$. Since $\overline{v(E^{\#})}$ is absolutely convex there is a $y' \in G'$ such that

$$\sup_{z \in v(E^{\#})} |\langle z, y' \rangle| < \langle y, y' \rangle$$

(Corollary 1.3.1.7). Thus

$$\alpha \|y'\| \leq \|v'y'\| = \sup_{x \in E^{\#}} |\langle x, v'y' \rangle| = \sup_{x \in E^{\#}} |\langle vx, y' \rangle| < \langle y, y' \rangle \leq \alpha \|y'\|$$

which is a contradiction. Hence

$$\alpha G^{\#} \subset \overline{v(E^{\#})}$$

and 0 is an interior point of $\overline{v(E^{\#})}$. By Proposition 1.4.2.2, 0 is an interior point of $v(E^{\#})$. Thus v is surjective. It follows that

$$\operatorname{Im} u = v(E) = G$$

and so $\operatorname{Im} u$ is closed in F. ■

Proposition 1.4.2.12 *Let E, F be Banach spaces and $\mathfrak{G}$ the set of closed vector subspaces of E. Take $u \in \mathcal{L}(E,F)$ and let $q : E \to E/\operatorname{Ker} u$ be the quotient map, v the factorization of u through $E/\operatorname{Ker} u$ and w the algebraic isomorphism associated to u. Then the followsing are equivalent:*

a) $u(G)$ *is closed for every* $G \in \mathfrak{G}$.

b) $q(G)$ *is closed for every* $G \in \mathfrak{G}$ *and* v *is bounded below.*

c) $q(G)$ *is closed for every* $G \in \mathfrak{G}$ *and* w *is an isomorphism.*

a $\Rightarrow$ b . Take $G \in \mathfrak{G}$. It follows from

$$q(G) = \overset{-1}{v}(u(G))$$

that $q(G)$ is closed. Since

$$\operatorname{Im} v = u(E)$$

it follows that $\operatorname{Im} v$ is closed. Thus, by Corollary 1.4.2.8 a $\Rightarrow$ b, v is bounded below.

b $\Rightarrow$ c . By Corollary 1.4.2.8 b $\Rightarrow$ a , $\operatorname{Im} v$ is closed. Hence $\operatorname{Im} u$ is closed and by the Principle of Inverse Operators, w is an isomorphism.

c $\Rightarrow$ a . Since w is an isomorphism, $\operatorname{Im} w$ is complete and

$$u(G) = w(q(G))$$

is closed in F for every $G \in \mathfrak{G}$. ■

Proposition 1.4.2.13 (0) *Let* E, F *be Banach spaces and take* $u \in \mathcal{L}(E,F)$. *Let* p *be a projection of* E *onto* $\operatorname{Ker} u$ *and* q *a projection of* F *onto* $\operatorname{Im} u$. *There is a* $v \in \mathcal{L}(F,E)$, *such that*

$$1_E - v \circ u = p\,, \quad u \circ v = q\,, \quad v \circ q = v\,.$$

Define

$$G := \operatorname{Ker} p$$

and

$$w : G \longrightarrow \operatorname{Im} u\,, \quad x \longmapsto ux\,.$$

Then w is obviously linear and continuous. For $x \in \operatorname{Ker} w$,

$$ux = wx = 0\,,$$

so that $x = 0$, i.e. w is injective. Take $y \in \operatorname{Im} u$. Then

$$y = ux\,,$$

for some $x \in E$. Then

$$x - px \in G\,,$$

$$w(x - px) = ux - upx = ux = y\,,$$

i.e. w is surjective. Since $\operatorname{Im} u$ is closed (Theorem 1.2.5.8 b $\Rightarrow$ a, Proposition 1.2.5.2 a $\Rightarrow$ b), it is complete. By the Principle of Inverse Operators, w is an isomorphism. Set

$$v : F \longrightarrow E\,, \quad y \longmapsto w^{-1}(qy)\,.$$

Then v is linear and continuous and

$$v \circ q = v\,.$$

For $x \in E$,

$$x - px \in G\,, \quad u(x - px) = ux\,,$$

$$(1_E - v \circ u)x = x - w^{-1}(qux) = x - w^{-1}(ux) =$$

$$= x - w^{-1}(u(x - px)) = x - (x - px) = px\,,$$

so that

$$1_E - v \circ u = p\,.$$

For $y \in F$,

$$(u \circ v)y = u(w^{-1}(qy)) = qy$$

and so

$$u \circ v = q\,. \qquad \blacksquare$$

Proposition 1.4.2.14 *Let E, F be Banach spaces and $u \in \mathcal{L}(E, F)$.*

a) *There is a $v \in \mathcal{L}(F, E)$ such that $u \circ v = 1_F$ iff u is surjective and* $\operatorname{Ker} u$ *is a complemented subspace of E.*

b) *There is a* $v \in \mathcal{L}(F, E)$ *such that* $v \circ u = 1_E$ *iff* u *is injective and* $\operatorname{Im} u$ *is a complemented subspace of* F.

a) Assume first that $u \circ v = 1_F$ for some $v \in \mathcal{L}(F, E)$. Then u is surjective and v is injective. Put

$$p := v \circ u .$$

Then

$$p \circ p = v \circ u \circ v \circ u = v \circ u = p ,$$

i.e. p is a projection in E. For $x \in E$,

$$(px = 0) \Longleftrightarrow (vux = 0) \Longleftrightarrow (ux = 0) ,$$

so that

$$\operatorname{Ker} u = \operatorname{Ker} p$$

and $\operatorname{Ker} u$ is a complemented subspace of E by Murray's Theorem (Theorem 1.2.5.8 b $\Rightarrow$ a).

The reverse implication follows from Proposition 1.4.2.13 and Murray's Theorem.

b) Assume first that $v \circ u = 1_E$ for some $v \in \mathcal{L}(F, E)$. Then u is injective. Put

$$p := u \circ v .$$

Then

$$p \circ p = u \circ v \circ u \circ v = u \circ v = p ,$$

i.e. p is a projection in F. If $y \in \operatorname{Im} u$ then there is an $x \in E$ with

$$y = ux .$$

It follows

$$y = ux = uvux = pux \in \operatorname{Im} p ,$$

so that

$$\operatorname{Im} u \subset \operatorname{Im} p .$$

The reverse inclusion is trivial, so that

$$\operatorname{Im} u = \operatorname{Im} p$$

and $\operatorname{Im} u$ is a complemented subspace of F (Theorem 1.2.5.8 b $\Rightarrow$ a).

The reverse implication follows from Proposition 1.4.2.13 and Murray's Theorem. ■

Proposition 1.4.2.15 *Let E be a Banach space and p be a projection of E'' onto $\operatorname{Im} j_E$. Put $F := \operatorname{Ker} p$ and*

$$u : E \longrightarrow ({}^\circ F)', \quad x \longmapsto (j_E x) \mid {}^\circ F .$$

a) *u is an operator and $\|u\| \leq 1$.*

b) *Given $y' \in ({}^\circ F)'$, there is an $x \in E$ such that*

$$ux = y', \quad \|x\| \leq \|p\| \, \|y'\| .$$

c) *u is a isomorphism iff F is closed in $E''_{E'}$.*

d) *If F is closed in $E''_{E'}$ and $\|p\| \leq 1$, then u is an isometry.*

a) u is linear and

$$\|ux\| = \|(j_E x) \mid {}^\circ F\| \leq \|j_E\| = \|x\|$$

for every $x \in E$ (Theorem 1.3.6.3 a)).

b) By the Hahn–Banach Theorem, there is an $x'' \in E''$ such that

$$x'' \mid {}^\circ F = y', \quad \|x''\| = \|y'\| .$$

Take $x \in E$ with

$$j_E x = p x'' .$$

Then

$$x'' - p x'' \in F \subset ({}^\circ F)^\circ ,$$

so that

$$\langle ux, x' \rangle = \langle j_E x, x' \rangle = \langle p x'', x' \rangle = \langle x'', x' \rangle = \langle y', x' \rangle$$

for every $x' \in {}^\circ F$ and thus

$$ux = y' .$$

Moreover,

$$\|x\| = \|j_E x\| = \|px''\| \le \|p\| \, \|x''\| = \|p\| \, \|y'\| .$$

c) Assume that u is an isomorphism and take $x'' \in ({}^\circ F)^\circ$. There is an $x \in E$ such that

$$j_E x = px'' .$$

Since

$$x'' - px'' \in F ,$$

it follows that

$$\langle ux, x' \rangle = \langle j_E x, x' \rangle = \langle px'', x' \rangle = \langle x'', x' \rangle = 0$$

for every $x' \in {}^\circ F$. Hence

$$ux = 0 , \quad x = 0 , \quad x'' \in F ,$$

$$F = ({}^\circ F)^\circ ,$$

and F is a closed subset of $E''_{E'}$.

Now suppose that F is closed in $E''_{E'}$. Take $x \in \operatorname{Ker} u$. Then

$$j_E x \in ({}^\circ F)^\circ = F$$

(Proposition 1.3.5.4) and so

$$j_E x = 0 , \quad x = 0$$

(Theorem 1.3.6.3 a)). Hence u is injective. By b), u is surjective. By the Principle of Inverse Operators, u is an isomorphism.

d) follows from a), b), and c). ■

Corollary 1.4.2.16 *Given a Banach space E, the following are equivalent:*

a) *E is a dual space.*

b) *There is a projection p of E'' onto $\operatorname{Im} j_E$ such that $\|p\| \le 1$ and $\operatorname{Ker} p$ is closed in $E''_{E'}$.*

a $\Rightarrow$ b follows from Proposition 1.3.6.19 b), c).

b $\Rightarrow$ a. Let $F := \operatorname{Ker} p$ and take

$$u : E \longrightarrow (^\circ F)', \quad x \longmapsto (j_E x) \mid {}^\circ F .$$

By Proposition 1.4.2.15 d) u is an isometry. ∎

Remark. Let T be a Hausdorff space. Let $\mu \neq 0$ be an atomfree Radon measure on T. Let j be the evaluation map of $L^1(\mu)$. Then there is a projection p of $L^1(\mu)''$ onto $\operatorname{Im} j$ such that $\|p\| \le 1$. But $L^1(\mu)$ is not a dual space (Example 1.3.1.15). This shows that we cannot omit the hypothesis "$\operatorname{Ker} p$ is closed in $E''_{E'}$" from b).

Proposition 1.4.2.17 *Let E be a Banach space and F, G, H closed vector subspaces of E such that*

$$E = F \oplus G, \quad H = F \cap H + G \cap H .$$

If q denotes the quotient map $E \to E/H$, and if $q(F)$, $q(G)$ are closed, then

$$E/H = q(F) \oplus q(G) .$$

Define

$$\varphi : q(F) \times q(G) \longrightarrow E/H , \quad (X, Y) \longmapsto X + Y .$$

Step 1 φ is injective

Take $(X, Y) \in q(F) \times q(G)$ with

$$X + Y = 0 .$$

Take $x \in F$, $y \in G$ with

$$X = qx , \quad Y = qy .$$

Then

$$q(x + y) = X + Y = 0 ,$$

so that

$$x + y \in H .$$

By hypothesis, there are $a \in F \cap H$ and $b \in G \cap H$ with

$$a + b = x + y\,.$$

We see that

$$a - x = y - b \in F \cap G$$

and therefore

$$a - x = y - b = 0\,,$$

$$x = a \in F \cap H\,, \quad y = b \in G \cap H\,,$$

$$X = qa = 0\,, \quad Y = qb = 0\,.$$

Hence φ is injective.

Step 2 φ is surjective

Take $z \in E$. There are $x \in F$ and $y \in G$ such that

$$z = x + y\,.$$

We get

$$qz = qx + qy = \varphi(qx, qy) \in \operatorname{Im} \varphi\,.$$

Hence, φ is surjective.

Step 3 $E/H = q(F) \oplus q(G)$

By the first two steps and Corollary 1.4.2.6

$$E/H = q(F) \oplus q(G)\,. \qquad \blacksquare$$

Definition 1.4.2.18 (0) *Let X, Y be sets and take $f : X \to Y$. The set*

$$\{(x, f(x)) \mid x \in X\}$$

is called ***the graph of*** f.

If X, Y are topological spaces, Y Hausdorff, and $f : X \to Y$ continuous, then the graph of f is closed. If E, F are vector spaces and $u : E \to F$ is a linear map, then the graph of u is a vector subspace of $E \times F$.

Theorem 1.4.2.19 (0) (The Closed Graph Theorem, Banach 1929) *Let E, F be Banach spaces and $u : E \to F$ a linear map. If the graph of u is closed, then u is continuous.*

Let p, q be the projections of $E \times F$ onto E and F, respectively. $E \times F$ is a Banach space (Proposition 1.1.5.1) and p, q are continuous. Let G be the graph of u. G is a vector subspace of $E \times F$. Being closed, it is a Banach space. Put

$$v := p|G\,.$$

Then v is bijective. By the Principle of Inverse Operators (Corollary 1.4.2.4), v^{-1} is continuous. From

$$u = q \circ v^{-1}$$

it follows that u is also continuous. ■

Corollary 1.4.2.20 *Let E, F be Banach spaces, $u : E \to F$ a linear map and $f : F' \to E'$ an arbitrary map such that*

$$\langle ux, y'\rangle = \langle x, f(y')\rangle$$

for every $(x, y') \in E \times F'$. Then u is continuous and $f = u'$.

Let G be the graph of u and take $(x, y) \in \overline{G}$. Then there is a sequence $((x_n, y_n))_{n\in\mathbb{N}}$ in G converging to (x, y). Take $y' \in F'$. Then

$$\langle x_n, f(y')\rangle = \langle ux_n, y'\rangle = \langle y_n, y'\rangle$$

for every $n \in \mathbb{N}$. Thus

$$\langle ux, y'\rangle = \langle x, f(y')\rangle = \lim_{n\to\infty} \langle x_n, f(y')\rangle = \lim_{n\to\infty} \langle y_n, y'\rangle = \langle y, y'\rangle\,.$$

Since y' is arbitrary,

$$ux = y$$

(Corollary 1.3.3.9). Hence $(x, y) \in G$ and the graph of u is closed. By the Closed Graph Theorem, u is continuous. By Theorem 1.3.4.2 a) $f = u'$. ■

Exercises

E 1.4.1 Take $p \in [1, \infty]$ and let $\mathbb{K}^{(\mathbb{N})}$ be endowed with the norm induced by ℓ^p. Given $x \in \mathbb{K}^{(\mathbb{N})}$ and $n \in \mathbb{N}$, set

$$x_n : \mathbb{N} \longrightarrow \mathbb{K}, \quad m \longmapsto \begin{cases} nx(n) & \text{if } m = n \\ 0 & \text{if } m \neq n \end{cases}$$

and

$$u_n : \mathbb{K}^{(\mathbb{N})} \longrightarrow \mathbb{K}^{(\mathbb{N})}, \quad x \longmapsto x_n .$$

Prove the following:

a) $u_n \in \mathcal{L}(\mathbb{K}^{(\mathbb{N})})$ for every $n \in \mathbb{N}$.

b) $\lim\limits_{n \to \infty} u_n x = 0$ for every $x \in \mathbb{K}^{(\mathbb{N})}$.

c) $\sup\limits_{n \in \mathbb{N}} \|u_n\| = \infty$ (i.e. the conclusion of the Banach–Steinhaus Theorem does not hold).

E 1.4.2 Let E, F be Banach spaces and $u : E \to F$ a surjective operator. Show that there is an $\alpha > 0$ such that

$$\|u'y'\| \geq \alpha \|y'\|$$

for every $y' \in F'$.

E 1.4.3 Let E be a vector space and let $p \leq q$ be complete norms on E. Show that p and q are equivalent.

1.5 Banach Categories

The set of operators on a Banach space forms a Banach algebra and the whole of Chapter II is devoted to the study of such algebras. Unfortunately this theory cannot be applied to the case of operators between two different Banach spaces. The corresponding general theory is the theory of Banach categories, which is the subject of this section. Note that a Banach algebra is no more than a Banach category with precisely one object. This theory is not developed further in this book and so the reader may choose to omit this paragraph.

1.5.1 Definitions

Definition 1.5.1.1 (1) (2) (3) *A* ***Banach system*** *is a class* Ω *and a map* $\mathcal{A}$ *defined on* Ω^2 *such that* $\mathcal{A}(E,F)$ *is a Banach space for every* $E,F \in \Omega$ *and*

$$(E,F) \neq (G,H) \Longrightarrow \mathcal{A}(E,F) \cap \mathcal{A}(G,H) = \emptyset$$

for every $E,F,G,H \in \Omega$. *We use the expressions "the Banach system* $(\Omega, \mathcal{A})$*" or "the Banach system* $\mathcal{A}$ *over* Ω*" or, simply, "the Banach system* $\mathcal{A}$*". The elements of* Ω *are called* ***the objects of the Banach system*** *and the elements of* $\mathcal{A}(E,F)$ *(for* $E,F \in \Omega$*) are called* ***the morphisms of the Banach system****. We put*

$$\mathcal{A}(E) := \mathcal{A}(E,E)$$

for $E \in \Omega$ *and*

$$E \xrightarrow{x} F :\Longleftrightarrow E \underset{\mathcal{A}}{\xrightarrow{x}} F : \Longleftrightarrow x \in \mathcal{A}(E,F)$$

for $E,F \in \Omega$.

Let $\mathcal{A},\mathcal{B},\mathcal{C}$ *be Banach systems over the same class* Ω. *An* $(\mathcal{A},\mathcal{B},\mathcal{C})$*-* ***multiplication*** *is a map* φ *defined on* Ω^3 *such that the following holds for every* $E,F,G \in \Omega$*:*

$\varphi(E,F,G)$ *is a bilinear map*

$$\mathcal{A}(E,F) \times \mathcal{B}(F,G) \longrightarrow \mathcal{C}(E,G)\,, \quad (x,y) \longmapsto yx$$

such that

$$\|yx\| \leq \|x\|\,\|y\|$$

for all $E \overset{x}{\to} F \overset{y}{\to} G$.

A ***left*** *(resp.* ***right)*** ***multiplication on*** $\mathcal{A}$ ***over*** $\mathcal{B}$ *is an* $(\mathcal{A}, \mathcal{B}, \mathcal{A})$ *(resp.* $(\mathcal{B}, \mathcal{A}, \mathcal{A})$*)–multiplication. A* ***unit*** *for such a multiplication is a map* 1, *defined on* Ω, *such that*

$$1_E \in \mathcal{B}(E),$$
$$1_E \neq 0 \Longrightarrow \|1_E\| = 1,$$
$$1_E x = x \ \ (\text{resp. } x 1_E = x)$$

for every $E, F \in \Omega$ *and* $F \underset{\mathcal{A}}{\overset{x}{\to}} E$ *(resp.* $E \underset{\mathcal{A}}{\overset{x}{\to}} F$*). A left and a right multiplication on* $\mathcal{A}$ *is called* ***compatible*** *if*

$$(ax)b = a(xb)$$

for every $E, F, G, H \in \Omega$ *and*

$$E \overset{b}{\to} F \underset{\mathcal{A}}{\overset{x}{\to}} G \overset{a}{\to} H.$$

An ***inner multiplication on*** $\mathcal{A}$ *is a left (or right) multiplication on* $\mathcal{A}$ *over* $\mathcal{A}$ *such that*

$$(xy)z = x(yz)$$

for every $E, F, G, H \in \Omega$ *and*

$$E \overset{z}{\to} F \overset{y}{\to} G \overset{x}{\to} H.$$

An inner multiplication ***has a unit*** *if its left and right multiplications have units.*

Let $(\Omega, \mathcal{A}), (\Omega, \mathcal{B})$ *be Banach systems. A left and a right multiplication on* $\mathcal{A}$ *over* $\mathcal{B}$ *are called* ***simultaneously compatible*** *with an inner multiplication on* $\mathcal{A}$ *if*

$$(xa)y = x(ay)$$

for every $E, F, G, H \in \Omega$ *and*

$$E \underset{\mathcal{A}}{\overset{y}{\to}} F \underset{\mathcal{B}}{\overset{a}{\to}} G \underset{\mathcal{A}}{\overset{x}{\to}} H.$$

A ***Banach category (unital Banach category)*** *is a Banach system endowed with an inner multiplication (which has a unit).*

If Ω *is a class of Banach spaces, then the map*

$$(E, F) \longmapsto \mathcal{L}(E, F)$$

defined on Ω^2 *is a Banach system (Theorem 1.2.1.9). It is a unital Banach category with the usual composition of the maps as multiplication (Corollary 1.2.1.5). We denote it by* $\mathcal{L}_\Omega$ *or, simply,* $\mathcal{L}$.

Example 1.5.1.2 *Take $p \in \{0\} \cup [1, \infty]$ and let q be the conjugate exponent of p. Let Ω be a class of sets. Then the Banach system*

$$(S,T) \longmapsto \ell^{p,q}(S,T)$$

on Ω with the multiplication defined in Proposition 1.2.3.5 is a Banach category.

The claim follows from Proposition 1.2.3.5. ■

Proposition 1.5.1.3 (2) *Let $\mathcal{A}, \mathcal{B}, \mathcal{C}$ be Banach systems on the same class Ω, φ an $(\mathcal{A}, \mathcal{B}, \mathcal{C})$-multiplication, and take $E, F, G \in \Omega$.*

a) *$\varphi(E, F, G)$ is continuous.*

b) *If $(x_\iota)_{\iota \in I}$ is a summable family in $\mathcal{A}(E, F)$ (in $\mathcal{B}(F, G)$), then*

$$x\Big(\sum_{\iota\in I} x_\iota\Big) = \sum_{\iota \in I} x x_\iota\,, \quad \Big(\Big(\sum_{\iota\in I} x_\iota\Big)x = \sum_{\iota\in I} x_\iota x\Big)$$

for every $x \in \mathcal{B}(F,G)$ $(x \in \mathcal{A}(E,F))$.

a) follows from Proposition 1.2.9.2 c $\Rightarrow$ a.
b) follows from a) and Proposition 1.2.1.16. ■

Proposition 1.5.1.4 (2) *Every unital Banach category admits a unique unit for the left multiplication and a unique unit for the right multiplication and they coincide; we call it* ***the unit*** *of the Banach category.*

Let (Ω, Λ) be a Banach category and 1 (resp. $1'$) be a unit of the left (resp. right) multiplication of Λ. Then

$$1_E = 1_E 1'_E = 1'_E$$

for every $E \in \Omega$. ■

Definition 1.5.1.5 (1) (2) (3) *Let (Ω, Λ) be a unital Banach category and take $E, F \in \Omega$. An element $x \in \Lambda(E,F)$ is called* ***left invertible (right invertible)*** *if there is an $y \in \Lambda(F,E)$ such that*

$$yx = 1_E \quad (xy = 1_F)\,.$$

x is called ***invertible*** *if it is both left and right invertible.*

The invertible morphisms of $\mathcal{L}$ are precisely the isomorphisms of Banach spaces.

Proposition 1.5.1.6 (2) *Let Λ be a unital Banach category, E, F objects of Λ, and x an invertible morphism of $\Lambda(E, F)$. Then there is a unique $y \in \Lambda(F, E)$ with*

$$yx = 1_E , \quad xy = 1_F .$$

Put

$$x^{-1} := y .$$

x^{-1} is called ***the inverse of*** *x.*

Since x is invertible, there are $y, z \in \Lambda(F, E)$ so that

$$yx = 1_E , \quad xz = 1_F .$$

Then

$$y = y1_F = y(xz) = (yx)z = 1_E z = z .$$ ∎

Corollary 1.5.1.7 (2) *Let (Ω, Λ) be a unital Banach category. Take $E, F, G \in \Omega$, and $E \xrightarrow{x} F \xrightarrow{y} G$. If x and y are invertible, then yx is invertible and*

$$(yx)^{-1} = x^{-1}y^{-1} .$$

We have

$$(x^{-1}y^{-1})(yx) = x^{-1}(y^{-1}y)x = x^{-1}x = 1_E ,$$

$$(yx)(x^{-1}y^{-1}) = y(xx^{-1})y^{-1} = yy^{-1} = 1_G .$$ ∎

Corollary 1.5.1.8 (2) *Let (Ω, Λ) be a unital Banach category. Take $E, F \in \Omega$, $E \xrightarrow{x} F$, and $F \xrightarrow{y} E$. Then x and y are invertible iff xy and yx are invertible.*

By Corollary 1.5.1.7, if x and y are invertible, then xy and yx are also invertible. Assume now that xy and yx are invertible. Put

$$u := xy , \quad v := yx .$$

Then

$$x(yu^{-1}) = (xy)u^{-1} = 1_F ,$$

$$(v^{-1}y)x = v^{-1}(yx) = 1_E ,$$

so that x is invertible. Hence, y is also invertible. ∎

Definition 1.5.1.9 (1) (2) *Let $(\Omega, \mathcal{A})$ be a Banach system. The map*

$$(E, F) \longmapsto \mathcal{A}(F, E)'$$

defined on Ω^2 is a Banach system. It is called ***the dual of*** $\mathcal{A}$ *and is denoted by $\mathcal{A}'$. The dual of $\mathcal{A}'$ is called* ***the bidual of*** $\mathcal{A}$ *and is denoted by $\mathcal{A}''$ and the dual of $\mathcal{A}''$ is called* ***the tridual of*** $\mathcal{A}$ *and is denoted by $\mathcal{A}'''$.*

Definition 1.5.1.10 (1) (2) (3) *Let Ω be a class and Λ a Banach category (unital Banach category) over Ω. A* ***left Λ–module (unital left Λ–module)*** *is a Banach system $\mathcal{A}$ over Ω together with a left multiplication over Λ such that*

$$(ab)x = a(bx)$$

for every $E, F, G, H \in \Omega$ and

$$E \underset{\mathcal{A}}{\overset{x}{\rightarrow}} F \underset{\Lambda}{\overset{b}{\rightarrow}} G \underset{\Lambda}{\overset{a}{\rightarrow}} H$$

(and such that the unit of Λ is the unit for the left multiplication).

The right Λ–module (unital right Λ–module) *is defined in a similar way.*

Λ is a left and a right Λ–module (unital left and unital right Λ–module) in a natural way.

Proposition 1.5.1.11 (1) (2) *Let (Ω, Λ) be a Banach category (unital Banach category), $\mathcal{A}$ a left Λ–module (unital left Λ–module), and H a Banach space. Given $E, F, G \in \Omega$, put*

$$\mathcal{A}_H(E, F) := \mathcal{L}(\mathcal{A}(F, E), H),$$

$$ua : \mathcal{A}(G, E) \longrightarrow H, \quad x \longmapsto u[ax]$$

for all $E \underset{\Lambda}{\overset{a}{\rightarrow}} F \underset{\mathcal{A}_H}{\overset{u}{\rightarrow}} G$. Then the Banach system $\mathcal{A}_H$ with the above multiplication is a right Λ–module (unital right Λ–module), and, similarly, if we interchange left and right. In particular, $\mathcal{A}_{\mathbb{K}} = \mathcal{A}'$ is a right Λ–module (unital right Λ–module).

It is easy to see that the maps defined form a multiplication. Take

$$E, F, G, D \in \Omega$$

and

$$E \underset{\Lambda}{\overset{a}{\to}} F \underset{\Lambda}{\overset{b}{\to}} G \underset{\mathcal{A}_H}{\overset{u}{\to}} D.$$

Then

$$(u(ba))[x] = u[(ba)x] = u[b(ax)] = (ub)[ax] = ((ub)a)[x]$$

for every $D \underset{\mathcal{A}}{\overset{a}{\to}} E$ (and

$$(u1_G)[y] = u[1_G y] = u[y]$$

for every $D \underset{\mathcal{A}}{\overset{a}{\to}} G$), so that

$$u(ba) = (ub)a \quad (u1_G = u).$$

■

Definition 1.5.1.12 (1) (2) (3) *Let (Ω, Λ), (Ω, Δ) be two Banach categories (unital Banach categories). A (Λ, Δ)*–***module (unital (Λ, Δ)–module)*** *is a Banach system over Ω endowed with the structure of a left Λ–module (unital left Λ–module) and right Δ–module (unital right Δ–module) such that the left and the right multiplications are compatible. A* ***Λ–module (unital Λ–module)*** *is a (Λ, Λ)–module (unital (Λ, Λ)–module).*

Every Banach category (unital Banach category) Λ is a Λ–module (unital Λ–module).

Corollary 1.5.1.13 (1) (2) *Let $(\Omega, \Lambda), (\Omega, \Delta)$ be two Banach categories (unital Banach categories) and $\mathcal{A}$ a (Λ, Δ)–module (unital (Λ, Δ)–module). Then $\mathcal{A}'$ is a (Δ, Λ)–module (unital (Δ, Λ)–module). If $\mathcal{A}$ is a Λ–module (unital Λ–module), then $\mathcal{A}'$ is also a Λ–module (unital Λ–module).*

By Proposition 1.5.1.11, $\mathcal{A}'$ is a left Δ–module (unital left Δ–module) and a right Λ–module (unital right Λ–module).

Take $E, F, G, H \in \Omega$ and

$$E \underset{\Lambda}{\overset{a}{\to}} F \underset{\mathcal{A}'}{\overset{x'}{\to}} G \underset{\Delta}{\overset{b}{\to}} H.$$

Then

$$\langle x, (bx')a\rangle = \langle ax, bx'\rangle = \langle (ax)b, x'\rangle = \langle a(xb), x'\rangle = \langle xb, x'a\rangle = \langle x, b(x'a)\rangle$$

for every $x \in \mathcal{A}(H, E)$. Thus

$$(bx')a = b(x'a).$$

■

Definition 1.5.1.14 (1) (2) (3) *Let Λ be a Banach category (unital Banach category). A Λ-**category (unital Λ-category)** is a Λ-module (unital Λ-module) $\mathcal{A}$ endowed with an inner multiplication (with a unit) such that each left multiplication on $\mathcal{A}$ is compatible with each right multiplication on $\mathcal{A}$ and that the left and the right multiplication on $\mathcal{A}$ over Λ are simultaneously compatible with the inner multiplication on $\mathcal{A}$.*

Λ is a Λ-category (unital Λ-category). Every Λ-category (unital Λ-category) is a Banach category (unital Banach category) with respect to the inner multiplication.

1.5.2 Functors

Definition 1.5.2.1 (1) (2) (3) *Let* $(\Omega, \mathcal{A}), (\Omega, \mathcal{B})$ *be two Banach systems. A* ***functor of*** $\mathcal{A}$ ***into*** $\mathcal{B}$ *is a map* f *defined on* Ω^2 *such that*

$$f(E, F) \in \mathcal{L}(\mathcal{A}(E, F), \mathcal{B}(E, F))$$

for every $E, F \in \Omega$. *The functor* f *is called* ***isometric*** *if* $f(E, F)$ *is an isometry for every* $E, F \in \Omega$. $\mathcal{A}$ *and* $\mathcal{B}$ *are called* ***isometric*** *if there is an isometric functor of* $\mathcal{A}$ *into* $\mathcal{B}$. *The map*

$$(E, F) \longmapsto 1_{\mathcal{A}(E,F)}$$

defined on Ω^2 *is an isometric functor of* $\mathcal{A}$ *into* $\mathcal{A}$. *It is called* ***the identity functor of*** $\mathcal{A}$.

Let $(\Omega, \mathcal{A}), (\Omega, \mathcal{B}), (\Omega, \mathcal{C})$ *be Banach systems,* f *a functor of* $\mathcal{A}$ *into* $\mathcal{B}$, *and* g *a functor of* $\mathcal{B}$ *into* $\mathcal{C}$. *The map*

$$(E, F) \longmapsto g(E, F) \circ f(E, F)$$

defined on Ω^2 *is a functor of* $\mathcal{A}$ *into* $\mathcal{C}$. *It is called* ***the composition of the functors*** f ***and*** g *and it is denoted by* $g \circ f$.

Let $(\Omega, \mathcal{A})$ *be a Banach system. Given* $E, F \in \Omega$, *let* j_{EF} *denote the evaluation on* $\mathcal{A}(E, F)$. *Then the map*

$$(E, F) \longmapsto j_{EF}$$

defined on Ω^2 *is a functor of* $\mathcal{A}$ *into its bidual* $\mathcal{A}''$. *It is called* ***the evaluation functor of*** $\mathcal{A}$. *Given* $E \in \Omega$, *put*

$$j_E := j_{EE}\,.$$

Let $(\Omega, \mathcal{A}), (\Omega, \mathcal{B})$ *be Banach categories (unital Banach categories). A* ***functor of Banach categories (unital Banach categories)*** *of* $\mathcal{A}$ *into* $\mathcal{B}$ *is a functor* f *of* $\mathcal{A}$ *into* $\mathcal{B}$ *such that*

$$f(xy) = f(x)f(y) \qquad (\text{and } f(1_E) = 1_E)$$

for every $E, F, G \in \Omega$ *and*

$$E \underset{\mathcal{A}}{\overset{y}{\to}} F \underset{\mathcal{A}}{\overset{x}{\to}} G\,.$$

Let (Ω, Λ) *be a Banach category and* $\mathcal{A}, \mathcal{B}$ *left (right)* Λ*-modules. A* ***functor of left (right)*** Λ***-modules*** *of* $\mathcal{A}$ *into* $\mathcal{B}$ *is a functor* f *of* $\mathcal{A}$ *into* $\mathcal{B}$ *such that*

$$f(ax) = af(x) \qquad (f(xa) = f(x)a)$$

for every $E, F, G \in \Omega$ *and*

$$E \underset{\mathcal{A}}{\overset{x}{\to}} F \underset{\Lambda}{\overset{a}{\to}} G. \qquad (E \underset{\Lambda}{\overset{a}{\to}} F \underset{\mathcal{A}}{\overset{x}{\to}} G).$$

A ***functor of*** Λ***–modules*** *is a functor of left and right* Λ*–modules.*

Let (Ω, Λ) *be a Banach category (unital Banach category) and let* $\mathcal{A}, \mathcal{B}$ *be two* Λ*–categories (unital* Λ*–categories). A* ***functor of (unital)*** Λ***–categories*** *of* $\mathcal{A}$ *into* $\mathcal{B}$ *is a functor of* Λ*–modules of* $\mathcal{A}$ *into* $\mathcal{B}$ *which is also a functor of Banch categories (unital Banach categories).*

Example 1.5.2.2 (2) *Let* Ω *be a class of Banach spaces and* $\mathcal{A}$ *the Banach system*

$$(E, F) \longmapsto \mathcal{L}(F', E')$$

over Ω*. Given* $E, F \in \Omega$*, put*

$$f(E, F) : \mathcal{L}(E, F) \longrightarrow \mathcal{A}(E, F), \quad u \longmapsto u'.$$

Given $E, F, G \in \Omega$ *and* $E \underset{\mathcal{A}}{\overset{u}{\to}} F \underset{\mathcal{A}}{\overset{v}{\to}} G$*, put*

$$vu := u \circ v.$$

Then $\mathcal{A}$ *with this multiplication is a unital Banach category and* f *is a functor of unital Banach categories of* $\mathcal{L}$ *into* $\mathcal{A}$*.* $\mathcal{A}$ *is called* ***the transpose unital category of*** $\mathcal{L}$ *and* f ***the transposition functor of*** $\mathcal{L}$*.*

The result follows from Theorem 1.3.4.2 and Corollaries 1.3.4.3, 1.3.4.4, and 1.3.4.5. ■

Proposition 1.5.2.3 (2) *Let* (Ω, Λ) *be a Banach category. Let* $\mathcal{A}, \mathcal{B}$ *be left (right)* Λ*–modules and let* f *be a functor of left (right)* Λ*–modules of* $\mathcal{A}$ *into* $\mathcal{B}$*. Given* $E, F \in \Omega$*, define*

$$f'(E, F) = f(F, E)'.$$

Then f' *is a functor of right (left)* Λ*–modules of* $\mathcal{B}'$ *into* $\mathcal{A}'$*, called* ***the transpose of*** f*.*

Given $E,F\in\Omega$,

$$f'(E,F)=f(F,E)'\in\mathcal{L}\Big(\mathcal{B}(F,E)',\,\mathcal{A}(F,E)'\Big)=\mathcal{L}\Big(\mathcal{B}'(E,F)\,,\,\mathcal{A}'(E,F)\Big).$$

Take $E,F,G\in\Omega$ and

$$E\underset{\Lambda}{\overset{a}{\to}}F\underset{\mathcal{B}'}{\overset{y'}{\to}}G.$$

Then

$$\langle x,f'(y'a)\rangle=\langle fx,y'a\rangle=\langle afx,y'\rangle=\langle f(ax),y'\rangle=\langle ax,f'y'\rangle=\langle x,(f'y')a\rangle$$

for every $x\in\mathcal{A}(G,E)$ (Theorem 1.3.4.2 a)). Thus

$$f'(y'a)=(f'y')a.$$ ■

Proposition 1.5.2.4 (1) (2) *Let (Ω,Λ) be a Banach category and $\mathcal{A}$ a left (right) Λ–module. Then the evaluation functor on $\mathcal{A}$ is a functor of left (right) Λ–modules into its bidual.*

Given $E,F\in\Omega$ and $x\in\mathcal{A}(E,F)$, set

$$\widetilde{x}=j_{\mathcal{A}}x.$$

Take $E,F,G\in\Omega$ and

$$E\underset{\mathcal{A}}{\overset{x}{\to}}F\underset{\Lambda}{\overset{a}{\to}}G.$$

Then

$$\langle\widetilde{ax},x'\rangle=\langle ax,x'\rangle=\langle x,x'a\rangle=\langle\widetilde{x},x'a\rangle=\langle a\widetilde{x},x'\rangle$$

for every $x'\in\mathcal{A}'(G,E)$, so

$$\widetilde{ax}=a\widetilde{x}.$$ ■

Definition 1.5.2.5 (1) (2) *Let (Ω,Λ) be a Banach category and $\mathcal{A}$ a left (right) Λ–module. Take $E,F,G\in\Omega$. Given*

$$E\underset{\mathcal{A}'}{\overset{x'}{\to}}F\underset{\mathcal{A}}{\overset{x}{\to}}G\quad\Big(E\underset{\mathcal{A}}{\overset{x}{\to}}F\underset{\mathcal{A}'}{\overset{x'}{\to}}G\Big),$$

set

$$xx':\Lambda(G,E)\longrightarrow\mathbb{K},\quad a\longmapsto\langle ax,x'\rangle,$$

$$(x'x:\Lambda(G,E)\longrightarrow\mathbb{K},\quad \longmapsto\langle xa,x'\rangle).$$

If $\mathcal{A} = \Lambda$, then it is easy to see that the above composition law coincides with the multiplication introduced in Proposition 1.5.1.11.

Proposition 1.5.2.6 (1) (2) *Let Λ be a Banach category and $\mathcal{A}$ a left (right) Λ-module. The composition law introduced in Definition 1.5.2.5 is an $(\mathcal{A}', \mathcal{A}, \Lambda')$-multiplication ($(\mathcal{A}, \mathcal{A}', \Lambda')$-multiplication) such that*

$$E \overset{x'}{\to} F \overset{x}{\to} G \overset{a}{\to} H \Longrightarrow (ax)x' = a(xx')$$

$$(\textit{resp. } E \overset{a}{\to} F \overset{x}{\to} G \overset{x'}{\to} H \Longrightarrow (x'x)a = x'(xa)\,),$$

$$E \overset{a}{\to} F \overset{x'}{\to} G \overset{x}{\to} H \Longrightarrow (xx')a = x(x'a)\,,$$

$$(\textit{resp. } E \overset{x}{\to} F \overset{x'}{\to} G \overset{a}{\to} H \Longrightarrow (ax')x = a(x'x)\ )\,,$$

where a is a morphism of Λ, x is a morphism of $\mathcal{A}$, and x' is a morphism of $\mathcal{A}'$. If $\mathcal{A}$ is a Λ-module, then

$$E \overset{x}{\to} F \overset{a}{\to} G \overset{x'}{\to} H \Longrightarrow (x'a)x = x'(ax)\,,$$

$$E \overset{x'}{\to} F \overset{a}{\to} G \overset{x}{\to} H \Longrightarrow (xa)x' = x(ax')\,,$$

with the same conventions as above.

The first assertion is easy to verify. We have

$$\langle b, (ax)x' \rangle = \langle b(ax), x' \rangle = \langle (ba)x, x' \rangle = \langle ba, xx' \rangle = \langle b, a(xx') \rangle$$

$$(\text{resp. } \langle b, (x'x)a \rangle = \langle ab, x'x \rangle = \langle x(ab), x' \rangle = \langle (xa)b, x' \rangle = \langle b, x'(xa) \rangle)\,,$$

$$\langle b, (xx')a \rangle = \langle ab, xx' \rangle = \langle (ab)x, x' \rangle = \langle a(bx), x' \rangle = \langle bx, x'a \rangle = \langle b, x(x'a) \rangle$$

$$(\text{resp. } \langle b, (ax')x \rangle = \langle xb, ax' \rangle = \langle (xb)a, x' \rangle =$$

$$= \langle x(ba), x' \rangle = \langle ba, x'x \rangle = \langle b, a(x'x) \rangle\,)$$

for every $H \underset{\Lambda}{\overset{b}{\to}} E$, which proves the relations. If $\mathcal{A}$ is a Λ-module, then

$$\langle b, (x'a)x \rangle = \langle xb, x'a \rangle = \langle a(xb), x' \rangle = \langle (ax)b, x' \rangle = \langle b, x'(ax) \rangle\,,$$

$$\langle b, (xa)x' \rangle = \langle b(xa), x' \rangle = \langle (bx)a, x' \rangle = \langle bx, ax' \rangle = \langle b, x(ax') \rangle$$

for every $H \underset{\Lambda}{\overset{b}{\to}} E$, which proves the last assertion. ∎

Corollary 1.5.2.7 (1) (2) *If Λ is a Banach category and $\mathcal{A}$ is a left (right) Λ–module, then*

$$(j_{FG}x)x' = xx' \quad (x'(j_{EF}x) = x'x)$$

for all objects E, F, G of Λ and

$$E \underset{\mathcal{A}'}{\overset{x'}{\to}} F \underset{\mathcal{A}}{\overset{x}{\to}} G \quad \left(E \underset{\mathcal{A}}{\overset{x}{\to}} F \underset{\mathcal{A}'}{\overset{x'}{\to}} G\right).$$

Given $a \in \Lambda(G, E)$

$$\langle a, (j_{FG}x)x'\rangle = \langle x'a, j_{FG}x\rangle = \langle x, x'a\rangle = \langle ax, x'\rangle = \langle a, xx'\rangle$$

$$\left(\langle a, x'(j_{EF}x)\rangle = \langle ax', j_{EF}x\rangle = \langle x, ax'\rangle = \langle xa, x'\rangle = \langle a, x'x\rangle\right).$$

Thus

$$(j_{FG}x)x' = xx' \quad (x'(j_{EF}x) = x'x). \qquad \blacksquare$$

Definition 1.5.2.8 (1) (2) *Let (Ω, Λ) be a Banach category and $\mathcal{A}$ a left (right) Λ–module. Define*

$$x'a'' : \mathcal{A}(G, E) \longrightarrow \mathbb{K}, \quad x \longmapsto \langle a'', xx'\rangle$$

for $E, F, G \in \Omega$ and

$$E \underset{\Lambda''}{\overset{a''}{\to}} F \underset{\mathcal{A}'}{\overset{x'}{\to}} G$$

$$\Big(a''x' : \mathcal{A}(G, E) \longrightarrow \mathbb{K}, \quad x \longmapsto \langle a'', x'x\rangle$$

for $E, F, G \in \Omega$ and

$$E \underset{\mathcal{A}'}{\overset{x'}{\to}} F \underset{\Lambda''}{\overset{a''}{\to}} G\Big).$$

It is easy to see that if $\mathcal{A} = \Lambda$, then the above definition coincides with Definition 1.5.2.5.

Proposition 1.5.2.9 (1) (2) *Let Λ be a Banach category, $\mathcal{A}$ a left (right) Λ–module, and E, F, G, H objects of Λ.*

a) *The composition law introduced above is a right (left) multiplication on $\mathcal{A}'$ over Λ'' with the property that*

$$E \underset{\Lambda''}{\overset{a''}{\to}} F \underset{\mathcal{A}'}{\overset{x'}{\to}} G \underset{\mathcal{A}}{\overset{x}{\to}} H \Longrightarrow x(x'a'') = (xx')a''$$

$$\left(E \underset{\mathcal{A}}{\overset{x}{\to}} F \underset{\mathcal{A}'}{\overset{x'}{\to}} G \underset{\Lambda''}{\overset{a''}{\to}} H \Longrightarrow a''(x'x) = (a''x')x\right).$$

b) *If* $F \underset{\mathcal{A}'}{\overset{x'}{\to}} G$ *and*

$$u : \mathcal{A}(G,H) \longrightarrow \Lambda'(F,H)\,, \quad x \longmapsto xx'$$

$$(u : \mathcal{A}(E,F) \longrightarrow \Lambda'(E,G)\,, \quad x \longmapsto x'x)\,,$$

then

$$H \underset{\Lambda''}{\overset{a''}{\to}} F \Longrightarrow u'a'' = x'a''$$

$$(G \underset{\Lambda''}{\overset{a''}{\to}} E \Longrightarrow u'a'' = a''x')\,.$$

a) The first assertion is easy to verify. By Proposition 1.5.2.6,

$$\langle a, x(x'a'')\rangle = \langle ax, x'a''\rangle = \langle a'', (ax)x'\rangle = \langle a'', a(xx')\rangle = \langle a, (xx')a''\rangle$$

$$\Big(\langle a, a''(x'x)\rangle = \langle a'', (x'x)a\rangle = \langle a'', x'(xa)\rangle = \langle xa, a''x'\rangle = \langle a, (a''x')x\rangle\Big)$$

for every $H \underset{\Lambda}{\overset{a}{\to}} E$.

b) We have

$$\langle x, u'a''\rangle = \langle ux, a''\rangle = \langle xx', a''\rangle = \langle x, x'a''\rangle$$

for every $G \underset{\mathcal{A}}{\overset{x}{\to}} H$

$$(\langle x, u'a''\rangle = \langle ux, a''\rangle = \langle x'x, a''\rangle = \langle x, a''x'\rangle$$

for every $E \underset{\mathcal{A}}{\overset{x}{\to}} F$). ■

Definition 1.5.2.10 (1) (2) *Let* (Ω, Λ) *be a Banach category. Define*

$$x'' \dashv y'' : \Lambda'(G,E) \longrightarrow \mathbb{K}\,, \quad x' \longmapsto \langle y'', x'x''\rangle\,,$$

$$x'' \vdash y'' : \Lambda'(G,E) \longrightarrow \mathbb{K}\,, \quad x' \longmapsto \langle x'', y''x'\rangle$$

for $E,F,G \in \Omega$, $E \underset{\Lambda''}{\overset{y''}{\to}} F \underset{\Lambda''}{\overset{x''}{\to}} G$. $\dashv$ *and* $\vdash$ *are called* ***the right*** *and* ***the left Arens multiplication****, respectively (1951).*

Remark. $\dashv$ and $\vdash$ may be different (see Remark to Example 1.5.2.15).

Proposition 1.5.2.11 (1) (2) *Let Λ be a Banach category, E, F, G, H objects of Λ, and $\mathcal{A}$ a left (right) Λ-module. Then*

$$E \underset{\Lambda''}{\overset{a''}{\to}} F \underset{\Lambda''}{\overset{b''}{\to}} G \underset{\mathcal{A}'}{\overset{x'}{\to}} H \Longrightarrow x'(b'' \dashv a'') = (x'b'')a''$$

$$\left(E \underset{\mathcal{A}'}{\overset{x'}{\to}} F \underset{\Lambda''}{\overset{a''}{\to}} G \underset{\Lambda''}{\overset{b''}{\to}} H \Longrightarrow (b'' \vdash a'')x' = b''(a''x')\right).$$

By Proposition 1.5.2.9 a)

$$\langle x, x'(b'' \dashv a'')\rangle = \langle b'' \dashv a'', xx'\rangle = \langle a'', (xx')b''\rangle = \langle a'', x(x'b'')\rangle = \langle x, (x'b'')a''\rangle$$

$$\Big(\langle x, (b'' \vdash a'')x'\rangle = \langle b'' \vdash a'', x'x\rangle = \langle b'', a''(x'x)\rangle = \langle b'', (a''x')x\rangle = \langle x, b''(a''x')\rangle\Big)$$

for every $H \underset{\mathcal{A}}{\overset{x}{\to}} E$. ■

Theorem 1.5.2.12 (1) (2) *Let Λ be a Banach category.*

a) *$\dashv$ and $\vdash$ are inner multiplications on Λ''.*

b) *Λ'' together with $\dashv$ (with $\vdash$) is a Λ-category. We denote this category by $\Lambda''_{\dashv}$ ($\Lambda''_{\vdash}$).*

c) *The evaluation functor of Λ into $\Lambda''_{\dashv}$ (into $\Lambda''_{\vdash}$) is a functor of Λ-categories.*

d) *If $\mathcal{A}$ is a left (right) Λ-module, then $\mathcal{A}'$ is a right $\Lambda''_{\dashv}$-module (left $\Lambda''_{\vdash}$-module).*

e) *Λ' is a left $\Lambda''_{\vdash}$-module and a right $\Lambda''_{\dashv}$-module.*

f) *If Λ' is a $(\Lambda''_{\vdash}, \Lambda''_{\dashv})$-module, then*

$$(z'' \dashv y'') \vdash x'' = z'' \dashv (y'' \vdash x'')$$

for all objects E, F, G, H of Λ and

$$E \underset{\Lambda''}{\overset{x''}{\to}} F \underset{\Lambda''}{\overset{y''}{\to}} G \underset{\Lambda''}{\overset{z''}{\to}} H.$$

g) *If Λ is a unital Banach category, then $\Lambda''_{\dashv}, \Lambda''_{\vdash}$ are unital Λ-categories. In this case, the evaluation functor of Λ into these unital categories is a functor of unital Λ-categories, and Λ' is a unital left $\Lambda''_{\dashv}$-module and a unital right $\Lambda''_{\dashv}$-module.*

h) *If j denotes the evaluation functor of Λ, then*

$$x''x = x'' \vdash (j_{EF}x) = x'' \dashv (j_{EF}x)\,,$$

$$yx'' = (j_{GH}y) \vdash x'' = (j_{GH}y) \dashv x''$$

for all objects E, F, G, H of Λ and

$$E \underset{\Lambda}{\overset{x}{\to}} F \underset{\Lambda''}{\overset{x''}{\to}} G \underset{\Lambda}{\overset{y}{\to}} H\,.$$

i) *Let E, F, G, H be objects of Λ and take $F \underset{\Lambda''}{\overset{x''}{\to}} G$. Then the maps*

$$\Lambda''(E,F) \longrightarrow \Lambda''(E,G)\,, \quad y'' \longmapsto x'' \dashv y''\,,$$

$$\Lambda''(G,H) \longrightarrow \Lambda''(E,G)\,, \quad y'' \longmapsto y'' \vdash x''$$

are continuous with respect to the topologies of pointwise convergence .

It is easy to see that $\dashv$ and $\vdash$ are $(\Lambda'', \Lambda'', \Lambda'')$–multiplications. Let E, F, G, H be objects of Λ. In order to simplify notation, we adopt the convention that the symbols x, y denote morphisms of Λ, the symbol x' denotes a morphism of Λ', and the symbols x'', y'', z'' denote morphisms of Λ''.

a) By Corollary 1.5.2.7, Λ'' is a Λ– module. Let

$$E \overset{x''}{\to} F \overset{x''}{\to} G \overset{z''}{\to} H\,.$$

By Proposition 1.5.2.11,

$$\langle (z'' \dashv y'') \dashv x'', x' \rangle = \langle x'', x'(z'' \dashv y'') \rangle = \langle x'', (x'z'')y'' \rangle =$$

$$= \langle y'' \dashv x'', x'z'' \rangle = \langle z'' \dashv (y'' \dashv x''), x' \rangle\,,$$

$$\langle (z'' \vdash y'') \vdash x'', x' \rangle = \langle z'' \vdash y'', x''x' \rangle = \langle z'', y''(x''x') \rangle = \langle z'', (y'' \vdash x'')x' \rangle =$$

$$= \langle z'' \vdash (y'' \vdash x''), x' \rangle\,,$$

for every $H \overset{x'}{\to} E$. Thus

$$(z'' \dashv y'') \dashv x'' = z'' \dashv (y'' \dashv x'')\,,$$

$$(z'' \vdash y'') \vdash x'' = z'' \vdash (y'' \vdash x'')\,.$$

b) Step 1 $E \overset{x''}{\to} F \overset{y''}{\to} F \overset{x}{\to} H \Longrightarrow \begin{cases} x(y'' \dashv x'') = (xy'') \dashv x'' \\ x(y'' \vdash x'') = (xy'') \vdash x'' \end{cases}$

By Proposition 1.5.2.9,

$$\langle x(y'' \dashv x''), x'\rangle = \langle y'' \dashv x'', x'x\rangle = \langle x'', (x'x)y''\rangle =$$

$$= \langle x'', x'(xy'')\rangle = \langle (xy'') \dashv x'', x'\rangle\,,$$

$$\langle x(y'' \vdash x''), x'\rangle = \langle y'' \vdash x'', x'x\rangle = \langle y'', x''(x'x)\rangle =$$

$$= \langle y'', (x''x')x\rangle = \langle xy'', x''x'\rangle = \langle (xy'') \vdash x'', x'\rangle$$

for every $H \overset{x'}{\to} E$.

Step 2 $E \overset{x}{\to} F \overset{x''}{\to} F \overset{y''}{\to} H \Longrightarrow \begin{cases} (y'' \dashv x'')x = y'' \dashv (x''x) \\ (y'' \vdash x'')x = y'' \vdash (x''x) \end{cases}$.

Given $H \overset{x'}{\to} E$,

$$\langle (y'' \dashv x'')x, x'\rangle = \langle y'' \dashv x'', xx'\rangle = \langle x'', (xx')y''\rangle =$$

$$= \langle x'', x(x'y'')\rangle = \langle x''x, x'y''\rangle = \langle y'' \dashv (x''x), x'\rangle\,,$$

$$\langle (y'' \vdash x'')x, x'\rangle = \langle y'' \vdash x'', xx'\rangle =$$

$$= \langle y'', x''(xx')\rangle = \langle y'', (x''x)x'\rangle = \langle y'' \vdash (x''x), x'\rangle$$

(Proposition 1.5.2.6).

Step 3 $E \overset{x''}{\to} F \overset{x}{\to} G \overset{y''}{\to} H \Longrightarrow \begin{cases} (y''x) \dashv x'' = y'' \dashv (xx'') \\ (y''x) \vdash x'' = y'' \vdash (xx'') \end{cases}$.

Given $H \overset{x'}{\to} E$,

$$\langle (y''x) \dashv x'', x'\rangle = \langle x'', x'(y''x)\rangle = \langle x'', (x'y'')x\rangle = \langle xx'', x'y''\rangle = \langle y'' \dashv (xx''), x'\rangle\,,$$

$$\langle (y''x) \vdash x'', x'\rangle = \langle y''x, x''x'\rangle = \langle y'', x(x''x')\rangle = \langle y'', (xx'')x'\rangle = \langle y'' \vdash (xx''), x'\rangle$$

(Proposition 1.5.2.6).

Step 4 b)

Follows from Corollary 1.5.1.13 and the preceding steps.

c) By Proposition 1.5.2.4, the evaluation functor j of Λ is a functor of Λ-modules. Take $E \xrightarrow{x} F \xrightarrow{y} G$. Then

$$\langle j(yx), x' \rangle = \langle y(jx), x' \rangle = \langle y, (jx)x' \rangle = \langle jy, (jx)x' \rangle = \langle (jy) \vdash (jx), x' \rangle ,$$

$$\langle j(yx), x' \rangle = \langle (jy)x, x' \rangle = \langle x, x'(jy) \rangle = \langle jx, x'(jy) \rangle = \langle (jy) \dashv (jx), x' \rangle ,$$

for every $G \xrightarrow{x'} E$, so that

$$j(yx) = (jy) \vdash (jx) = (jy) \dashv (jx) .$$

d) follows from b) and Proposition 1.5.2.11.

e) follows from Corollary 1.5.1.13 and d).

f) Given $H \xrightarrow{x'} E$,

$$\langle (z'' \dashv y'') \vdash x'', x' \rangle = \langle z'' \dashv y'', x''x' \rangle = \langle y'', (x''x')z'' \rangle =$$

$$= \langle y'', x''(x'z'') \rangle = \langle y'' \vdash x'', x'z'' \rangle = \langle z'' \dashv (y'' \vdash x''), x' \rangle .$$

g) Take $E \xrightarrow{x''} F$. Then

$$\langle x'' \vdash (j_E 1_E), x' \rangle = \langle x'', (j_E 1_E)x' \rangle = \langle x'', 1_E x' \rangle = \langle x'', x' \rangle ,$$

$$\langle (j_F 1_F) \vdash x'', x' \rangle = \langle j_F 1_F, x''x' \rangle = \langle 1_F, x''x' \rangle = \langle x'', x'1_F \rangle = \langle x'', x' \rangle ,$$

$$\langle x'' \dashv (j_E 1_E), x' \rangle = \langle j_E 1_E, x'x'' \rangle = \langle 1_E, x'x'' \rangle = \langle x'', 1_E x' \rangle = \langle x'', x' \rangle ,$$

$$\langle (j_F 1_F) \dashv x'', x' \rangle = \langle x'', x'(j_F 1_F) \rangle = \langle x'', x'1_F \rangle = \langle x'', x' \rangle$$

for every $F \xrightarrow{x'} E$ (Corollary 1.5.2.7, Corollary 1.5.1.13). Thus

$$x'' \vdash (j_E 1_E) = x'' , \qquad (j_F 1_F) \vdash x'' = x'' ,$$

$$x'' \dashv (j_E 1_E) = x'' , \qquad (j_F 1_F) \dashv x'' = x'' .$$

Hence $\Lambda''_\vdash$ and $\Lambda''_\dashv$ are unital Λ-categories and the evaluation functor of Λ into these categories is a functor of unital Λ-categories.

h) We have

$$\langle x'' \vdash (j_{EF}x), x' \rangle = \langle x'', (j_{EF}x)x' \rangle = \langle x'', xx' \rangle = \langle x''x, x' \rangle \,,$$

$$\langle x'' \dashv (j_{EF}x), x' \rangle = \langle j_{EF}x, x'x'' \rangle = \langle x, x'x'' \rangle = \langle x''x, x' \rangle$$

for every $G \overset{x'}{\to} E$ and

$$\langle (j_{GH}y) \vdash x'', x' \rangle = \langle j_{GH}y, x''x' \rangle = \langle y, x''x' \rangle = \langle yx'', x' \rangle \,,$$

$$\langle (j_{GH}y) \dashv x'', x' \rangle = \langle x'', x'(j_{GH}y) \rangle = \langle x'', x'y \rangle = \langle yx'', x' \rangle$$

for every $H \overset{x'}{\to} F$ (Corollary 1.5.2.7).

i) We have

$$\lim_{z'' \to y''} \langle x'' \dashv z'', x' \rangle = \lim_{z'' \to y''} \langle z'', x'x'' \rangle = \langle y'', x'x'' \rangle = \langle x'' \dashv y'', x' \rangle$$

for every $E \overset{y''}{\to} F$, $G \overset{x'}{\to} E$, and

$$\lim_{z'' \to y''} \langle z'' \vdash x'', x' \rangle = \lim_{z'' \to y''} \langle z'', x''x' \rangle = \langle y'', x''x' \rangle = \langle y'' \vdash x'', x' \rangle$$

for every $G \overset{y''}{\to} H$, $H \overset{x'}{\to} F$. ■

Proposition 1.5.2.13 *Let Λ be a Banach category and E, F, G, H objects of Λ. Take $F \underset{\Lambda''}{\overset{x''}{\to}} G$. Let j denote the evaluation functor of Λ' and put*

$$u_{x'} : \Lambda(F, G) \longrightarrow \Lambda'(E, G) \,, \quad x \longmapsto xx'$$

$$(\textit{resp. } u_{x'} : \Lambda(F, G) \longrightarrow \Lambda'(F, H) \,, \quad x \longmapsto x'x)$$

for every $E \underset{\Lambda'}{\overset{x'}{\to}} F$ (resp. $G \underset{\Lambda'}{\overset{x'}{\to}} H$). Then the following are equivalent:

a) $E \underset{\Lambda''}{\overset{x''}{\to}} F \Rightarrow x'' \vdash y'' = x'' \dashv y''$
(resp. $G \underset{\Lambda''}{\overset{x''}{\to}} H \Rightarrow y'' \vdash x'' = y'' \dashv x''$).

b) *The map*

$$\Lambda''(E, F) \longrightarrow \Lambda''(E, G) \,, \quad y'' \longmapsto x'' \vdash y''$$

$$(\textit{resp.} \Lambda''(G, H) \longrightarrow \Lambda''(F, G) \,, \quad y'' \longmapsto y'' \dashv x'')$$

is continuous with respect to the topologies of pointwise convergence.

c) $E \underset{\Lambda'}{\overset{x'}{\rightarrow}} F \Rightarrow u''_{x'}x'' = j_{EG}(x''x')$
(resp. $G \underset{\Lambda'}{\overset{x'}{\rightarrow}} H \Rightarrow u''_{x'}x'' = j_{FH}(x'x'')$ *).*

d) $E \underset{\Lambda'}{\overset{x'}{\rightarrow}} F \Rightarrow u''_{x'}x'' \in j_{EG}(\Lambda'(E,G))$
(resp. $G \underset{\Lambda'}{\overset{x'}{\rightarrow}} H \Rightarrow u''_{x'}x'' \in j_{FH}(\Lambda'(F,H))$ *).*

a $\Rightarrow$ b follows from Theorem 1.5.2.12 i).

b $\Rightarrow$ a. Let i be the evaluation functor on Λ. Then $i_{EF}(\Lambda(E,F))$ (resp. $i_{GH}(\Lambda(G,H))$) is dense in $\Lambda''(E,F)$ (resp. in $\Lambda''(G,H)$) with respect to the topology of pointwise convergence (Corollary 1.3.6.5). By Theorem 1.5.2.12 h),

$$x'' \vdash y'' = x'' \dashv y'' \qquad \text{(resp. } y'' \vdash x'' = y'' \dashv x''\text{)}$$

for every $y'' \in i_{EF}(\Lambda(E,F))$ (resp. $y'' \in i_{GH}(\Lambda(G,H))$). By continuity, (b) and Theorem 1.5.2.12 i)),

$$x'' \vdash y'' = x'' \dashv y'' \quad \text{(resp. } y'' \vdash x'' = y'' \dashv x''\text{)}$$

for every $E \underset{\Lambda''}{\overset{y''}{\rightarrow}} F$ (resp. $G \underset{\Lambda''}{\overset{y''}{\rightarrow}} H$).

a $\Rightarrow$ c. By Theorem 1.5.2.12 h) and Proposition 1.5.2.9 b),

$$\langle y'', u''_{x'}x''\rangle = \langle x'', u'_{x'}y''\rangle = \langle x'', x'y''\rangle = \langle y'' \dashv x'', x'\rangle =$$

$$= \langle y'' \vdash x'', x'\rangle = \langle y'', x''x'\rangle = \langle y'', j_{EG}(x''x')\rangle$$

for every $G \underset{\Lambda''}{\overset{y''}{\rightarrow}} E$ (resp.

$$\langle y'', u''_{x'}x''\rangle = \langle x'', u'_{x'}y\rangle = \langle x'', y''x'\rangle = \langle x'' \vdash y'', x'\rangle =$$

$$= \langle x'' \dashv y'', x'\rangle = \langle y'', x'x''\rangle = \langle y'', j_{FH}(x'x'')\rangle$$

for every $H \underset{\Lambda''}{\overset{y''}{\rightarrow}} F$), so

$$u''_{x'}x'' = j_{EG}(x''x') \qquad \text{(resp. } u''_{x'}x'' = j_{FH}(x'x'')\text{)}.$$

c $\Rightarrow$ d is trivial.

d $\Rightarrow$ b. By Proposition 1.5.2.9 b),

$$\langle x'' \vdash y'', x'\rangle = \langle x'', y''x'\rangle = \langle x'', u'_{x'}y''\rangle = \langle u''_{x'}x'', y''\rangle$$

for every $E \underset{\Lambda''}{\overset{y''}{\rightarrow}} F$ and $G \underset{\Lambda'}{\overset{x'}{\rightarrow}} E$

$$(\text{resp. } \langle y'' \dashv x'', x' \rangle = \langle x'', x'y'' \rangle = \langle x'', u'_{x'}, y'' \rangle = \langle u''_{x'}x'', y'' \rangle$$

for every $G \underset{\Lambda''}{\overset{y''}{\to}} F$ and $H \underset{\Lambda'}{\overset{x'}{\to}} F$). By d), the map

$$\Lambda''(E,F) \longrightarrow \Lambda''(E,G), \quad y'' \longmapsto x'' \vdash y''$$

$$(\text{resp. } \Lambda''(G,H) \longrightarrow \Lambda''(F,G), \quad y'' \longmapsto y'' \dashv x'')$$

is continuous with respect to the topologies of pointwise convergence. ■

Proposition 1.5.2.14 *Let $(\Omega, \mathcal{A}), (\Omega, \mathcal{B})$ be two Banach categories and let f be a functor of Banach categories of $\mathcal{A}$ into $\mathcal{B}$.*

a) $E \underset{\mathcal{A}}{\overset{x}{\to}} F \underset{\mathcal{B}'}{\overset{y'}{\to}} G \Rightarrow (f'_{FG}y')x = f'_{EG}(y' f_{EF}x)$,
$E \underset{\mathcal{B}'}{\overset{y'}{\to}} F \underset{\mathcal{A}}{\overset{x}{\to}} G \Rightarrow x(f'_{EF}y') = f'_{EG}((f_{FG}x)y')$.

b) $E \underset{\mathcal{A}''}{\overset{x''}{\to}} F \underset{\mathcal{B}'}{\overset{y'}{\to}} G \Rightarrow (f'_{FG}y')x'' = f'_{EG}(y' f''_{EF}x'')$,
$E \underset{\mathcal{B}'}{\overset{y'}{\to}} F \underset{\mathcal{A}''}{\overset{x''}{\to}} G \Rightarrow x''(f'_{EF}y') = f'_{EG}((f''_{FG}x'')y')$.

c) $E \underset{\mathcal{A}''}{\overset{x''}{\to}} F \underset{\mathcal{A}''}{\overset{y''}{\to}} G \Rightarrow \begin{cases} f''_{EG}(y'' \vdash x'') = (f''_{FG}y'') \vdash (f''_{EF}x'') \\ f''_{EG}(y'' \dashv x'') = (f''_{FG}y'') \dashv (f''_{EF}x''), \end{cases}$

i.e. f'' is a functor of Banach categories of $\mathcal{A}''_\vdash$ into $\mathcal{B}''_\vdash$ (of $\mathcal{A}''_\dashv$ into $\mathcal{B}''_\dashv$).

a) Take $G \underset{\mathcal{A}}{\overset{a}{\to}} E$. Then

$$\langle a, (f'_{FG}y')x \rangle = \langle xa, f'_{FG}y' \rangle = \langle f_{GF}(xa), y' \rangle = \langle (f_{EF}x)(f_{GE}a), y' \rangle =$$

$$= \langle f_{GE}a, y' f_{EF}x \rangle = \langle a, f'_{EG}(y' f_{EF}x) \rangle$$

in the first case and

$$\langle a, x(f'_{EF}y') \rangle = \langle ax, f'_{EF}y' \rangle = \langle f_{FE}(ax), y' \rangle = \langle (f_{GE}a)(f_{FG}x), y' \rangle =$$

$$= \langle f_{GE}a, (f_{FG}x)y' \rangle = \langle a, f'_{EG}((f_{FG}x)y') \rangle$$

in the second case.

b) Let $G \underset{\mathcal{A}}{\overset{x}{\to}} E$. By a),

$$\langle x, (f'_{FG}y')x'' \rangle = \langle x'', x f_{FG}y' \rangle = \langle x'', f'_{FE}((f_{GE}x)y') \rangle =$$

$$= \langle f''_{EF}x'', (f_{GE}x)y'\rangle = \langle f_{GE}x, y'f''_{EF}x''\rangle = \langle x, f'_{EG}(y'f''_{EF}x'')\rangle$$

in the first case and

$$\langle x, x''f'_{EF}y'\rangle = \langle x'', (f'_{EF}y')x\rangle = \langle x'', f'_{GF}(y'f_{GE}x)\rangle =$$

$$= \langle f''_{FG}x'', y'f_{GE}x\rangle = \langle f_{GE}x, (f''_{FG}x'')y'\rangle = \langle x, f'_{EG}((f''_{FG}x'')y')\rangle$$

in the second case.

c) Let $G \overset{x'}{\underset{\mathcal{A}'}{\rightarrow}} E$. By b),

$$\langle f''_{EG}(y'' \vdash x''), x'\rangle = \langle y'' \vdash x'', f'_{GE}x'\rangle = \langle y'', x''f'_{GE}x'\rangle =$$

$$= \langle y'', f'_{GF}((f''_{EF}x'')x')\rangle = \langle f''_{FG}y'', (f''_{EF}x'')x'\rangle = \langle (f''_{FG}y'') \vdash (f''_{EF}x''), x'\rangle,$$

$$\langle f''_{EG}(y'' \dashv x''), x'\rangle = \langle y'' \dashv x'', f'_{GE}x'\rangle = \langle x'', (f'_{GE}x')y''\rangle =$$

$$= \langle x'', f'_{FE}(x'f''_{FG}y'')\rangle = \langle f''_{EF}x'', x'f''_{FG}y''\rangle = \langle (f''_{FG}y'') \dashv (f''_{EF}x''), x'\rangle . \qquad \blacksquare$$

Example 1.5.2.15 *Let Ω be a class of sets and Λ the Banach system*

$$(S,T) \longmapsto \ell^{0,1}(S,T)$$

defined on Ω^2 endowed with the multiplication defined in Proposition 1.2.3.5. Take $P, R, S, T \in \Omega$.

a) *Λ is a Banach category. For convenience, we settle on the following notations: 1) h, k denote morphisms of Λ, k' denotes a morphism of Λ' and h'', k'', ℓ'' denote morphisms of Λ''; 2) p, r, s, t denote points of P, R, S, T, respectively.*

b) *Λ' may be identified with the Λ-module*

$$(S,T) \longmapsto \ell^{1,\infty}(S,T),$$

$$kk' : T \times R \longrightarrow \mathbb{K}, \quad (t,r) \longmapsto \sum_s k(s,t)k'(s,r)$$

for $R \overset{k'}{\rightarrow} S \overset{k}{\rightarrow} T$, and

$$k'k : T \times R \longrightarrow \mathbb{K}, \quad (t,r) \longmapsto \sum_s k(r,s)k'(t,s)$$

for $R \overset{k}{\rightarrow} S \overset{k'}{\rightarrow} T$.

c) Λ'' *may be identified with the map*

$$(S,T) \longmapsto E(S,T) \ ,$$

where $E(S,T)$ *is the Banach space introduced in Proposition 1.2.3.7* a) *for* $p = 1$ *and*

c₁) $R \overset{k''}{\to} S \overset{k}{\to} T \Rightarrow (kk'')_r = \int k(s,\cdot)dk''_r(s) = \sum_s k(s,\cdot)k''_r(\{s\})\,,$

c₂) $R \overset{k}{\to} S \overset{k''}{\to} T \Rightarrow (k''k)_r = \sum_s k(r,s)k''_s\,,$

c₃) $R \overset{k''}{\to} S \overset{k'}{\to} T \Rightarrow (k'k'')[t,r] = \int k'(t,\cdot)dk''_r\,,$

c₄) $R \overset{k'}{\to} S \overset{k''}{\to} T \Rightarrow (k''k')[t,r] = \sum_s k'(s,r)k''_s(\{t\})\,,$

c₅) $R \overset{k''}{\to} S \overset{h''}{\to} T \Rightarrow \begin{cases} (h'' \dashv k'')_r = \int h''_r dk''_r(s) \\ (h'' \vdash k'')_r = \sum_s k''_r(\{s\})h''_s\,. \end{cases}$

d) $\Lambda_{\dashv}$ *is a unital Banach category. For* $t \in T, (1_T)_t$ *is Dirac measure at* t *and* Λ' *is a unital left* $\Lambda''_{\vdash}$*–module.*

e) Λ' *is a* $(\Lambda''_{\vdash}, \Lambda''_{\dashv})$*–module.*

f) $P \overset{\ell''}{\to} R \overset{k''}{\to} S \overset{h''}{\to} T \Rightarrow (h'' \dashv k'') \vdash \ell'' = h'' \dashv (k'' \vdash \ell'')\,.$

a) is easy to see.

b) The identification follows from Proposition 1.2.3.6 c). Given $T \overset{h}{\to} R$,

$$\langle h, kk' \rangle = \langle hk, k' \rangle = \sum_{s,r} (hk)[s,r]k'(s,r) =$$

$$= \sum_{s,r} \Big(\sum_t k(s,t)h(t,r) \Big) k'(s,r) = \sum_{t,r} h(t,r) \Big(\sum_s k(s,t)k'(s,r) \Big)\,,$$

$$\langle h, k'k \rangle = \langle kh, k' \rangle = \sum_{t,s} (kh)[t,s]k'(t,s) =$$

$$= \sum_{t,s} \Big(\sum_r h(t,r)k(r,s) \Big) k'(t,s) = \sum_{t,r} h(t,r) \Big(\sum_s k(r,s)k'(t,s) \Big)\,.$$

c) The identification follows from Proposition 1.2.3.7 b) . We have:

c$_1$) $T \xrightarrow{k'} R \Longrightarrow \langle kk'', k'\rangle = \langle k'', k'k\rangle = \sum_r \int (k'k)(r,s)dk''_r(s) =$

$$= \sum_r \int \Big(\sum_t k(s,t)k'(r,t)\Big) dk''_r(s) =$$

$$= \sum_{r,t} k'(r,t)\Big(\int k(s,t)dk''_r(s)\Big) =$$

$$= \sum_{r,t} k'(r,t)\Big(\sum_s k(s,t)k''_r(\{s\})\Big) .$$

c$_2$) $T \xrightarrow{k'} R \Longrightarrow \langle k''k, k'\rangle = \langle k''kk'\rangle = \sum_s \int (kk')[s,t]dk''_s(t) =$

$$= \sum_s \int \Big(\sum_r k(r,s)k'(r,t)\Big) dk''_s(t) =$$

$$= \sum_r \int k'(r,t)d\Big(\sum_s k(r,s)dk''_s(t)\Big) .$$

c$_3$) $T \xrightarrow{k} R \Longrightarrow \langle k, k'k''\rangle = \langle k'', kk'\rangle = \sum_s \int (kk')[r,s]dk''_r(s) =$

$$= \sum_s \int \Big(\sum_t k(t,r)k'(t,s)\Big) dk''_r(s) =$$

$$= \sum_{t,r} k(t,r)\int k'(t,s)dk''_r(s) .$$

c$_4$) $T \xrightarrow{k} R \Longrightarrow \langle k, k''k'\rangle = \langle k'', k'k\rangle = \sum_s \int (k'k)[s,t]dk''_s(t) =$

$$= \sum_s \int \Big(\sum_r k(t,r)k'(s,r)\Big) dk''_s(t) =$$

$$= \sum_{s,r} k'(s,r)\Big(\int k(t,r)dk''_s(t)\Big) =$$

$$= \sum_{s,r} k'(s,r)\Big(\sum_t k(t,r)k''_s(\{t\})\Big) =$$

$$= \sum_{t,r} k(t,r)\Big(\sum_s k'(s,r)k''_s(\{t\})\Big) .$$

c$_5$) $T \xrightarrow{k'} R \Longrightarrow \langle h'' \dashv k'', k'\rangle = \langle k'', k'h''\rangle = \sum_r \int (k'h'')[r,s]dk''_r(s) =$

$$= \sum_r \int \Big(\int k'(r,t)dh''_s(t)\Big) dk''_r(s) = \sum_r \int k'(r,\cdot)d\Big(\int h''_s dk''_r(s)\Big) ,$$

$$= \langle h'' \vdash k'', k'\rangle = \langle h'', k''k'\rangle = \sum_s \int (k''k')[s,t]dh''_s(t) =$$

$$= \sum_s \int \Big(\sum_r k'(r,t)k''_r(\{s\})dh''_s(t)\Big) =$$

$$= \sum_r \int k'(r,\cdot)d\Big(\sum_s k''_r(\{s\})h''_s\Big) .$$

d) We have

$$(1_T \dashv k'')_s = \int (1_T)_t dk''_s(t) = k''_s,$$

$$(h'' \dashv 1_T)_t = \int h''_{t'} d(1_T)_t[t'] = h''_t$$

for $S \overset{k''}{\to} T \overset{h''}{\to} R$ and $S \overset{k'}{\to} T$.

e) $P \overset{h''}{\to} R \overset{k'}{\to} S \overset{k''}{\to} T \Longrightarrow ((k''k')h'')[t,p] = \int (k''k')[t,r]dh''_p(r) =$

$$= \int \Big(\sum_s k'(s,r)k''_s(\{t\})\Big) dh''_p(r) = \sum_s \Big(\int k'(s,r)dh''_p(r)\Big) k''_s(\{t\}) =$$
$$= \sum_s (k'h'')[s,p]k''_s(\{t\}) = (k''(k'h''))[t,p].$$

f) follows from e) and Theorem 1.5.2.12 f). ■

Remark. If $k''_r(\{s\}) = 0$ for every $(r,s) \in R \times S$ in c_5), then

$$h'' \vdash k'' = 0$$

for every h''. Take $S \overset{h''}{\to} S$ and for $s \in S$, let h''_s be the Dirac measure at s. Then

$$h'' \dashv k'' = k''$$

by d). Hence $\vdash$ and $\dashv$ may be different.

Definition 1.5.2.16 (2) *Let (Ω, Λ) be a Banach category and $\mathcal{A}$ a Λ–module (Λ–category). A Λ–**submodule** (Λ–**subcategory**) of $\mathcal{A}$ is a Λ–module (Λ–category) $\mathcal{B}$ such that $\mathcal{B}(E,F)$ is a Banach subspace of $\mathcal{A}(E,F)$ for every $E,F \in \Omega$ and such that the multiplications of $\mathcal{B}$ are the restrictions of the multiplications of $\mathcal{A}$. The map defined on Ω^2 such that $f(E,F)$ is the inclusion map*

$$\mathcal{B}(E,F) \longrightarrow \mathcal{A}(E,F)$$

*for every $E,F \in \Omega^2$ is called **the inclusion functor of $\mathcal{B}$ into $\mathcal{A}$**.*

Proposition 1.5.2.17 (2) *Let (Ω, Λ) be a Banach category. Let $\mathcal{A}$ be a Λ–module (Λ–category) and $\mathcal{B}$ a Λ–submodule (Λ–subcategory) of $\mathcal{A}$. Put*

$$\mathcal{C}(E,F) := \mathcal{A}(E,F)/\mathcal{B}(E,F)$$

and let $f(E,F)$ *denote the quotient map*

$$\mathcal{A}(E,F) \longrightarrow \mathcal{C}(E,F)$$

for $E,F \in \Omega$. *Define multiplications on* $\mathcal{C}$ *by taking the factorizations. Then* $\mathcal{C}$ *is a* Λ*-module (*Λ*-category) and* f *is a functor of* Λ*-modules (*Λ*-categories).* $\mathcal{C}$ *is called the* **quotient Λ-module (quotient Λ-category) of $\mathcal{A}$ by $\mathcal{B}$** *and is denoted by* $\mathcal{A}/\mathcal{B}$; f *is called* **the quotient functor of $\mathcal{A}$ onto $\mathcal{A}/\mathcal{B}$**.

Let (Ω,Λ) *be a unital Banach category. If* $\mathcal{A}$ *is a unital* Λ*-module, then* $\mathcal{A}/\mathcal{B}$ *is a unital* Λ*-module. If* $\mathcal{A}$ *is a unital* Λ*-category, then* $\mathcal{A}/\mathcal{B}$ *is a unital* Λ*-category and the quotient functor of* $\mathcal{A}$ *onto* $\mathcal{A}/\mathcal{B}$ *is a functor of unital* Λ*-categories.*

The proof is a long verification. ■

Proposition 1.5.2.18 (2) *Let* (Ω,Λ) *be a Banach category,* $\mathcal{A}$ *a* Λ*-module, and* $\mathcal{B}$ *a* Λ*-submodule of* $\mathcal{A}$. *Given* $E,F \in \Omega$, *set*

$$\mathcal{B}^\circ(E,F) := \mathcal{B}(F,E)^\circ .$$

Then $\mathcal{B}^\circ$ *is a* Λ*-submodule of* $\mathcal{A}'$. *For* $E,F \in \Omega$, *let* $f(E,F)$ *denote the canonical isometry*

$$(\mathcal{A}'/\mathcal{B}^\circ)(E,F) \longrightarrow \mathcal{B}'(E,F)$$

(Proposition 1.3.5.2). Then f *is an isometric functor of* Λ*-modules of* $\mathcal{A}'/\mathcal{B}^\circ$ *into* $\mathcal{B}'$.

Take $E,F,G,H \in \Omega$ and consider

$$E \underset{\Lambda}{\overset{a}{\rightarrow}} F \underset{\mathcal{B}^\circ}{\overset{x'}{\rightarrow}} G \underset{\Lambda}{\overset{b}{\rightarrow}} H .$$

Then

$$\langle x, x'a\rangle = \langle ax, x'\rangle = 0 ,$$

$$\langle y, bx'\rangle = \langle yb, x'\rangle = 0$$

for every

$$G \underset{\mathcal{B}}{\overset{x}{\rightarrow}} E , \quad H \underset{\mathcal{B}}{\overset{y}{\rightarrow}} F ,$$

so that

$$x'a \in \mathcal{B}^\circ(E,G)\,, \quad bx' \in \mathcal{B}^\circ(F,H)\,.$$

Hence $\mathcal{B}^\circ$ is a Λ-submodule of $\mathcal{A}'$.

Take $E,F,G,H \in \Omega$,

$$E \underset{\Lambda}{\overset{a}{\to}} F \underset{\mathcal{A}'}{\overset{x'}{\to}} G \underset{\Lambda}{\overset{b}{\to}} H\,,$$

and let q be the quotient functor of $\mathcal{A}'$ onto $\mathcal{A}'/\mathcal{B}^\circ$. We have

$$\langle x, f((qx')a)\rangle = \langle x, (qx')a\rangle = \langle ax, qx'\rangle = \langle ax, f(qx')\rangle = \langle x, (f(qx'))a\rangle\,,$$

$$\langle y, f(b(qx'))\rangle = \langle y, b(qx')\rangle = \langle yb, qx'\rangle = \langle yb, f(qx')\rangle = \langle y, bf(qx')\rangle$$

for every

$$G \underset{\mathcal{B}}{\overset{x}{\to}} E\,, \quad H \underset{\mathcal{B}}{\overset{y}{\to}} F\,.$$

This proves the last assertion. ■

Proposition 1.5.2.19 (2) *Let $(\Omega, \mathcal{A})$ be a Banach system, i the evaluation functor of $\mathcal{A}$, and*

$$(i(\mathcal{A})^\circ)(E,F) := (i(\mathcal{A})(F,E))^\circ\,.$$

If $\mathcal{A}'$ is a Banach category and $i(\mathcal{A})$ is a submodule of $\mathcal{A}''$, then $i(\mathcal{A})^\circ$ is an $\mathcal{A}'''_\vdash$-submodule and an $\mathcal{A}'''_\dashv$-submodule.

We denote by j the evaluation functor of $\mathcal{A}'$.

Step 1 $E \underset{i(\mathcal{A})}{\overset{x''}{\to}} F \underset{i(\mathcal{A})^\circ}{\overset{x'''}{\to}} G \underset{i(\mathcal{A})}{\overset{y''}{\to}} H \Longrightarrow x'''x'' = 0\,,\ y''x''' = 0\,.$

We have

$$\langle x', x'''x''\rangle = \langle x''', x''x'\rangle = 0\,, \quad \langle y', y''x'''\rangle = \langle x''', y'y''\rangle = 0$$

for every

$$G \underset{\mathcal{A}'}{\overset{x'}{\to}} E\,, \quad H \underset{\mathcal{A}'}{\overset{y'}{\to}} F\,.$$

Step 2 $E \underset{i(\mathcal{A})^\circ}{\overset{x'''}{\to}} F \underset{i(\mathcal{A})^\circ}{\overset{y'''}{\to}} G \Longrightarrow y''' \vdash x''', y''' \dashv x''' \in (i(\mathcal{A})^\circ)(E,G)\,.$

For $G \underset{i(\mathcal{A})}{\overset{x''}{\to}} E$,

$$\langle y''' \vdash x''', x''\rangle = \langle y''', x'''x''\rangle = 0\,,$$

$$\langle y''' \dashv x''', x''\rangle = \langle x''', x''y'''\rangle = 0$$

by Step 1.

Step 3 $E \underset{j(\mathcal{A}')}{\overset{x'''}{\rightarrow}} F \underset{i(\mathcal{A})^\circ}{\overset{y'''}{\rightarrow}} G \underset{j(\mathcal{A}')}{\overset{z'''}{\rightarrow}} H \Longrightarrow$

$$\Longrightarrow \begin{cases} y''' \vdash x''', y''' \dashv x''', z''' \vdash y''', z''' \dashv y''' \\ \text{are morphisms of} \quad i(\mathcal{A})^\circ\,. \end{cases}$$

Take

$$E \underset{\mathcal{A}'}{\overset{x'}{\rightarrow}} F\,, \quad G \underset{\mathcal{A}'}{\overset{z'}{\rightarrow}} H$$

with

$$x''' = j_{EF}x'\,, \quad z''' = j_{GH}z'\,.$$

Then

$$\langle y''' \dashv x''', x''\rangle = \langle x''', x''y'''\rangle = 0\,,$$

$$\langle y''' \vdash x''', x''\rangle = \langle y''', x'''x''\rangle = \langle y''', (j_{EF}x')x''\rangle = \langle y''', x'x''\rangle = 0\,,$$

$$\langle z''' \vdash y''', y''\rangle = \langle z''', y'''y''\rangle = 0\,,$$

$$\langle z''' \dashv y''', y''\rangle = \langle y''', y''z'''\rangle = \langle y''', y''(j_{GH}z')\rangle = \langle y''', y''z'\rangle = 0$$

for every

$$G \underset{i(\mathcal{A})}{\overset{x''}{\rightarrow}} E\,, \quad H \underset{i(\mathcal{A})}{\overset{y''}{\rightarrow}} F$$

by Step 1 and Corollary 1.5.2.7.

Step 4 $i(\mathcal{A})^\circ$ is an $\mathcal{A}'''_\vdash$–submodule and an $\mathcal{A}'''_\dashv$–submodule.

The assertion follows from Steps 2 and 3 together with Proposition 1.3.6.19 d).

■

1.6 Nuclear Maps

Several classes of compact operators on Hilbert spaces are known. These classes have some connection or other to the ℓ^p–spaces $(p \in \{0\} \cup [1, \infty[)$. The class of nuclear operators, which are also called trace operators, arises when $p = 1$. They are the subject of this section. Their strong properties make the theory also applicable to Banach spaces, as shown by Alexander Grothendieck. Since this theory is not pursued further in this book, the reader may skip this section.

1.6.1 General Results

Definition 1.6.1.1 (0) (Grothendieck, 1952) *Let E, F be normed spaces. A map $u : E \to F$ is called* **nuclear** *if there is a family $((x'_\iota, y_\iota))_{\iota \in I}$ in $E' \times F$ such that*

$$\sum_{\iota \in I} \|x'_\iota\| \, \|y_\iota\| < \infty$$

and

$$ux = \sum_{\iota \in I} \langle x, x'_\iota \rangle y_\iota$$

for every $x \in E$. We write $\mathcal{L}^1(E, F)$ for the set of nuclear maps of E into F and define

$$\|u\|_1 := \inf \sum_{\iota \in I} \|x'_\iota\| \, \|y_\iota\|$$

for $u \in \mathcal{L}^1(E, F)$, where the infimum is taken over all families $((x'_\iota, y_\iota))_{\iota \in I}$ in $E' \times F$ with the above properties. We set

$$\mathcal{L}^1(E) := \mathcal{L}^1(E, F) .$$

We can replace the indexing set I by $\mathbb{N}$ in the above definition, since

$$\{\iota \in I \mid \|x'_\iota\| \, \|y_\iota\| \neq 0\}$$

is countable.

Example 1.6.1.2 (3) *Let T be a set, $E := c_0(T)$ (resp. $E := \ell^\infty(T)$), $F := \ell^p(T)$, $p \in [1, \infty]$, $y \in F$, and take*

$$u : E \longrightarrow F , \quad x \longmapsto xy .$$

Then $u \in \mathcal{L}^1(E)$ iff $y \in \ell^1(T)$ and in this case

$$\|u\|_1 = \|y\|_1 .$$

Assume that $y \in \ell^1(T)$. Given $t \in T$, set

$$x'_t : E \longrightarrow \mathbb{K}, \quad x \longmapsto x(t)y(t).$$

Then

$$\sum_{t \in T} \|x'_t\| \, \|e_t\| = \sum_{t \in T} |y(t)| = \|y\|_1 < \infty,$$

$$\sum_{t \in T} \langle x, x'_t \rangle e_t = \sum_{t \in T} x(t)y(t)e_t = ux$$

for every $x \in E$. Hence u is nuclear and

$$\|u\| \leq \|y\|_1 .$$

Assume that u is nuclear. There is a family $((x'_\iota, y_\iota))_{\iota \in I}$ in $E' \times F$, such that

$$\sum_{\iota \in I} \|x'_\iota\| \, \|y_\iota\| < \infty$$

and

$$ux = \sum_{\iota \in I} \langle x, x'_\iota \rangle y_\iota$$

for every $x \in E$. Then

$$y(t)e_t = ue_t = \sum_{\iota \in I} \langle e_t, x'_\iota \rangle y_\iota ,$$

so that

$$|y(t)| = \left\| \sum_{\iota \in I} \langle e_t, x'_\iota \rangle y_\iota \right\| \leq \sum_{\iota \in I} |\langle e_t, x'_\iota \rangle| \, \|y_\iota\|$$

for every $t \in T$. Hence

$$\sum_{t \in T} |y(t)| \leq \sum_{t \in T} \sum_{\iota \in I} |\langle e_t, x'_\iota \rangle| \, \|y_\iota\| =$$

$$= \sum_{\iota \in I} \|y_\iota\| \sum_{t \in T} |\langle e_t, x'_\iota \rangle| \leq \sum_{\iota \in I} \|y_\iota\| \, \|x'_\iota\| ,$$

so that $y \in \ell^1(T)$ and

$$\|y\|_1 \leq \|u\|_1 .$$

■

Theorem 1.6.1.3 (0) *Let E, F be normed spaces.*

a) *$\mathcal{L}^1(E,F)$ is a vector subspace of $\overline{\mathcal{L}_f(E,F)}$ and*

$$\|u\| \leq \|u\|_1$$

for every $u \in \mathcal{L}^1(E,F)$.

b) *The map*

$$\mathcal{L}^1(E,F) \longrightarrow \mathbb{R}_+ , \quad u \longmapsto \|u\|_1$$

is a norm. Take $\mathcal{L}^1(E,F)$ with this norm.

c) *$\mathcal{L}_f(E,F)$ is a dense set of $\mathcal{L}^1(E,F)$.*

d) *$\mathcal{L}^1(E,F)$ is complete whenever F is complete.*

e) *$u \in \mathcal{L}^1(E,F) \Rightarrow u' \in \mathcal{L}^1(F',E')$, $\|u'\|_1 \leq \|u\|_1$.*

a), b), and c). Take $u \in \mathcal{L}^1(E,F)$. Let $((x'_\iota, y_\iota))_{\iota\in I}$ be a family in $E' \times F$ with

$$\sum_{\iota\in I} \|x'_\iota\| \, \|y_\iota\| < \infty$$

and such that

$$ux = \sum_{\iota\in I} \langle x, x'_\iota\rangle y_\iota$$

for every $x \in E$. u is linear and

$$\|ux\| \leq \sum_{\iota\in I} \|\langle x, x'_\iota\rangle y_\iota\| = \sum_{\iota\in I} |\langle x, x'_\iota\rangle| \, \|y_\iota\| \leq$$

$$\leq \sum_{\iota\in I} \|x\| \, \|x'_\iota\| \, \|y_\iota\| = \|x\| \sum_{\iota\in I} \|x'_\iota\| \, \|y_\iota\|$$

for every $x \in E$. Hence u is continuous and

$$\|u\| \leq \|u\|_1 .$$

Given $u, v \in \mathcal{L}^1(E,F)$ and $\alpha \in \mathbb{K}$, there are families $((x'_\iota, x_\iota))_{\iota\in I}$, $((y'_\lambda, y_\lambda))_{\lambda\in L}$ in $E' \times F$ such that

$$\sum_{\iota\in I} \|x'_\iota\| \, \|x_\iota\| < \infty , \quad \sum_{\lambda\in L} \|y'_\lambda\| \, \|y_\lambda\| < \infty$$

and

$$ux = \sum_{\iota \in I} \langle x, x'_\iota \rangle x_\iota \,, \quad vx = \sum_{\lambda \in L} \langle x, y'_\lambda \rangle y_\lambda$$

for every $x \in E$. Then

$$(u + v)x = \sum_{\iota \in I} \langle x, x'_\iota \rangle x_\iota + \sum_{\lambda \in L} \langle x, y'_\lambda \rangle y_\lambda \,, \quad (\alpha u)x = \sum_{\iota \in I} \langle x, x'_\iota \rangle \alpha x_\iota$$

for every $x \in E$. Hence $u + v$, $\alpha u \in \mathcal{L}^1(E, F)$ and

$$\|u + v\|_1 \leq \|u\|_1 + \|v\|_1 \,, \quad \|\alpha u\|_1 = |\alpha| \, \|u\|_1 \,.$$

From

$$\|u\|_1 = 0 \,,$$

it follows that

$$\|u\| = 0$$

and so

$$u = 0 \,.$$

Hence $\mathcal{L}^1(E, F)$ is a vector subspace of $\mathcal{L}(E, F)$ and the map

$$\mathcal{L}^1(E, F) \longrightarrow \mathbb{R}_+ \,, \quad u \longmapsto \|u\|_1$$

is a norm.

Given $u \in \mathcal{L}^1(E, F)$ and $\varepsilon > 0$, there is a family $((x'_\iota, y_\iota))_{\iota \in I}$ in $E' \times F$ such that

$$\sum_{\iota \in I} \|x'_\iota\| \, \|y_\iota\| < \infty$$

and

$$ux = \sum_{\iota \in I} \langle x, x'_\iota \rangle y_\iota$$

for every $x \in E$. There is a finite subset J of I with

$$\sum_{\iota \in I \setminus J} \|x'_\iota\| \, \|y_\iota\| < \varepsilon \,.$$

Put

$$w : E \longrightarrow F\,, \quad x \longmapsto \sum_{\iota\in J} \langle x, x'_\iota\rangle y_\iota\,.$$

Then

$$(u-w)x = \sum_{\iota\in I\setminus J} \langle x, x'_\iota\rangle y_\iota$$

for every $x \in E$. It follows that $u-w$ is nuclear and

$$\|u-w\| \le \|u-w\|_1 \le \sum_{\iota\in I\setminus J} \|x'_\iota\|\,\|y_\iota\| < \varepsilon\,.$$

Since $w \in \mathcal{L}_f(E,F)$, $u \in \overline{\mathcal{L}_f(E,F)}$ and $\mathcal{L}_f(E,F)$ is a dense set of $\mathcal{L}^1(E,F)$.

d) Let $(u_n)_{n\in\mathbb{N}}$ be a Cauchy sequence in $\mathcal{L}^1(E,F)$. By a) and b), $(u_n)_{n\in\mathbb{N}}$ is a Cauchy sequence in $\mathcal{L}(E,F)$. Put

$$u := \lim_{n\to\infty} u_n$$

(Theorem 1.2.1.9 b)). We may assume that

$$\|u_n - u_{n+1}\|_1 < \frac{1}{2^n}$$

for every $n \in \mathbb{N}$. Given $n \in \mathbb{N}$, there is a family $((x'_{n\iota}, y_{n\iota}))_{\iota\in I_n}$ in $E' \times F$ such that

$$\sum_{\iota\in I_n} \|x'_{n\iota}\|\,\|y_{n\iota}\| < \frac{1}{2^n}$$

and

$$(u_n - u_{n+1})x = \sum_{\iota\in I_n} \langle x, x'_{n\iota}\rangle y_{n\iota}$$

for every $x \in E$. Hence,

$$u_n x - u_p x = \sum_{q=n}^{p-1} \sum_{\iota\in I_q} \langle x, x'_{q\iota}\rangle y_{q\iota}$$

for every $n, p \in \mathbb{N}$, $n < p$, and $x \in E$. Then

$$\sum_{q=n}^{\infty} \sum_{\iota\in I_q} \|x'_{q\iota}\|\,\|y_{q\iota}\| < \sum_{q=n}^{\infty} \frac{1}{2^q} = \frac{1}{2^{n-1}}$$

and

$$(u_n - u)x = u_n x - ux = \lim_{p\to\infty}(u_n x - u_p x) = \sum_{q=n}^{\infty}\sum_{\iota\in I_q}\langle x, x'_{q\iota}\rangle y_{q\iota}$$

for $n \in \mathbb{N}$ and $x \in E$. Hence $u_n - u \in \mathcal{L}^1(E,F)$ and

$$\|u_n - u\|_1 < \frac{1}{2^{n-1}}$$

for every $n \in \mathbb{N}$. By a) and b), u is nuclear and $(u_n)_{n\in\mathbb{N}}$ converges to u in $\mathcal{L}^1(E,F)$. Hence $\mathcal{L}^1(E,F)$ is complete.

e) Let $((x'_\iota, y_\iota))_{\iota\in I}$ be a family in $E' \times F$ such that

$$\sum_{\iota\in I}\|x'_\iota\|\,\|y_\iota\| < \infty$$

and

$$ux = \sum_{\iota\in I}\langle x, x'_\iota\rangle y_\iota$$

for every $x \in E$. Then

$$\langle x, u'y'\rangle = \langle ux, y'\rangle = \sum_{\iota\in I}\langle x, x'_\iota\rangle\langle y_\iota, y'\rangle = \langle x, \sum_{\iota\in I}\langle y_\iota, y'\rangle x'_\iota\rangle$$

for every $(x,y') \in E \times F'$ (Theorem 1.3.4.2 a), Proposition 1.2.1.16, Corollary 1.2.1.10, Corollary 1.1.6.10). Thus

$$u'y' = \sum_{\iota\in I}\langle y_\iota, y'\rangle x'_\iota$$

for every $y' \in F'$ and the assertion now follows. ■

Corollary 1.6.1.4 (0) *If E, F are normed spaces, then*

$$\langle\cdot, x'\rangle y \in \mathcal{L}^1(E,F),$$

$$\|\langle\cdot, x'\rangle y\|_1 = \|\langle\cdot, x'\rangle y\| = \|x'\|\,\|y\|$$

for every $(x', y) \in E' \times F$.

Now

$$\|\langle\cdot, x'\rangle y\| = \sup_{x\in E^\#}\|\langle x, x'\rangle y\| = \|y\|\sup_{x\in E^\#}|\langle x, x'\rangle| = \|y\|\,\|x'\|.$$

By Theorem 1.6.1.3 a),

$$\|\langle\cdot, x'\rangle y\|_1 \geq \|y\|\,\|x'\|,$$

and the reverse inequality is trivial. ■

Proposition 1.6.1.5 (0) *Let E, F, G, H be normed spaces. Take $u \in \mathcal{L}(E,F)$, $v \in \mathcal{L}^1(F,G)$, and $w \in \mathcal{L}(G,H)$. Then $w \circ v \circ u \in \mathcal{L}^1(E,H)$ and*

$$\|w \circ v \circ u\|_1 \leq \|w\| \, \|v\|_1 \|u\| .$$

Let $((y'_\iota, z_\iota))_{\iota \in I}$ be a family in $F' \times G$ such that

$$\sum_{\iota \in I} \|y'_\iota\| \, \|z_\iota\| < \infty$$

and

$$vy = \sum_{\iota \in I} \langle y, y'_\iota \rangle z_\iota$$

for every $y \in F$. Then

$$\sum_{\iota \in I} \|u' y'_1\| \, \|w z_\iota\| \leq \sum_{\iota \in I} \|u'\| \, \|y'_\iota\| \, \|w\| \, \|z_\iota\| =$$

$$= \|u\| \, \|w\| \sum_{\iota \in I} \|y'_\iota\| \, \|z_\iota\| < \infty$$

(Theorem 1.3.4.2 b)) and

$$(w \circ v \circ u)x = w\Big(\sum_{\iota \in I} \langle ux, y'_\iota \rangle z_\iota\Big) = \sum_{\iota \in I} \langle x, u' y'_\iota \rangle w z_\iota$$

for every $x \in E$ (Proposition 1.2.1.16, Theorem 1.3.4.2 a)). Hence $w \circ v \circ u$ is nuclear and

$$\|w \circ v \circ u\|_1 \leq \|w\| \, \|v\|_1 \|u\| . \qquad \blacksquare$$

Corollary 1.6.1.6 *Let Ω be a class of Banach spaces. The map on Ω^2 defined by*

$$(E, F) \longmapsto \mathcal{L}^1(E, F) ,$$

together with the usual composition of maps as multiplication, is an $\mathcal{L}_\Omega$-category. We denote it by $\mathcal{L}^1_\Omega$, or simply, by $\mathcal{L}^1$.

The assertion follows immediately from Theorem 1.6.1.3 a), b), d), and Proposition 1.6.1.5. ■

Corollary 1.6.1.7 (3) *Let E, F be normed spaces, G a subspace of F, p a projection of F onto G, and j the inclusion map $G \to F$. Then*

$$\|j \circ u\|_1 \leq \|u\|_1 \leq \|p\| \, \|j \circ u\|_1$$

for every $u \in \mathcal{L}^1(E, G)$.

We have

$$u = p \circ j \circ u\,,$$

so that

$$\|j \circ u\|_1 \le \|u\|_1 \le \|p\|\,\|j \circ u\|_1$$

by Proposition 1.6.1.5. ■

Corollary 1.6.1.8 (3) *Let E, F be normed spaces and take $u \in \mathcal{L}^1(E,F)$. If F is a dual space, then*

$$\|u\|_1 = \|u'\|_1\,.$$

By Theorem 1.6.1.3 e), u' and u'' are nuclear and

$$\|u''\|_1 \le \|u'\|_1 \le \|u\|_1\,.$$

By Proposition 1.3.6.19 b), there is a projection p of F'' onto $\operatorname{Im} j_F$ with $\|p\| \le 1$, so that

$$\|u\|_1 = \|j_F \circ u\|_1$$

(Corollary 1.6.1.7). Since

$$j_F \circ u = u'' \circ j_E$$

(Proposition 1.3.6.16),

$$\|u\|_1 = \|j_F \circ u\|_1 = \|u'' \circ j_E\|_1 \le \|u''\|_1 \|j_F\| = \|u''\|_1 \le \|u'\|_1 \le \|u\|_1$$

(Proposition 1.6.1.5). Hence

$$\|u\|_1 = \|u'\|_1\,.$$ ■

Corollary 1.6.1.9 (3) *Let E, F be normed spaces and take $u \in \mathcal{L}^1(E,F)$. If F is reflexive, then $u' \in \mathcal{L}^1(F', E')$ and*

$$\|u'\|_1 = \|u\|_1\,.$$ ■

Corollary 1.6.1.10 (3) *Let E, F be normed spaces and take $u \in \mathcal{L}^1(E,F)$. Then u' and u'' are nuclear and*

$$\|u'\|_1 = \|u''\|_1\,.$$

By Theorem 1.6.1.3 e), u' and u'' are nuclear and by Corollary 1.6.1.8,

$$\|u'\|_1 = \|u''\|_1 .$$ ■

Proposition 1.6.1.11 *Let E, F be normed spaces and G a subspace of E. Then, given $u \in \mathcal{L}^1(G, F)$, there is a $v \in \mathcal{L}^1(E, F)$ such that*

$$v|G = u\,, \quad \|v\|_1 = \|u\|_1 .$$

Let $((x'_\iota, y_\iota))_{\iota \in I}$ be a family in $G' \times F$ such that

$$\sum_{\iota \in I} \|x'_\iota\| \, \|y_\iota\| < \infty$$

and

$$ux = \sum_{\iota \in I} \langle x, x'_\iota \rangle y_\iota$$

for every $x \in G$. By the Hahn–Banach Theorem, given $\iota \in I$, there is a $z'_\iota \in E'$ such that

$$z'_\iota | G = x'_\iota\,, \quad \|z'_\iota\| = \|x_\iota\| .$$

Set

$$v : E \longrightarrow F\,, \quad x \longmapsto \sum_{\iota \in I} \langle x, z'_\iota \rangle y_\iota .$$

Then $v \in \mathcal{L}^1(E, F)$,

$$v|G = u\,, \quad \|v\|_1 \le \sum_{\iota \in I} \|z'_\iota\| \, \|y_\iota\| = \sum_{\iota \in I} \|x'_\iota\| \, \|y_\iota\| .$$

Hence

$$\|v\|_1 \le \|u\|_1 .$$

By Proposition 1.6.1.5,

$$\|u\|_1 \le \|v\|_1 ,$$

so that

$$\|u\|_1 = \|v\|_1 .$$ ■

Theorem 1.6.1.12 (3) *Let E, F be normed spaces.*

a) *Given* $u \in \mathcal{L}(E, F'')$, *there is a unique* $\widetilde{u} \in \mathcal{L}^1(F, E)'$ *such that*

$$\widetilde{u}(\langle \cdot, y' \rangle x) = \langle ux, y' \rangle$$

for every $(x, y') \in E \times F'$.

b) *The map*

$$\mathcal{L}(E, F'') \longrightarrow \mathcal{L}^1(F, E)', \quad u \longmapsto \widetilde{u}$$

is an isometry.

Step 1 a) and $\|\widetilde{u}\| \le \|u\|$

Let $((x_\iota, y'_\iota))_{\iota \in I}$ be a finite family in $E \times F'$ such that

$$\sum_{\iota \in I} \langle \cdot, y'_\iota \rangle x_\iota = 0 \,.$$

We show that

$$\sum_{\iota \in I} \langle ux_\iota, x'_\iota \rangle = 0 \,.$$

Since the map

$$E \longrightarrow \mathbb{K}, \quad x \longmapsto \langle ux, y' \rangle$$

is linear for every $y' \in F'$, we may assume that $(x_\iota)_{\iota \in I}$ is linearly independent. Given $y \in F$,

$$\sum_{\iota \in I} \langle y, y'_\iota \rangle x_\iota = 0$$

so that

$$\langle y, y'_\iota \rangle = 0$$

for every $y \in F$ and $\iota \in I$. Hence $y'_\iota = 0$ for every $\iota \in I$ and so

$$\sum_{\iota \in I} \langle ux_\iota, y'_\iota \rangle = 0 \,.$$

By the above proof (and Corollary 1.3.3.4), the map

$$\dot{u} : \mathcal{L}_f(F, E) \longrightarrow \mathbb{K}, \quad \sum_{\iota \in I} \langle \cdot, y'_\iota \rangle x_\iota \longmapsto \sum_{\iota \in I} \langle ux_\iota, y'_\iota \rangle$$

is well–defined. It is obviously linear. Now

$$\left|\dot{u}\Big(\sum_{\iota\in I}\langle\cdot,y'_\iota\rangle x_\iota\Big)\right| = \left|\sum_{\iota\in I}\langle ux_\iota,y'_\iota\rangle\right| \leq \sum_{\iota\in I}|\langle ux_\iota,y'_\iota\rangle| \leq \sum_{\iota\in I}\|ux_\iota\|\,\|y'_\iota\| \leq$$

$$\leq \sum_{\iota\in I}\|u\|\,\|x_\iota\|\,\|y'_\iota\| = \|u\|\sum_{\iota\in I}\|x_\iota\|\,\|y'_\iota\|\,,$$

so that

$$\left|\dot{u}\Big(\sum_{\iota\in I}\langle\cdot,y'_\iota\rangle x_\iota\Big)\right| \leq \|u\|\left\|\sum_{\iota\in I}\langle\cdot,y'_\iota\rangle x_\iota\right\|_1$$

for every finite family $((x_\iota,y'_\iota))_{\iota\in I}$ in $E\times F'$. Hence $\dot{u}$ is continuous with respect to the norm $\|\cdot\|_1$ on $\mathcal{L}_f(F,E)$ and $\|\dot{u}\|\leq\|u\|$. By Theorem 1.6.1.3 c) and Proposition 1.2.1.13, there is a unique $\widetilde{u}\in\mathcal{L}^1(F,E)'$, such that

$$\widetilde{u}=\dot{u}$$

on $\mathcal{L}_f(F,E)$ and

$$\|\widetilde{u}\|=\|\dot{u}\|\leq\|u\|\,.$$

Step 2 b)

Take $\theta\in\mathcal{L}^1(F,E)'$ and $x\in E$. Set

$$ux: F'\longrightarrow\mathbb{K}\,,\quad y'\longmapsto\theta(\langle\cdot,y'\rangle x)\,.$$

Then

$$|\langle ux,y'\rangle| = |\theta(\langle\cdot,y'\rangle x)| \leq \|\theta\|\,\|\langle\cdot,y'\rangle x\|_1 = \|\theta\|\,\|y'\|\,\|x\|$$

for every $y'\in F'$ (Corollary 1.6.1.4), so that

$$ux\in F''\,,\quad \|ux\|\leq\|\theta\|\,\|x\|\,.$$

Hence

$$u\in\mathcal{L}(E,F'')\,,\quad \|u\|\leq\|\theta\|\,.$$

By the definition of u, $\widetilde{u}=\theta$ on $\mathcal{L}_f(F,E)$ (Corollary 1.3.3.4). By Step 1 and Theorem 1.6.1.3 c), $\widetilde{u}=\theta$ and

$$\|\theta\|=\|\widetilde{u}\|\leq\|u\|\leq\|\theta\|\,,$$

$$\|\widetilde{u}\|=\|u\|\,.$$

■

Corollary 1.6.1.13 *Let Ω be a class of reflexive Banach spaces. Given $E, F \in \Omega$, let $f(E,F)$ denote the isometry*

$$\mathcal{L}(E,F) \longrightarrow \mathcal{L}^1(F,E)'$$

defined in Theorem 1.6.1.12 b). Then f is an isometric functor of the unital $\mathcal{L}_\Omega$–module $\mathcal{L}_\Omega$ into the dual of the unital $\mathcal{L}_\Omega$–modulus $\mathcal{L}^1_\Omega$.

Given $E, F \in \Omega$ and $u \in \mathcal{L}(E,F)$, put

$$\widetilde{u} := f_{EF} u \,.$$

Take $E, F, G \in \Omega$, $u \in \mathcal{L}(E,F)$, and $v \in \mathcal{L}(F,G)$. Then

$$\widetilde{vu}\Big(\sum_{\iota \in I} \langle \cdot, z'_\iota \rangle x_\iota\Big) = \sum_{\iota \in I} \langle vux_\iota, z'_\iota \rangle = \sum_{\iota \in I} \langle ux_\iota, v'z'_\iota \rangle =$$

$$= \widetilde{u}\Big(\sum_{\iota \in I} \langle \cdot, v'z'_\iota \rangle x_\iota\Big) = \widetilde{u}\Big(\Big(\sum_{\iota \in I} \langle \cdot, z'_\iota \rangle x_\iota\Big) v\Big) = v\widetilde{u}\Big(\sum_{\iota \in I} \langle \cdot, z'_\iota \rangle x_\iota\Big),$$

$$\widetilde{vu}\Big(\sum_{\iota \in I} \langle \cdot, z'_\iota \rangle x_\iota\Big) = \sum_{\iota \in I} \langle vux_\iota, z'_\iota \rangle = \widetilde{v}\Big(\sum_{\iota \in I} \langle \cdot, z'_\iota \rangle ux_\iota\Big) =$$

$$= \widetilde{v}\Big(u\Big(\sum_{\iota \in I} \langle \cdot, z'_\iota \rangle x_\iota\Big)\Big) = \widetilde{v}u\Big(\sum_{\iota \in I} \langle \cdot, z'_\iota \rangle x_\iota\Big)$$

for every finite family $((x_\iota, z'_\iota))_{\iota \in I}$ in $E \times G'$ (Corollary 1.3.4.8). By Theorem 1.6.1.3 c),

$$f(vu) = \widetilde{vu} = v\widetilde{u} = vf(u)\,, \qquad f(vu) = \widetilde{vu} = \widetilde{v}u = f(v)u\,,$$

which proves the assertion. ■

Proposition 1.6.1.14 (3) *Let E, F be normed spaces and take $u \in \mathcal{L}^1(F, E'')$. There is a unique $\widetilde{u} \in \mathcal{L}_f(E,F)'$ such that*

$$\widetilde{u}(\langle \cdot, x' \rangle y) = \langle uy, x' \rangle$$

for every $(x', y) \in E' \times F$. Moreover,

$$\|\widetilde{u}\| \le \|u\|_1 \,.$$

Let $((x'_\iota, y_\iota))_{\iota\in I}$ be a finite family in $E' \times F$ such that

$$\sum_{\iota\in I}\langle\cdot, x'_\iota\rangle y_\iota = 0\,.$$

We show that

$$\sum_{\iota\in I}\langle uy_\iota, x'_\iota\rangle = 0\,.$$

Since the map

$$F \longrightarrow \mathbb{K}\,, \quad y \longmapsto \langle uy, x'\rangle$$

is linear for every $x' \in E'$, we may assume $(y_\iota)_{\iota\in I}$ to be linearly independent. Since

$$\sum_{\iota\in I}\langle x, x'_\iota\rangle y_\iota = 0$$

for every $x \in E$, it follows

$$\langle x, x'_\iota\rangle = 0$$

for every $x \in E$ and $\iota \in I$. Hence $x'_\iota = 0$ for every $\iota \in I$ and

$$\sum_{\iota\in I}\langle uy_\iota, x'_\iota\rangle = 0\,.$$

By the above proof (and Corollary 1.3.3.4), the map

$$\tilde{u} : \mathcal{L}_f(E,F) \longrightarrow \mathbb{K}\,, \quad \sum_{\iota\in I}\langle\cdot, x'_\iota\rangle y_\iota \longmapsto \sum_{\iota\in I}\langle uy_\iota, x'_\iota\rangle$$

is well–defined; it is obviously linear. Let $((x'', y'_\lambda))_{\lambda\in L}$ be a family in $E'' \times F'$ such that

$$\sum_{\lambda\in L}\|x''_\lambda\|\,\|y'_\lambda\| < \infty$$

and

$$uy = \sum_{\lambda\in L}\langle y, y'_\lambda\rangle x''$$

for every $y \in F$. Let $((x'_\iota, y_\iota))_{\iota\in I}$ be a finite family in $E' \times F$. Then

$$\left|\tilde{u}\Big(\sum_{\iota\in I}\langle\cdot, x'_\iota\rangle y_\iota\Big)\right| = \left|\sum_{\iota\in I}\langle uy_\iota, x'_\iota\rangle\right| =$$

$$= \left| \sum_{\iota \in I} \langle \sum_{\lambda \in L} \langle y_\iota, y'_\lambda \rangle x''_\lambda, x'_\iota \rangle \right| = \left| \sum_{\iota \in I} \sum_{\lambda \in L} \langle y_\iota, y'_\lambda \rangle \langle x''_\lambda, x'_\iota \rangle \right| =$$

$$= \left| \sum_{\lambda \in L} \sum_{\iota \in I} \langle y_\iota, y'_\lambda \rangle \langle x''_\lambda, x'_\iota \rangle \right| = \left| \sum_{\lambda \in L} \left\langle x''_\lambda, \sum_{\iota \in I} \langle y_\iota, y'_\lambda \rangle x'_\iota \right\rangle \right| \leq$$

$$\leq \sum_{\lambda \in L} \left| \langle x''_\lambda, \sum_{\iota \in I} \langle y_\iota, y'_\lambda \rangle x'_\iota \rangle \right| \leq \sum_{\lambda \in L} \|x''_\lambda\| \left\| \sum_{\iota \in I} \langle y_\iota, y'_\lambda \rangle x'_\iota \right\| =$$

$$= \sum_{\lambda \in L} \|x'_\lambda\| \left\| \sum_{\iota \in I} \langle y'_\lambda, j_F y_\iota \rangle x'_\iota \right\| =$$

$$= \sum_{\lambda \in L} \|x'_\lambda\| \left\| \Big(\sum_{\iota \in I} \langle \cdot, j_F y_\iota \rangle x'_\iota \Big) y'_\lambda \right\| \leq$$

$$\leq \sum_{\lambda \in L} \|x''_\lambda\| \, \|y'_\lambda\| \, \| \sum_{\iota \in I} \langle \cdot, j_F y_\iota \rangle x'_\iota \| = \sum_{\lambda \in L} \|x''_\lambda\| \, \|y'_\lambda\| \left\| \sum_{\iota \in I} \langle \cdot, x'_\iota \rangle y_\iota \right\|$$

by Corollary 1.3.6.6 and Theorem 1.3.4.2 b). Hence $\widetilde{u} \in \mathcal{L}_f(E,F)'$ and

$$\|\widetilde{u}\| \leq \sum_{\lambda \in L} \|x''\| \, \|y'_\lambda\| \, .$$

It follows that

$$\|\widetilde{u}\| \leq \|u\|_1 \, . \qquad \blacksquare$$

1.6.2 Examples

Example 1.6.2.1 *Take* $\mathbb{K}^2$ *with the Euclidean norm. Take* $u \in \mathcal{L}(\mathbb{K}^2)$ *and let*

$$\begin{bmatrix} \alpha & \gamma \\ \beta & \delta \end{bmatrix}$$

be the matrix associated to u *. Let*

$$\Lambda^2 := |\alpha|^2 + |\beta|^2 + |\gamma|^2 + |\delta|^2 ,$$

and

$$\Delta := |\alpha\delta - \beta\gamma| .$$

Then

$$\|u\|_1^2 = \Lambda^2 + 2\Delta .$$

Assume that $\Delta \neq 0$ *and choose* $\theta \in \mathbb{R}$ *such that*

$$\alpha\delta - \beta\gamma = \Delta e^{i\theta} .$$

Let v *be the operator on* $\mathbb{K}^2$ *associated to the matrix*

$$\frac{1}{\sqrt{\Lambda^2 + 2\Delta}} \begin{bmatrix} \overline{\alpha} + \delta e^{-i\theta} & \overline{\beta} - \delta e^{-i\theta} \\ \overline{\gamma} - \beta e^{-i\theta} & \overline{\delta} + \alpha e^{-i\theta} \end{bmatrix} .$$

Then

$$\|v\| = 1$$

and the following are equivalent for every $w \in \mathcal{L}(\mathbb{K}^2)^\#$ *:*

a) $w = v$.

b) *If* $\widetilde{w}$ *denotes the element of* $\mathcal{L}^1(\mathbb{K}^2)'$ *defined in Theorem 1.6.1.12 a), then*

$$\widetilde{w}(u) = \|u\|_1 .$$

Case 1 $\Delta = 0$

In this case $\operatorname{Im} u$ is one–dimensional and

$$\|u\|_1^2 = \|u\|^2 = \frac{1}{2}\left(\Lambda^2 + \sqrt{\Lambda^4 - 4\Delta^2}\right) = \Lambda^2 = \Lambda^2 + 2\Delta$$

(Corollary 1.6.1.4, Example 1.2.2.8).

Case 2 $\quad \Delta \neq 0$

By Theorem 1.6.1.12 b) (and Corollary 1.3.3.8 b), Corollary 1.2.1.6),

$$\|u\|_1 = \sup \{\mathrm{re}\langle u, w\rangle \mid w \in \mathcal{L}(\mathbb{K}^2),\ \|w\| = 1\},$$

where $\mathcal{L}(\mathbb{K}^2)$ is identified with $\mathcal{L}^1(\mathbb{K}^2)'$. Take $w \in \mathcal{L}(\mathbb{K}^2)$ with $\|w\| = 1$, and let

$$\begin{bmatrix} a & b \\ c & d \end{bmatrix}$$

by the matrix associated to w. Set

$$A^2 := |a|^2 + |b|^2 + |c|^2 + |d|^2, \quad D = ad - bc.$$

Then

$$1 = \frac{1}{2}\left(A^2 + \sqrt{A^4 - 4|D|^2}\right)$$

(Example 1.2.2.8) and this is equivalent to

$$A^2 - |D|^2 = 1.$$

Since

$$\langle u, w\rangle = \alpha a + \beta b + \gamma c + \delta d,$$

we have to find the supremum of

$$\mathrm{re}\,(\alpha a + \beta b + \gamma c + \delta d) = \frac{1}{2}(\alpha a + \beta b + \gamma c + \delta d + \overline{\alpha a} + \overline{\beta b} + \overline{\gamma c} + \overline{\delta d})$$

in the variables a, b, c, d on

$$A^2 - |D|^2 = 1.$$

We first show that the supremum is not attained at a regular point (a, b, c, d) of

$$A^2 - |D|^2 = 1.$$

Assume the contrary. Then there is a $\lambda \in \mathbb{R}$ with

$$\alpha = \frac{\lambda}{2}\frac{\partial(A^2 - |D|^2)}{\partial a} = \lambda(\overline{a} - \overline{D}d)\,,$$

$$\beta = \frac{\lambda}{2}\frac{\partial(A^2 - |D|^2)}{\partial b} = \lambda(\overline{b} + \overline{D}c)\,,$$

$$\gamma = \frac{\lambda}{2}\frac{\partial(A^2 - |D|^2)}{\partial c} = \lambda(\overline{c} + \overline{D}b)\,,$$

$$\delta = \frac{\lambda}{2}\frac{\partial(A^2 - |D|^2)}{\partial d} = \lambda(\overline{d} - \overline{D}a)\,,$$

and we get the contradiction that

$$0 \neq \alpha\delta - \beta\gamma = \lambda^2\Big(\overline{ad} - \overline{D}|a|^2 - \overline{D}|d|^2 + \overline{D}^2 ad - \overline{bc} - \overline{D}|b|^2 - \overline{D}|c|^2 - \overline{D}^2 bc\Big) =$$

$$= \lambda^2\Big(\overline{D} - \overline{D}A^2 + \overline{D}^2 D\Big) = \lambda^2\overline{D}\Big(1 - A^2 + |D|^2\Big) = 0\,.$$

Hence the supremum is attained at a singular point (a, b, c, d) of

$$A^2 - |D|^2 = 1\,.$$

Now

$$0 = \frac{\partial(A^2 - |D|^2)}{\partial a} = \overline{a} - \overline{D}d\,,$$

$$0 = \frac{\partial(A^2 - |D|^2)}{\partial b} = \overline{b} + \overline{D}c\,,$$

$$0 = \frac{\partial(A^2 - |D|^2)}{\partial c} = \overline{c} + \overline{D}b\,,$$

$$0 = \frac{\partial(A^2 - |D|^2)}{\partial d} = \overline{d} - \overline{D}a\,,$$

so that

$$\overline{D} = \overline{ad} - \overline{bc} = \overline{D}^2(ad - bc) = \overline{D}|D|^2\,.$$

Since obviously $D \neq 0$,

$$|D| = 1\,, \quad A^2 = 1 + |D|^2 = 2\,.$$

Let

$$D = e^{i\tau}$$

with $\tau \in \mathbb{R}$. Then

$$\overline{c} = -be^{-i\tau}\,, \quad \overline{d} = ae^{-i\tau}\,,$$

$$|a|^2 + |b|^2 = \frac{1}{2}A^2 = 1\,.$$

Hence we have to find the supremum of

$$M := \mathrm{re}\,(\alpha a + \beta b + \gamma c + \delta d) = \frac{1}{2}\,(\alpha a + \beta b + \gamma c + \delta d + \overline{\alpha a} + \overline{\beta b} + \overline{\gamma c} + \overline{\delta d}) =$$

$$= \frac{1}{2}\,(\alpha a + \beta b - \gamma \overline{b} e^{i\tau} + \delta \overline{a} e^{i\tau} + \overline{\alpha a} + \overline{\beta b} - \overline{\gamma} b e^{-i\tau} + \overline{\delta} a e^{-i\tau})$$

on

$$|a|^2 + |b|^2 = 1\,.$$

There is a $\lambda \in \mathbb{R}$ so that

$$\lambda \overline{a} = \lambda \frac{\partial(|a|^2 + |b|^2)}{\partial a} = 2\frac{\partial M}{\partial a} = \alpha + \overline{\delta} e^{-i\tau}$$

$$\lambda \overline{b} = \lambda \frac{\partial(|a|^2 + |b|^2)}{\partial b} = 2\frac{\partial M}{\partial b} = \beta - \overline{\gamma} e^{-i\tau}\,.$$

Hence

$$\lambda^2 = \lambda^2(|a|^2 + |b|^2) = |\alpha|^2 + |\delta|^2 + 2\mathrm{re}\,\alpha\delta e^{i\tau} + |\beta|^2 + |\gamma|^2 - 2\mathrm{re}\,\beta\delta e^{i\tau} =$$

$$= \Lambda^2 + 2\mathrm{re}\,\Delta e^{i(\theta+\tau)}\,,$$

$$\lambda M = \mathrm{re}\left(\alpha(\overline{\alpha} + \delta e^{i\tau}) + \beta(\overline{\beta} - \gamma e^{i\tau}) - \gamma(\beta - \overline{\gamma} e^{-i\tau})e^{i\tau} + \delta(\alpha + \overline{\delta} e^{-i\tau})e^{i\tau}\right) =$$

$$= \mathrm{re}\left(|\alpha|^2 + \alpha\delta e^{i\tau} + |\beta|^2 - \beta\gamma e^{i\tau} - \gamma\beta e^{i\tau} + |\gamma|^2 + \delta\alpha e^{i\tau} + |\delta|^2\right) =$$

$$= \mathrm{re}\,(\Lambda^2 + 2\Delta e^{i(\theta+\tau)}) = \lambda^2\,,$$

$$M = \lambda\,.$$

Hence

$$\tau = -\theta\,.$$

$$\|u\|_1^2 = \Lambda^2 + 2\Delta\,,$$

$$w = \frac{1}{(\Lambda^2+2\Delta)^{\frac{1}{2}}}\begin{bmatrix} \overline{\alpha}+\delta e^{-i\theta} & \overline{\beta}-\gamma e^{-i\theta} \\ \overline{\gamma}-\beta e^{-i\theta} & \overline{\delta}+\alpha e^{-i\theta} \end{bmatrix} = v\,. \qquad \blacksquare$$

Proposition 1.6.2.2 (1) (3) *Let* S, T *be sets and take* $p \in [1,\infty] \cup \{0\}$. *Given* $k \in \ell^{1,p}(T,S)$, *set*

$$\overset{\cup}{k} : \ell^\infty(T) \longrightarrow \ell^p(S)\,, \quad x \longmapsto \overset{\cup}{k}x$$

(Proposition 1.2.3.8 a)). Then

$$\overset{\cup}{k} \in \mathcal{L}^1(\ell^\infty(T), \ell^p(S))\,, \quad \|\overset{\cup}{k} \mid c_0(T)\|_1 = \|\overset{\cup}{k}\|_1 = \|k\|$$

for every $k \in \ell^{1,p}(T,S)$ *and the map*

$$\ell^{1,p}(T,S) \longrightarrow \mathcal{L}^1\Big(c_0(T), \ell^p(S)\Big)\,, \quad k \longrightarrow \overset{\cup}{k} \mid c_0(T)$$

is an isometry.

$$\text{Step 1} \quad u \in \mathcal{L}^1\Big(c_0(T)\,,\, \ell^p(S)\Big) \Longrightarrow \begin{cases} \exists! k \in \ell^{1,p}(T,S)\,, \\ \overset{\cup}{k} \mid c_0(T) = u\,,\ \|k\| \le \|u\|_1\,. \end{cases}$$

Let $((x'_\iota, y_\iota))_{\iota\in I}$ be a family in $c_0(T)' \times \ell^p(S)$ such that

$$\sum_{\iota\in I} \|x'_\iota\|\,\|y_\iota\| < \infty$$

and

$$ux = \sum_{\iota\in I} \langle x, x'_\iota\rangle y_\iota$$

for every $x \in c_0(T)$. We identify $c_0(T)'$ canonically with $\ell^1(T)$ (Example 1.2.2.3 e)) and set

$$k : T \times S \longrightarrow \mathbb{K}\,, \quad (t,s) \longmapsto (ue_t^T)(s)\,.$$

Let q be the conjugate exponent of p. Take $t \in T$. Then

$$\Big|\sum_{s\in S} y'(s)k(t,s)\Big| = \Big|\sum_{s\in S} y'(s)(ue_t^T)(s)\Big| = \Big|\sum_{s\in S} y'(s)\Big(\sum_{\iota\in I} x'_\iota(t)y_\iota(s)\Big)\Big| =$$

$$= \Big|\sum_{\iota\in I} x'_\iota(t)\Big(\sum_{s\in S} y_\iota(s)y'(s)\Big)\Big| \le \sum_{\iota\in I} |x'_\iota(t)|\,\Big|\sum_{s\in S} y_\iota(s)y'(s)\Big| \le$$

$$\le \sum_{\iota\in I} |x'_\iota(t)|\,\|y_\iota\|\,\|y'\|_q = \|y'\|_q \sum_{\iota\in I} |x'_\iota(t)|\|y_\iota\|$$

for every $y' \in \mathbb{K}^{(S)}$ so that $k(t,\cdot) \in \ell^p(S)$ and

$$\|k(t,\cdot)\|_p \le \sum_{\iota\in I} |x'_\iota(t)|\,\|y_\iota\|$$

(Proposition 1.2.2.2). Now

$$\sum_{t\in T} \|k(t,\cdot)\|_p \le \sum_{t\in T}\Big(\sum_{\iota\in I} |x'_\iota(t)|\,\|y_\iota\|\Big) =$$

$$= \sum_{\iota\in I} \|y_\iota\|\Big(\sum_{t\in T} |x'_\iota(t)|\Big) = \sum_{\iota\in I} \|y_\iota\|\,\|x'_\iota\|\,.$$

Hence $k \in \ell^{1,p}(T,S)$ and

$$\|k\| \le \|u\|_1\,.$$

Given $t \in T$,

$$\overset{\cup}{k} e_t^T = k(t,\cdot) = ue_t^T\,,$$

so that

$$\overset{\cup}{k} x = ux$$

for every $x \in \mathbb{K}^{(T)}$. Since $\mathbb{K}^{(T)}$ is dense in $c_0(T)$ (Proposition 1.1.2.6 c)),

$$\overset{\cup}{k}|c_0(T) = u\,.$$

The uniqueness of k is obvious.

Step 2 $\quad k \in \ell^{1,p}(T,S) \Longrightarrow \begin{cases} \overset{\cup}{k} \in \mathcal{L}^1(1^\infty(T), \ell^p(S))\,, \\ \|\overset{\cup}{k}|c_0(T)\|_1 = \|\overset{\cup}{k}\|_1 = \|k\|\,. \end{cases}$

Given $t \in T$, take

$$x'_t : \ell^\infty(T) \longrightarrow \mathbb{K}\,, \quad x \longmapsto x(t)\,,$$

$$y_t := k(t, \cdot)\,.$$

Then $x'_t \in \ell^\infty(T)'$ and $y_t \in \ell^p(S)$ for every $t \in T$,

$$\sum_{t \in T} \|x'_t\| \, \|y_t\|_p = \|k\| < \infty\,,$$

and

$$\overset{\cup}{k} x = \sum_{t \in T} k(t, \cdot) x(t) = \sum_{t \in T} \langle x, x'_t \rangle y_t$$

for every $x \in \ell^\infty(T)$. Hence

$$\overset{\cup}{k} \in \mathcal{L}^1(\ell^\infty(T), \ell^p(S))$$

and

$$\|\overset{\cup}{k}\|_1 \le \|k\|\,.$$

By Step 1 (and Proposition 1.6.1.5),

$$\|k\| \le \|\overset{\cup}{k} \mid c_0(T)\|_1 \le \|\overset{\cup}{k}\|_1\,.$$

Hence

$$\|\overset{\cup}{k} | c_0(T)\|_1 = \|\overset{\cup}{k}\|_1 = \|k\|\,.$$ ■

Corollary 1.6.2.3 *Let S, T be sets. Take $p \in [1, \infty[$ and let q be the conjugate exponent of p.*

a) *$(h(t,s)(ue_t^T)(s))_{(t,s) \in T \times S}$ is summable for every $h \in \ell^{\infty,q}(T,S)$ and $u \in \mathcal{L}^1(c_0(T), \ell^p(S))$. Given $h \in \ell^{\infty,q}(T,S)$, let*

$$\widetilde{h} : \mathcal{L}^1(c_0(T)\,, \ell^p(S)) \longrightarrow \mathbb{K}\,, \quad u \longmapsto \sum_{(t,s) \in T \times S} h(t,s)(ue_t^T)(s)\,.$$

b) *$\widetilde{h} \in \mathcal{L}^1(c_0(T)\,, \ell^p(S))'$ for every $h \in \ell^{\infty,q}(T,S)$ and the map*

$$\ell^{\infty,q}(T,S) \longrightarrow \mathcal{L}^1(c_0(T)\,, \ell^p(S))'\,, \quad h \longmapsto \widetilde{h}$$

is an isometry.

By Proposition 1.6.2.2, the map

$$\overset{\cup}{k} : c_0(T) \longrightarrow \ell^p(S)\,, \quad x \longmapsto \overset{\cup}{k}x$$

is in $\mathcal{L}^1(c_0(T)\,,\,\ell^p(S))$ for every $k \in \ell^{1,p}(T,S)$ and the map

$$\ell^{1,p}(T,S) \longrightarrow \mathcal{L}^1(c_0(T)\,,\,\ell^p(S))\,, \quad k \longmapsto \overset{\cup}{k}$$

is an isometry. We identify $\ell^{1,p}(T,S)$ with $\mathcal{L}^1(c_0(T)\,,\,\ell^p(S))$ by means of this isometry.

By Proposition 1.2.3.6, $(h(t,s)k(t,s))_{(t,s)\in T\times S}$ is summable for every $k \in \ell^{1,p}(T,S)$ and $h \in \ell^{\infty,q}(T,S)$, the map

$$\widetilde{h} : \ell^{1,p}(T,S) \longrightarrow \mathbb{K}\,, \quad k \longmapsto \sum_{(t,s)\in T\times S} h(t,s)k(t,s)$$

belongs to $\ell^{1,p}(T,S)'$ for every $h \in \ell^{\infty,p}(T,S)$, and the map

$$\ell^{\infty,q}(T,S) \longrightarrow \ell^{1,p}(T,S)'\,, \quad h \longmapsto \widetilde{h}$$

is an isometry. This proves the corollary. ■

Proposition 1.6.2.4 (1) (3) *Let S,T be sets and let p,q be conjugate exponents. Given $k \in \ell^{1,q}(S,T)$, let*

$$\overset{\cap}{k} : \ell^p(T) \longrightarrow \ell^1(S)\,, \quad x \longmapsto \overset{\cap}{k}x$$

(Proposition 1.2.3.4 a)).

a) *If $p \neq \infty$, then $\overset{\cap}{k} \in \mathcal{L}^1(\ell^p(T),\ell^1(S))$ for every $k \in \ell^{1,q}(S,T)$ and the map*

$$\ell^{1,q}(S,T) \longrightarrow \mathcal{L}^1(\ell^p(T),\ell^1(S))\,, \quad k \longmapsto \overset{\cap}{k}$$

is an isometry.

b) *If $p = \infty$, then $\overset{\cap}{k} \in \mathcal{L}^1(\ell^\infty(T)\,,\,\ell^1(S))$,*

$$\|\overset{\cap}{k}|c_0(T)\|_1 = \|\overset{\cap}{k}\|_1 = \|k\|$$

for every $k \in \ell^{1,1}(S,T)$ and the map

$$\ell^{1,1}(S,T) \longrightarrow \mathcal{L}^1(c_0(T)\,,\,\ell^1(S))\,, \quad k \longmapsto \overset{\cap}{k} \mid c_0(T)$$

is an isometry.

Put

$$E := \begin{cases} \ell^p(T) & \text{if} \quad p \neq \infty\,, \\ c_0(T) & \text{if} \quad p = \infty\,. \end{cases}$$

Step 1 $\quad u \in \mathcal{L}^1(E, \ell^1(S)) \Longrightarrow \begin{cases} \exists\, k \in \ell^{1,q}(S,T)\,, \\ \overset{\frown}{k}|E = u\,, \quad \|k\| \leq \|u\|_1\,. \end{cases}$

Let $((x'_\iota, y_\iota))_{\iota \in I}$ be a family in $E' \times \ell^1(S)$ such that

$$\sum_{\iota \in I} \|x'_\iota\|\, \|y_\iota\| < \infty$$

and

$$ux = \sum_{\iota \in I} \langle x, x'_\iota \rangle y_\iota$$

for every $x \in E$. Identify E' with $\ell^q(T)$ (Example 1.2.2.3 d), e)) and set

$$k : S \times T \longrightarrow \mathbb{K}\,, \quad (s,t) \longmapsto (ue_t^T)(s)\,.$$

Then

$$\left|\sum_{t \in T} k(s,t)x(t)\right| = \left|\sum_{t \in T} (ue_t^T)(s)x(t)\right| = |(ux)(s)| =$$

$$= \left|\sum_{\iota \in I} \langle x, x'_\iota \rangle y_\iota(s)\right| \leq \sum_{\iota \in I} |\langle x, x'_\iota \rangle|\, |y_\iota(s)| \leq$$

$$\leq \sum_{\iota \in I} \|x\|_p \|x'_\iota\|\, |y_\iota(s)| =$$

$$= \|x\|_p \sum_{\iota \in I} \|x'_\iota\|\, |y_\iota(s)|$$

for every $x \in \mathbb{K}^{(T)}$ and $s \in S$. By Proposition 1.2.2.2, $k(s,\cdot) \in \ell^q(T)$ and

$$\|k(s,\cdot)\|_q \leq \sum_{\iota \in I} \|x'_\iota\|\, |y_\iota(s)|$$

for every $s \in S$. Hence $k \in \ell^{1,q}(S,T)$ and

$$\|k\| \leq \|u\|_1\,.$$

We have

$$\overset{\cap}{k}e_t^T = k(\cdot,t) = ue_t^T$$

for every $t \in T$ and so

$$\overset{\cap}{k}x = ux$$

for every $x \in \mathbb{K}^{(T)}$. Since $\mathbb{K}^{(T)}$ is dense in E (Proposition 1.1.2.6 c)), $\overset{\cap}{k}|E = u$.

The uniqueness of k is obvious.

$$\text{Step 2} \qquad k \in \ell^{1,q}(S,T) \Longrightarrow \begin{cases} \overset{\cap}{k} \in \mathcal{L}^1(\ell^p(T)\,\ell^1(S)), \\ \|\overset{\cap}{k}|E\|_1 = \|\overset{\cap}{k}\|_1 = \|k\|. \end{cases}$$

Given $s \in S$, let

$$x'_s := k(s,\cdot), \quad y_s := e_s^S .$$

Then $x'_s \in \ell^p(T)'$ and $y_s \in \ell^1(S)$ for every $s \in S$,

$$\sum_{s\in S} \|x'_s\|\,\|y_s\| = \sum_{s\in S} \|k(s,\cdot)\|_q = \|k\| < \infty$$

(Example 1.2.2.3 b)), and

$$\sum_{s\in S} \langle x, x'_s\rangle y_s = \sum_{s\in S}\Big(\sum_{t\in T} k(s,t)x(t)\Big)e_s^S = \sum_{s\in S}(\overset{\cap}{k}x)(s)e_s^S = \overset{\cap}{k}x$$

for every $x \in \ell^p(T)$ (Example 1.1.6.16). Hence

$$\overset{\cap}{k} \in \mathcal{L}^1\Big(\ell^p(T)\,,\,\ell^1(S)\Big)$$

and

$$\|\overset{\cap}{k}\|_1 \le \|k\| .$$

By Step 1 (and Proposition 1.6.1.5),

$$\|k\| \le \|\overset{\cap}{k} \mid E\|_1 \le \|\overset{\cap}{k}\|_1 .$$

Thus

$$\|\overset{\cap}{k} \mid E\|_1 = \|\overset{\cap}{k}\|_1 = \|k\| .$$

■

Corollary 1.6.2.5 (1) (3) *Let S,T be sets. Take $p \in]1,\infty[$ and let q be the conjugate exponent of p.*

a) *$(h(s,t)(ue_t^T)(s))_{(s,t)\in S\times T}$ is summable for every $h \in \ell^{\infty,q}(S,T)$ and $u \in \mathcal{L}^1(\ell^q(T),\ell^1(S))$. Given $h \in \ell^{\infty,q}(S,T)$, let*

$$\widetilde{h} : \mathcal{L}^1(\ell^q(T),\ell^1(S)) \longrightarrow \mathbb{K}, \quad u \longmapsto \sum_{(s,t)\in S\times T} h(s,t)(ue_t^T)(s).$$

b) *$\widetilde{h} \in \mathcal{L}^1(\ell^q(T), \ell^1(S))'$ for every $h \in \ell^{\infty,q}(S,T)$ and the map*

$$\ell^{\infty,q}(S,T) \longrightarrow \mathcal{L}^1(\ell^q(T),\ell^1(S))', \quad h \longmapsto \widetilde{h}$$

is an isometry.

By Proposition 1.6.2.4, the map

$$\overset{\cap}{k} : \ell^q(T) \longrightarrow \ell^1(S), \quad x \longmapsto \overset{\cap}{k}x$$

is in $\mathcal{L}^1(\ell^q(T),\ell^1(S))$ for every $k \in \ell^{1,p}(S,T)$ and the map

$$\ell^{1,p}(S,T) \longrightarrow \mathcal{L}^1(\ell^q(T),\ell^1(S)), \quad k \longmapsto \overset{\cap}{k}$$

is an isometry. We identify $\ell^{1,p}(S,T)$ with $\mathcal{L}^1(\ell^q(T),\ell^1(S))$ via this isometry.

By Proposition 1.2.3.6, $(h(s,t)k(s,t))_{(s,t)\in S\times T}$ is summable for every $k \in \ell^{1,p}(S,T)$ and $h \in \ell^{\infty,q}(S,T)$, the map

$$\widetilde{h} : \ell^{1,p}(S,T) \longrightarrow \mathbb{K}, \quad k \longmapsto \sum_{(s,t)\in S\times T} h(s,t)k(s,t)$$

belongs to $\ell^{1,p}(S,T)'$ for every $h \in \ell^{\infty,q}(S,T)$, and the map

$$\ell^{\infty,q}(S,T) \longrightarrow \ell^{1,p}(S,T)', \quad h \longmapsto \widetilde{h}$$

is an isometry. This proves the corollary. ■

Example 1.6.2.6 (3) *Let $(a)_{\iota\in I}$ be an absolutely summable family in the Banach space E and define*

$$u : \ell^\infty(I) \longrightarrow E, \quad x \longmapsto \sum_{\iota\in I} x(\iota)a_\iota.$$

Then u is nuclear and

$$\|u\|_1 \le \sum_{\iota\in I} \|a_\iota\|.$$

Given $\iota \in I$, define

$$x'_\iota : \ell^\infty(I) \longrightarrow \mathbb{K}\,, \quad x \longmapsto x(\iota)\ .$$

Then

$$\sum_{\iota \in I} \|x'_\iota\|\, \|a_\iota\| = \sum_{\iota \in I} \|a_\iota\|$$

and

$$ux = \sum_{\iota \in I} \langle x, x'_\iota \rangle a_\iota$$

for every $x \in \ell^\infty(I)$. ■

1.7 Ordered Banach spaces

C^*–algebras are equipped with an intrinsic order relation with respect to which they become ordered Banach spaces. Properties of ordered Banach spaces which are used later in the book are presented in this section. Because of its importance in the later study of W^*–algebras, order continuity is also studied here.

1.7.1 Ordered normed spaces

Definition 1.7.1.1 (0) *An **ordered vector space** is a vector space E with an order relation $\leq$ such that given any $x, y, z \in E$ and $\alpha \in \mathbb{R}_+$*

$$x + z \leq y + z \text{ and } \alpha x \leq \alpha y$$

whenever $x \leq y$. Defining

$$E_+ := \{x \in E \mid x \geq 0\},$$

*we call the elements of E_+ **positive** and the elements of $-E_+$ **negative**.*

Proposition 1.7.1.2 (0) *If E is an ordered vector space then E_+ is a sharp convex cone in E.*

Given $x, y \in E_+$ and $\alpha \in \mathbb{R}_+$,

$$x + y,\ \alpha x \in E_+,$$

so that E_+ is a convex cone in E. From

$$x \in E_+ \cap (-E_+),$$

it follows successively that

$$-x \in E_+, \qquad 0 \leq -x, \quad x \leq 0 \leq x, \qquad x = 0,$$

i.e. E_+ is sharp. ∎

Proposition 1.7.1.3 (0) *Let E be a vector space and C a sharp convex cone in E. Then there is a unique order relation $\leq$ on E, with respect to which E is an ordered vector space with*

$$E_+ = C.$$

Define the relation $\leq$ by

$$x \leq y :\Longleftrightarrow y - x \in C$$

for $x, y \in E$. Then, given $x, y, z \in E$ and $\alpha \in \mathbb{R}_+$,

$$x - x = 0 \in C\,, \quad x \leq x\,,$$

$$x \leq y\,,\, y \leq z \Longrightarrow y - x\,,\, z - y \in C \Longrightarrow$$

$$\Longrightarrow z - x = (z - y) + (y - x) \in C \Longrightarrow x \leq z\,,$$

$$x \leq y\,,\, y \leq x \Longrightarrow y - x\,,\, x - y \in C \Longrightarrow$$

$$\Longrightarrow y - x \in C \cap (-C) \Longrightarrow y - x = 0 \Longrightarrow x = y\,,$$

$$x \leq y \Longrightarrow (y + z) - (x + z) = y - x \in C \Longrightarrow x + z \leq y + z\,,$$

$$x \leq y \Longrightarrow y - x \in C \Longrightarrow \alpha y - \alpha x = \alpha(y - x) \in C \Longrightarrow \alpha x \leq \alpha y\,,$$

$$x \in C \Longleftrightarrow x - 0 \in C \Longleftrightarrow 0 \leq x\,.$$

Hence $\leq$ is an order relation on E with respect to which E is an ordered vector space such that

$$E_+ = C\,.$$

Let $\preceq$ be an order relation on E which renders E an ordered vector space with

$$E_+ = C\,.$$

Then

$$x \preceq y \Longleftrightarrow 0 \preceq y - x \Longleftrightarrow y - x \in C \Longleftrightarrow x \leq y$$

for every $x, y \in E$ and so $\leq$ coincides with $\preceq$. ■

Definition 1.7.1.4 (0) *An **ordered normed space** is an ordered vector space E endowed with a norm with respect to which E_+ is closed. In this case we put*

$$E_+^\# := E^\# \cap E_+ .$$

*An **ordered Banach space** is an ordered normed space with a complete norm.*

If T is a topological space, then $\mathcal{C}(T)$ endowed with the usual order is an ordered Banach space.

Proposition 1.7.1.5 (0) *Let E be an ordered normed space and let $(x_\iota)_{\iota\in I}$, $(y_\iota)_{\iota\in I}$ be families in E such that $x_\iota \le y_\iota$ for every $\iota \in I$.*

a) *If $\mathfrak{F}$ is a filter on I such that $\lim\limits_{\iota,\mathfrak{F}} x_\iota$ and $\lim\limits_{\iota,\mathfrak{F}} y_\iota$ exist, then*

$$\lim_{\iota,\mathfrak{F}} x_\iota \le \lim_{\iota,\mathfrak{F}} y_\iota .$$

b) *If $(x_\iota)_{\iota\in I}$ and $(y_\iota)_{\iota\in I}$ are summable then*

$$\sum_{\iota\in I} x_\iota \le \sum_{\iota\in I} y_\iota .$$

c) *If $(z_\iota)_{\iota\in I}$ is a summable family in E_+ then*

$$0 \le \sum_{\iota\in J} z_\iota \le \sum_{\iota\in I} z_\iota \in E_+$$

for every finite subset J of I.

a) $\lim\limits_{\iota,\mathfrak{F}} y_\iota - \lim\limits_{\iota,\mathfrak{F}} x_\iota = \lim\limits_{\iota,\mathfrak{F}} (y_\iota - x_\iota) \in E_+$.

b) By a),

$$\sum_{\iota\in I} x_\iota = \lim_{J,\mathfrak{F}_I} \sum_{\iota\in J} x_\iota \le \lim_{J,\mathfrak{F}_I} \sum_{\iota\in J} y_\iota = \sum_{\iota\in I} y_\iota .$$

c) follows from b). ■

Corollary 1.7.1.6 (0) *Let E be an ordered Banach space such that*

$$x, y \in E_+ ,\ x \le y \Longrightarrow \|x\| \le \|y\| .$$

a) *For all $x, y \in E$,*

$$-y \le x \le y \Longrightarrow \|x\| \le 3\|y\| .$$

b) *If* $(x_\iota)_{\iota\in I}$, $(y_\iota)_{\iota\in I}$ *are famlies in* E *such that* $(y_\iota)_{\iota\in I}$ *is summable and*

$$-y_\iota \leq x_\iota \leq y_\iota$$

for every $\iota \in I$ *then* $(x_\iota)_{\iota\in I}$ *is also summable and*

$$-\sum_{\iota\in I} y_\iota \leq \sum_{\iota\in I} x_\iota \leq \sum_{\iota\in I} y_\iota \,.$$

From

$$0 \leq x + y \leq 2y$$

it follows

$$\|x\| = \|x+y-y\| \leq \|x+y\| + \|y\| \leq \|2y\| + \|y\| = 3\|y\| \,.$$

b) Take $\varepsilon > 0$. There is a $J \in \mathfrak{P}_f(I)$ such that

$$\left\| \sum_{\iota\in K} y_\iota \right\| < \frac{\varepsilon}{3}$$

for every $K \in \mathfrak{P}_f(I\backslash J)$ (Proposition 1.1.6.6). By a),

$$\left\| \sum_{\iota\in K} x_\iota \right\| \leq 3 \left\| \sum_{\iota\in K} y_\iota \right\| < \varepsilon$$

for every $K \in \mathfrak{P}_f(I\backslash J)$. By Proposition 1.1.6.6), $(x_\iota)_{\iota\in I}$ is summable and by Proposition 1.7.1.5 b),

$$-\sum_{\iota\in I} y_\iota \leq \sum_{\iota\in I} x_\iota \leq \sum_{\iota\in I} y_\iota \,. \qquad \blacksquare$$

Corollary 1.7.1.7 (0) *Let* E *be a finite-dimensional ordered Banach space such that*

$$x, y \in E_+ \,,\; x \leq y \Longrightarrow \|x\| \leq \|y\| \,.$$

Then there is an $\alpha \in \mathbb{R}_+$ *such that*

$$\sum_{\iota\in I} \|x_\iota\| \leq \alpha \left\| \sum_{\iota\in I} x_\iota \right\|$$

for every summable family $(x_\iota)_{\iota\in I}$ *in* E_+.

Assume the contrary. Then for every $n \in \mathbb{N}$ there is a finite family $(x_{n,\iota})_{\iota \in I_n}$ in E_+ such that

$$\left\| \sum_{\iota \in I_n} x_{n,\iota} \right\| < \frac{1}{n^2}, \quad \sum_{\iota \in I_n} \|x_{n,\iota}\| \geq 1 .$$

Consider the family $(x_{n,\iota})_{n \in \mathbb{N}, \iota \in I_n}$. Take $\varepsilon > 0$ and $n_0 \in \mathbb{N}$ such that

$$\sum_{n=n_0}^{\infty} \frac{1}{n^2} < \varepsilon .$$

Let J be a finite subset of $\{(n,\iota) | n > n_0 , \iota \in I_n\}$ and put

$$p := \sup_{(n,\iota) \in J} n .$$

By Proposition 1.7.1.5 c),

$$0 \leq \sum_{(n,\iota) \in J} x_{n,\iota} \leq \sum_{n=n_0}^{p} \sum_{\iota \in I_n} x_{n,\iota} .$$

By our hypothesis about E,

$$\left\| \sum_{(n,\iota) \in J} x_{n,\iota} \right\| \leq \left\| \sum_{n=n_0}^{p} \sum_{\iota \in I_n} x_{n,\iota} \right\| \leq \sum_{n=n_0}^{p} \left\| \sum_{\iota \in I_n} x_{n,\iota} \right\| < \sum_{n=n_0}^{p} \frac{1}{n^2} < \varepsilon .$$

By Proposition 1.1.6.6, the family $(x_{n,\iota})_{n \in \mathbb{N}, \iota \in I_n}$ is summable. By Proposition 1.1.6.14, the family $(x_{n,\iota})_{n \in \mathbb{N}, \iota \in I_n}$ is absolutely summable, and we get the contradiction

$$\infty = \sum_{n \in \mathbb{N}} \sum_{\iota \in I_n} \|x_{n,\iota}\| < \infty .$$

■

Corollary 1.7.1.8 (0) *Let x' be a linear form on the ordered Banach space E such that $x'(E_+) \subset \mathbb{R}_+$.*

a) *$x'|E_+$ is continuous at 0.*

b) *If the absolutely convex hull of $E_+^{\#}$ is a 0–neighbourhood in E, then x' is continuous.*

a) Assume the contrary. Then for each $n \in \mathbb{N}$ there is an $x_n \in E_+$ with

$$\|x_n\| \leq \frac{1}{2^n}, \quad x'(x_n) \geq 1 .$$

Put

$$x := \sum_{n\in\mathbb{N}} x_n$$

(Corollary 1.1.6.10 a $\Rightarrow$ c). Then

$$\sum_{n=1}^{p} x_n \le x$$

(Proposition 1.7.1.5 c), so that

$$p \le \sum_{n=1}^{p} x'(x_n) = x'\Big(\sum_{n=1}^{p} x_n\Big) \le x'(x)$$

for every $p \in \mathbb{N}$ and this is a contradiction.

b) By a), x' is bounded on $E_+^\#$. Therefore it is bounded on the absolutely convex hull of $E_+^\#$. Hence x' is bounded on a 0–neighbourhood in E and so it must be continuous. ■

Proposition 1.7.1.9 *Let E, F be ordered normed spaces such that the vector subspace of E generated by E_+ is dense in E. Then there is a unique order relation on $\mathcal{L}(E,F)$ with respect to which $\mathcal{L}(E,F)$ is an ordered vector space and*

$$\mathcal{L}(E,F)_+ = \{u \in \mathcal{L}(E,F) \mid u(E_+) \subset F_+\} .$$

*$\mathcal{L}(E,F)_+$ is closed in the topology of pointwise convergence and so $\mathcal{L}(E,F)$ is an ordered normed space. The elements of $\mathcal{L}(E,\mathbb{K})_+$ will be called **positive linear forms on E**.*

Put

$$\mathcal{F} := \{u \in \mathcal{L}(E,F) \mid u(E_+) \subset F_+\} .$$

$\mathcal{F}$ is clearly a convex cone in $\mathcal{L}(E,F)$. Take

$$u \in \mathcal{F} \cap (-\mathcal{F}) .$$

Then

$$u(E_+) \subset F_+ \cap (-F_+) = \{0\} .$$

Thus u vanishes on the vector subspace of E generated by E_+. By hypothesis, $u = 0$ and $\mathcal{F}$ is sharp. The first claim now follows from Proposition 1.7.1.3.

The second claim is easy to prove. ■

1.7.2 Order Continuity

Definition 1.7.2.1 (0) *Let* A *be a subset of the ordered set* E. *An* ***upper (lower) bound for*** A ***in*** E *is an element* $x \in E$ *such that*

$$y \leq x \qquad (x \leq y)$$

for every $y \in A$. *If there is a least upper bound (greatest lower bound) for* A *in* E *then it is called* ***supremum (infimum)*** *of* A *in* E *and it is denoted by* $\overset{E}{\bigvee\limits_{x\in A}} x$ *or* $\bigvee\limits_{x\in A} x$ *(by* $\overset{E}{\bigwedge\limits_{x\in A}} x$ *or* $\bigwedge\limits_{x\in A} x$ *). If* $A = \{x, y\}$, *then we put*

$$x \vee y := x \overset{E}{\vee} y := \overset{E}{\bigvee_{z\in A}} z\,, \quad x \wedge y := x \overset{E}{\wedge} y := \overset{E}{\bigwedge_{z\in A}} z\,.$$

E *is called* ***complete*** *if every nonempty upward (downward) directed set of* E, *which is bounded above (below) has a supremum (infimum) in* E. E *is called* σ***–complete*** *if the previous conditions are fulfilled for countable subsets of* E. *If in addition* E *is a normed space, then we will use the expression* ***order complete*** *and* ***order*** σ***–complete*** *instead of complete and* σ*–complete, respectively, in order to avoid confusion.*

Let $(x_\iota)_{\iota\in I}$ *be a family in* E *and put*

$$A := \{x_\iota \mid \iota \in I\}\,.$$

We extend the above notions for $(x_\iota)_{\iota\in I}$ *by replacing it with* A *and define*

$$\bigvee_{\iota\in I} x_\iota := \overset{E}{\bigvee_{\iota\in I}} x_\iota := \overset{E}{\bigvee_{y\in A}} y\,, \quad \bigwedge_{\iota\in I} x_\iota := \overset{E}{\bigwedge_{\iota\in I}} x_\iota := \overset{E}{\bigwedge_{y\in A}} y\,.$$

E *is called a* ***lattice*** *(Dedekind, 1897) if any two elements of* E *have both a supremum and an infimum (or equivalently, if every finite subset of* E *has both a supremum and an infimum). A* ***vector lattice*** *is an ordered vector space which is a lattice. A* ***band*** *of an order complete vector lattice* E *is a vector subspace* F *of* E *such that:*

1) $x, y \in F \Rightarrow x \overset{E}{\vee} y \in F$.

2) $x \in E\,,\ y \in F\,,\ 0 \leq x \leq y \Rightarrow x \in F$.

3) *If* $(x_\iota)_{\iota\in I}$ *is an upward directed family in* F *then* $\overset{E}{\bigvee\limits_{\iota\in I}} x_\iota$ *belongs to* F *if it exists.*

Remark. An ordered set E is σ–complete iff every bounded increasing (decreasing) sequence in E has a supremum (infimum).

Proposition 1.7.2.2 (0) *Let E be an ordered Hausdorff vector space such that E_+ is closed and that the translations and multiplications with -1 are continuous (for example let E be an ordered normed space). Let A be a nonempty upward directed set of E, $\mathfrak{F}$ the upper section filter of A, and x a point of adherence of $\mathfrak{F}$. Then x is the supremum of A.*

Take $y \in A$. Then $\{z \in A \mid z \geq y\} \in \mathfrak{F}$ so that

$$x \in \overline{\{z \in A \mid z \geq y\}},$$

$$x - y \in \overline{\{z - y \mid z \in A,\, z \geq y\}} \subset \overline{E}_+ = E_+,$$

$$y \leq x.$$

Thus x is an upper bound for A.

Let y be an upper bound for A. Then

$$y - z \in E_+$$

for every $z \in A$ and so

$$y - x \in \overline{\{y - z \mid z \in A\}} \subset \overline{E}_+ = E_+.$$

$$x \leq y.$$

Hence x is the supremum of A. ∎

Definition 1.7.2.3 (0) *Let $u : E \to F$ be a linear map from the ordered vector space E to the normed space F. Let $\mathfrak{A}$ be the set of all downward directed (countable) sets of E whose infimum is 0. For each $A \in \mathfrak{A}$, let $\mathfrak{H}_A$ denote the lower section filter of A. Then u is said to be* ***order continuous (order σ–continuous)*** *if $u(\mathfrak{H}_A)$ converges to 0 for each $A \in \mathfrak{A}$. The vector subspace of $\mathbb{K}^E$ generated by the order continuous (order σ–continuous) positive linear forms on E is denoted by E^π (resp. E^σ).*

Every linear form of E^π (of E^σ) is order continuous (order σ–continuous).

Proposition 1.7.2.4 (0) *Let E be an ordered Banach space, F a normed space, and $u : E \to F$ an order σ–continuous linear map.*

a) *$u|E_+$ is continuous at 0.*

b) *If the absolutely convex hull of $E_+^\#$ is a 0–neighbourhood in E, then u is continuous.*

a) Assume that $u|E_+$ is not continuous at 0. Then for every $n \in \mathbb{N}$ there is an $x_n \in E_+$ with

$$\|x_n\| \leq \frac{1}{2^n}, \quad \|ux_n\| \geq 1 .$$

Given $p \in \mathbb{N}$, put

$$y_p := \sum_{n=p}^{\infty} x_n$$

(Corollary 1.1.6.10 a $\Rightarrow$ c). Then, for $p \in \mathbb{N}$,

$$\|y_p\| \leq \sum_{n=p}^{\infty} \|x_n\| < \sum_{n=p}^{\infty} \frac{1}{2^n} = \frac{1}{2^{p-1}} ,$$

$$1 \leq \|ux_p\| = \|uy_p - uy_{p+1}\| \leq \|uy_p\| + \|uy_{p+1}\| .$$

By Proposition 1.7.2.2

$$y_1 = \bigvee_{p \in \mathbb{N}} \left(\sum_{n=1}^{p} x_n \right) ,$$

so

$$\bigwedge_{p \in \mathbb{N}} y_p = 0 .$$

Since u is order σ–continuous, $(uy_p)_{p \in \mathbb{N}}$ converges to 0. This contradicts the above inequalities. Hence $u|E_+$ is continuous at 0.

b) By a), u is bounded on $E_+^\#$. By the hypothesis of b), u is bounded on a 0–neighbourhood in E. Hence u is continuous. ■

Proposition 1.7.2.5 (0) *Let E be an ordered Banach space such that*

$$\|x + y\| = \|x\| + \|y\|$$

for every $x, y \in E_+$. Let A be a nonempty upward directed subset of E_+. Then the following are equivalent:

a) *A is bounded above.*

b) *A is bounded in norm.*

c) *The upper section filter of A converges.*

d) *There is a point of adherence for the upper section filter of A.*

e) *A has a supremum.*

If these conditions are fulfilled, then the upper section filter of A converges to the supremum of A.

In particular, E is order complete and every operator defined on E is order continuous.

a $\Rightarrow$ b. Let y be an upper bound of A. By hypothesis

$$\|x\| \le \|y\|$$

for every $x \in A$, so that

$$\sup_{x \in A} \|x\| \le \|y\| .$$

b $\Rightarrow$ c. We set

$$\alpha := \sup_{x \in A} \|x\| < \infty .$$

Take $a \in A$ and suppose $x, y \in A$, $a \le x$, $a \le y$. Since A is upward directed, there is a $z \in A$ with

$$x \le z , \quad y \le z.$$

Then

$$\|x - y\| \le \|x - z\| + \|z - y\| = \|z\| - \|x\| + \|z\| - \|y\| \le 2(\alpha - \|a\|) .$$

It follows that the upper section filter of A is a Cauchy filter and it therefore converges.

d $\Rightarrow$ e and the last assertion follow from Proposition 1.7.2.2.

c $\Rightarrow$ d and e $\Rightarrow$ a are trivial. ∎

Proposition 1.7.2.6 (0) *Let E be a normed space. Assume that E' is an ordered vector space such that E'_+ is closed in E'_E. Then every norm bounded upward directed nonempty set A' of E'_+ has a supremum x' and the upper section filter of A' converges to x' in E'_E.*

We may assume that $A' \subset E'^{\#}$. Since $E'^{\#}_{E}$ is compact (Alaoglu–Bourbaki Theorem), the upper section filter $\mathfrak{F}$ of A' has a point of adherence x' in $E^{\#}_{E'}$. By Proposition 1.7.2.2, x' is the supremum of A'. In particular, x' is the unique point of adherence of $\mathfrak{F}$ in $E^{\#}_{E'}$. Using once again the fact that $E^{\#}_{E'}$ is compact, we deduce that $\mathfrak{F}$ converges to x' in E'_E. ∎

Proposition 1.7.2.7 *Let E be an ordered normed space whose norm is monotonic increasing (i.e. $\|x\| \le \|y\|$ whenever $x, y \in E_+$ and $x \le y$). Let F be a normed space and define $\mathcal{F}$ by*

$$\mathcal{F} := \{u \in \mathcal{L}(E,F) \mid u \textit{ is order continuous (order } \sigma\textit{-continuous)}\}.$$

Then $\mathcal{F}$ is a closed vector subspace of $\mathcal{L}(E,F)$.

It is easy to see that $\mathcal{F}$ is a vector subspace of $\mathcal{L}(E,F)$. Take $u \in \overline{\mathcal{F}}$ and let A be a downward directed (countable) subset of E with infimum 0. Take $\varepsilon > 0$ and $x_0 \in A$. There is a $v \in \mathcal{F}$ such that

$$\|u - v\| < \frac{\varepsilon}{2(1 + \|x_0\|)}.$$

Further there is an $x \in A$ such that

$$\|vy\| < \frac{\varepsilon}{2}$$

for every $y \in A$ with $y \le x$. Then

$$\|uy\| \le \|(u-v)y\| + \|vy\| \le \frac{\varepsilon \|y\|}{2(1+\|x_0\|)} + \frac{\varepsilon}{2} < \varepsilon$$

for $y \in A$ with $y \le x$ and $y \le x_0$. Hence $u(\mathfrak{G}_A)$ converges to 0 and $u \in \mathcal{F}$. ∎

Example 1.7.2.8 *Let T be a set and take $p \in \{0\} \cup [1, \infty]$.*

a) *There is a unique order relation on $\ell^p(T)$ such that $\ell^p(T)$ is an ordered vector space and*

$$\ell^p(T)_+ = \{x \in \ell^p(T) \mid x(T) \subset \mathbb{R}_+\}.$$

b) *$\ell^p(T)$ is an order complete vector lattice and an ordered Banach space.*

c) *Every order σ-continuous linear map of $\ell^p(T)$ into a normed space is continuous.*

d) *If $x, y \in \ell^1(T)_+$, then*

$$\|x + y\|_1 = \|x\|_1 + \|y\|_1.$$

a) follows from Proposition 1.7.1.3.

b) and d) are easy to see.

c) follows from Proposition 1.7.2.4 b). ■

Example 1.7.2.9 *Let T be a set. Take $p \in \{0\} \cup [1, \infty]$, and let q be the conjugate exponent of p. Given $x \in \ell^q(T)$, define*

$$\widetilde{x} : \ell^p(T) \longrightarrow \mathbb{K}, \quad y \longmapsto \sum_{t \in T} x(t)y(t)$$

(Example 1.2.2.3 a)).

a) *Given $x \in \ell^q(T)$, the linear form $\widetilde{x}$ is positive iff x is positive (Example 1.7.2.8 a)).*

b) *$\widetilde{x} \in \ell^p(T)^\pi$, whenever $x \in \ell^q(T)$.*

c) *$\ell^p(T)^\pi = \{x' \mid x'$ is an order continuous linear form on $\ell^p(T)\}$,*

$\ell^p(T)^\sigma = \{x' \mid x'$ is an order σ-continuous linear form on $\ell^p(T)\}$.

d) *The map*

$$\ell^q(T) \longrightarrow \ell^p(T)^\pi, \quad x \longmapsto \widetilde{x}$$

is an isometry of ordered Banach spaces (Proposition 1.7.1.9, Proposition 1.7.2.7).

a) is easy to see.

b) Since every element of $\ell^q(T)$ is a linear combination of positive elements it is sufficient to show that $\widetilde{x}$ is order continuous for every $x \in \ell^q(T)$.

Case 1 $\quad p = 1$

The assertion follows from the last assertion of Proposition 1.7.2.5 and Example 1.7.2.8 d)).

Case 2 $\quad p \neq 1$

It is obvious that $\widetilde{x}$ is order continuous if $x \in \mathbb{K}^{(T)}$. Since $\mathbb{K}^{(T)}$ is a dense set of $\ell^q(T)$ (Proposition 1.1.2.6 c)), it follows from Proposition 1.7.2.7 (and Example 1.2.2.3 b)), that $\widetilde{x}$ is order continuous for $x \in \ell^q(T)$.

c) follows from the fact that $\ell^p(T)$ is a vector lattice.

d) Case 1 $\quad p \neq \infty$

By Example 1.2.2.3 d), the map

$$\ell^q(T) \longrightarrow \ell^p(T)', \quad x \longmapsto \widetilde{x}$$

is an isometry of Banach spaces and so the assertion follows from a) and b).

Case 2 $\quad p = \infty$

Let x' be a positive order continuous linear form on $\ell^\infty(T)$. We define

$$x : T \longrightarrow \mathbb{K}, \quad t \longmapsto x'(e_t) .$$

Then

$$\sum_{t\in A} x(t) = \sum_{t\in A} x'(e_t) = x(e_A) \leq x(e_T)$$

for every $A \in \mathfrak{P}_f(T)$ and so $x \in \ell^1(T)$. Take $y \in \ell^\infty(T)_+$. Then

$$\{ye_{T\setminus A} \mid A \in \mathfrak{P}_f(T)\}$$

is a downward directed set of $\ell^\infty(T)$ with infimum 0. Thus

$$0 = \inf_{A\in\mathfrak{P}_f(T)} x'(ye_{T\setminus A}) = x'(y) - \sup_{A\in\mathfrak{P}_f(T)} \widetilde{x}(ye_A),$$

$$x'(y) = \sup_{A\in\mathfrak{P}_f(T)} x'(ye_A) = \sup_{A\in\mathfrak{P}_f(T)} \widetilde{x}(ye_A) = \widetilde{x}(y) .$$

Since y is arbitrary and every element of $\ell^\infty(T)$ is a linear combination of positive elements of $\ell^\infty(T)$, we see that $x' = \widetilde{x}$. Hence the map

$$\ell^1(T) \longrightarrow \ell^\infty(T)^\pi, \quad x \longmapsto \widetilde{x}$$

is surjective. By a) and Example 1.2.2.3 b), the above map is an isometry of ordered Banach spaces. ■

Remark. Let $\mathfrak{F}$ be a free ultrafilter on T closed under countable intersections. Then the linear map

$$\ell^\infty(T) \longrightarrow \mathbb{K}, \quad x \longmapsto \lim x(\mathfrak{F})$$

is positive and order σ–continuous but not order continuous.

Let ω_1 be the first uncountable ordinal number endowed with the usual topology and let $\mathfrak{F}$ be the filter on ω_1 of the compelements of the countable subsets of ω_1. Then the map

$$\mathcal{C}(\omega_1) \longrightarrow \mathbb{K}, \quad x \longmapsto \lim x(\mathfrak{F})$$

is positive and order σ–continuous but not order continuous.

Proposition 1.7.2.10 *Let E be a σ-complete ordered vector space. The countable family $(x_\iota)_{\iota\in I}$ in E_+ is called* ***order summable*** *if*

$$\{\sum_{\iota\in J} x_\iota \mid J \in \mathfrak{P}_f(I)\}$$

is bounded above. In this case we define

$$\sum_{\iota\in I}^{\leq} x_\iota := \bigvee\{\sum_{\iota\in J} x_\iota \mid J \in \mathfrak{P}_f(I)\} .$$

Then, given a linear form x' on E, the following are equivalent:

a) *x' is order σ-continuous.*

b) *given an order summable countable family $(x_\iota)_{\iota\in I}$ in E_+, the family $(x'(x_\iota))_{\iota\in I}$ is summable and*

$$\sum_{\iota\in I} x'(x_\iota) = x'\Big(\sum_{\iota\in I}^{\leq} x_\iota\Big) .$$

a $\Rightarrow$ b. We may assume I to be infinite. Let $\varphi : \mathbb{N} \to I$ be a bijective map. Then $\Big(\sum_{\iota\in I}^{\leq} x_\iota - \sum_{k=1}^{n} x_{\varphi(k)}\Big)_{n\in\mathbb{N}}$ is a decreasing sequence in E with infimum 0. Thus

$$0 = \lim_{n\to\infty} x'\Big(\sum_{\iota\in I}^{\leq} x_\iota - \sum_{k=1}^{n} x_{\varphi(k)}\Big) = x'\Big(\sum_{\iota\in I}^{\leq} x_\iota\Big) - \lim_{n\to\infty} \sum_{k=1}^{n} x'\Big(x_{\varphi(k)}\Big) ,$$

$$\lim_{n\to\infty} \sum_{k=1}^{n} x'\Big(x_{\varphi(k)}\Big) = x'\Big(\sum_{\iota\in I}^{\leq} x_\iota\Big)$$

and the assertion follows from Proposition 1.1.6.14 c $\Rightarrow$ a.

b $\Rightarrow$ a. Let A be a downward directed countable set of E with infimum 0 and let $\mathfrak{F}$ be its lower section filter. Assume that $x'(\mathfrak{F})$ does not converge to 0. Then there is an $\varepsilon > 0$ such that for every $x \in A$ there is some $y \in A$ such that $y \leq x$ and

$$|x'(y)| > \varepsilon .$$

Thus there is a decreasing sequence $(y_n)_{n\in\mathbb{N}}$ in A with infimum 0 such that

$$|x'(y_n)| > \varepsilon$$

for every $n \in \mathbb{N}$. Given $n \in \mathbb{N}$, define

$$x_n := y_n - y_{n+1} .$$

Then

$$y_n = \sum_{\substack{k\in\mathbb{N}\\ k\geq n}}^{\leq} x_k$$

for every $n \in \mathbb{N}$. By b),

$$x'(y_n) = \sum_{\substack{k\in\mathbb{N}\\ k\geq n}} x'(x_k)$$

for every $n \in \mathbb{N}$. We deduce the contradiction that

$$0 = \lim_{n\to\infty} \left|\sum_{\substack{k\in\mathbb{N}\\ k\geq n}} x'(x_k)\right| = \lim_{n\to\infty} |x'(y_n)| \geq \varepsilon \,. \qquad \blacksquare$$

Proposition 1.7.2.11 *Let E be a σ-complete ordered vector space and $(x'_n)_{n\in\mathbb{N}}$ a sequence of order continuous (order σ-continuous) linear forms on E such that $(x'_n(x))_{n\in\mathbb{N}}$ converges for every $x \in E$. Then*

$$x' : E \longrightarrow \mathbb{K}\,, \quad x \longmapsto \lim_{n\to\infty} x'_n(x)$$

is order continuous (order σ-continuous).

Let $(x_\iota)_{\iota\in I}$ be an order summable countable family in E_+. Given a linear form y' on E, define

$$\widetilde{y'} : I \longrightarrow \mathbb{K}\,, \quad \iota \longmapsto y'(x_\iota) \,.$$

Let $(\alpha_\iota)_{\iota\in I}$ be a bounded family in $\mathbb{R}_+$. By Proposition 1.7.2.10 a $\Rightarrow$ b, $\widetilde{x'_n} \in \ell^1(I)$ and

$$\sum_{\iota\in I} \alpha_\iota \widetilde{x'_n}(\iota) = \sum_{\iota\in I} \alpha_\iota x'_\iota(x_\iota) = x'_n\left(\sum_{\iota\in I}^{\leq} \alpha_\iota x_\iota\right)$$

for every $n \in \mathbb{N}$. Hence

$$\lim_{n\to\infty} \sum_{\iota\in I} \alpha_\iota \widetilde{x'_n}(\iota) = x'\left(\sum_{\iota\in I}^{\leq} \alpha_\iota x_\iota\right) .$$

It follows that $(\widetilde{x'_n})_{n\in\mathbb{N}}$ is a weak Cauchy sequence in $\ell^1(I)$. By Schur's Theorem (Theorem 1.3.6.11), $\widetilde{x'} \in \ell^1(I)$ and

$$\lim_{n\to\infty} \|\widetilde{x'_n} - \widetilde{x'}\|_1 = 0 \,.$$

Thus

$$x'\left(\sum_{\iota\in I}^{\leq} x_\iota\right) = \lim_{n\to\infty}\sum_{\iota\in I}\widetilde{x'_n}(\iota) = \sum_{\iota\in I}\widetilde{x'}(\iota) = \sum_{\iota\in I} x'(x_\iota)\,.$$

By Proposition 1.7.2.10 b $\Rightarrow$ a , x' is order σ–continuous.

Now suppose that each x'_n $(n \in \mathbb{N})$ is order–continuous and that x' is not order continuous. Then there are a downward directed set A of E with infimum 0 and an $\varepsilon > 0$ such that for every $x \in A$, there is a $y \in A$ with $y \leq x$ and

$$|x'(y)| \geq \varepsilon\,.$$

We may construct a decreasing sequence $(x_n)_{n\in\mathbb{N}}$ in A inductively such that for every $n \in \mathbb{N}$

$$|x'(x_n)| \geq \varepsilon$$

and

$$k \in \mathbb{N}_n \Longrightarrow |x'_k(x_n)| \leq \frac{1}{n}\,.$$

We put

$$x := \bigwedge_{n\in\mathbb{N}} x_n\,.$$

Then

$$x'_k(x) = \lim_{n\to\infty} x'_k(x_n) = 0$$

for every $k \in \mathbb{N}$ and

$$|x'(x)| = \lim_{n\to\infty} |x'(x_n)| \geq \varepsilon\,.$$

This leads to the contradiction that

$$\varepsilon \leq |x'(x)| = \lim_{k\to\infty} |x'_k(x)| = 0\,. \qquad \blacksquare$$

Definition 1.7.2.12 (0) *Let T be a locally compact space. The open set U of T is called an* ***exact set of*** *T if it is of the form*

$$U = \{x \neq 0\}$$

for some $x \in \mathcal{C}(T)$.

Let $\mathfrak{T}$ be the σ-algebra on T generated by the exact sets of T. The elements of $\mathfrak{T}$ are called **Baire sets** *and the $\mathfrak{T}$-measurable functions on T are called* **Baire functions**.

T is called a **Stone space** *(σ-**Stone space**) if the closure of any open (exact) set of T is open. A* **hyperstonian space** *is a Stone space T in which*

$$\bigcup_{\mu \in \mathcal{C}_0(T)^\pi} \operatorname{Supp} \mu$$

is dense.

By Urysohn's Theorem, every open σ-compact set of T is exact. The intersection of a finite family of exact sets is exact, as is the union of a countable family of exact sets. If T is metrizable, then every open set of T is exact, so that every Borel function on T is a Baire function.

Proposition 1.7.2.13 (0) *Let T be a locally compact space. Then the following are equivalent:*

a) *T is a Stone (σ-Stone) space.*

b) *If x is a bounded Borel (Baire) function on T, then there is a $y \in \mathcal{C}(T)$ such that $\{x \neq y\}$ is meager.*

c) *Every nonempty (countable) family $(x_\iota)_{\iota \in I}$ in $\mathcal{C}(T)_+$ has an infimum y in $\mathcal{C}(T)$ and*

$$\{t \in T \mid y(t) \neq \inf_{\iota \in I} x_\iota(t)\}$$

is meager.

d) *$\mathcal{C}(T)$ is order complete (order σ-complete).*

e) *$\mathcal{C}_0(T)$ is order complete (order σ-complete).*

The function y in b) *is unique and*

$$y(T) \subset \overline{x(T)}\ .$$

a $\Rightarrow$ b. Let $\mathfrak{R}$ be the set of subsets A of T for which there is a clopen set U of T such that

$$(A \setminus U) \cup (U \setminus A)$$

is meager. By a), $\mathfrak{R}$ contains the open (exact) sets of T. It is easy to see that $\mathfrak{R}$ is a σ-algebra. Hence, every Borel (Baire) set of T belongs to $\mathfrak{R}$.

By the definition of $\mathfrak{R}$, for each $A \in \mathfrak{R}$ there is some $y \in \mathcal{C}(T)$ for which $\{e_A \neq y\}$ is meager. It follows that b) holds for step functions on T with respect to $\mathfrak{R}$. Since x is a bounded Borel (Baire) function on T, there is a sequence $(x_n)_{n\in\mathbb{N}}$ of step functions on T with respect to $\mathfrak{R}$ converging uniformly to x. For every $n \in \mathbb{N}$, there is a $y_n \in \mathcal{C}(T)$ for which $\{x_n \neq y_n\}$ is meager. Since T is a Baire space,

$$\|y_m - y_n\|_\infty \leq \|x_m - x_n\|_\infty$$

for $m, n \in \mathbb{N}$. Hence $(y_n)_{n\in\mathbb{N}}$ is a Cauchy sequence in $\mathcal{C}(T)$. Set

$$y := \lim_{n\to\infty} y_n .$$

Then

$$\{x \neq y\} \subset \bigcup_{n\in\mathbb{N}} \{x_n \neq y_n\} .$$

Hence $\{x \neq y\}$ is meager.

b $\Rightarrow$ c. Define

$$y : T \longrightarrow \mathbb{C}, \quad t \longmapsto \inf_{\iota\in I} x_\iota(t) .$$

y is a bounded Borel (Baire) function. By b), there is an $x \in \mathcal{C}(T)$ for which $\{x \neq y\}$ is meager. Since T is a Baire space, x is the infimum in $\mathcal{C}(T)$ of $(x_\iota)_{\iota\in I}$.

c $\Rightarrow$ d $\Rightarrow$ e is easy to see.

e $\Rightarrow$ a. Let U be an open (exact) set of T. We are required to prove that $\overline{U}$ is open. We may assume U to be relatively compact.

First assume that U is open and put

$$\mathcal{F} := \{x \in \mathcal{C}_0(T)_+ \mid x \leq e_U\} ,$$

$$y := \bigvee_{x\in\mathcal{F}} x \in \mathcal{C}_0(T) .$$

By Urysohn's Lemma, y is 1 on U and 0 on $T\setminus\overline{U}$. Hence

$$\overline{U} = \overset{-1}{y}(]0, \infty[)$$

is open.

Now let U be an exact set of T and take $x \in \mathcal{C}_0(T)$ such that $U = \{x \neq 0\}$. Let

$$y := \bigvee_{n\in\mathbb{N}} (e_T \wedge (n|x|)) \in \mathcal{C}_0(T) .$$

Then y is 1 on U and 0 on $T\setminus\overline{U}$. Hence

$$\overline{U} = \overset{-1}{y}(]0,\infty[)$$

is open.

The last assertion follows from the fact that T is a Baire space. ∎

Example 1.7.2.14 (0) *Let T be a locally compact space. Take $\mu \in \mathcal{M}_b(T)$ and let*

$$F := \operatorname{Supp} \mu \ .$$

We identify $\mathcal{M}_b(T)$ canonically with $\mathcal{C}_0(T)'$.

a) *$\mu \in \mathcal{C}_0(T)^\pi$ iff every meager set of T is a μ–null set.*

b) *$\mu \in \mathcal{C}_0(T)^\pi \Rightarrow F = \overset{0}{\overline{F}}$.*

c) *If T is Stonian and $\mu \in \mathcal{C}_0(T)^\pi$, then F is open and there is a unique $x \in \mathcal{C}_0(T)$ with*

$$\mu = x \cdot |\mu| \, , \quad |x| \le e_F \ .$$

In this case $|x| = e_F$. If in addition μ is real, then x is also real and

$$\mu^+ = x^+ \cdot |\mu| \, , \quad \mu^- = x^- \cdot |\mu| \ .$$

d) *T is hyperstonian iff $\mathcal{C}_0(T)_{\mathcal{C}_0(T)^\pi}$ is Hausdorff and in this case for every open nonempty set U of T there is a $\nu \in \mathcal{C}_0(T)^\pi_+$ with $\operatorname{Supp} \nu$ compact, nonempty, and contained in U.*

a) Take $\mu \in \mathcal{C}_0(T)^\pi$. Let K be a nowhere dense compact set of T. Let

$$\mathcal{F} := \{x \in \mathcal{C}_0(T) \mid e_K \le x\}$$

and denote by $\mathfrak{F}$ the lower section filter of $\mathcal{F}$. Since the infimum of $\mathcal{F}$ is 0,

$$\mu(K) = \lim_{x,\mathfrak{F}} \int x d\mu = 0 \ .$$

Hence every meager set of T is a μ–null set.

Now suppose that every meager set of T is a μ–null set. Let $\mathcal{F}$ be a downward directed nonempty set in $\mathcal{C}_0(T)_+$ with infimum 0. Let $\mathfrak{F}$ be its lower section filter and put

$$A := \left\{ t \in T \,\middle|\, \inf_{x \in \mathcal{F}} x(t) \neq 0 \right\} \ .$$

Then A is a μ–null set. Hence

$$\lim_{x,\mathfrak{F}} \int x d\mu = \mu(A) = 0 \ ,$$

so that $\mu \in \mathcal{C}_0(T)^\pi$.

b) Since $F \backslash \overset{\circ}{\overline{F}}$ is nowhere dense, it is, by a), a μ–null set. Hence

$$F \backslash \overset{\circ}{\overline{F}} = \emptyset \ , \quad F = \overset{\circ}{\overline{F}} \ .$$

c) By b), F is open. There is a Borel function f on T such that

$$\mu = f \cdot |\mu| \ , \quad |f| = e_F \ .$$

By Proposition 1.7.2.13 a $\Rightarrow$ b, there is an $x \in \mathcal{C}(T)$ for which $\{x \neq f\}$ is meager. By a),

$$x = f \quad \mu\text{–a.e.} \ ,$$

so that

$$\mu = x \cdot |\mu| \ .$$

Moreover, since T is a Baire space,

$$|x| = e_F \ .$$

The other assertions are easy to see.

d) Let F_0 denote the closure of

$$\bigcup_{\nu \in \mathcal{C}_0(T)^\pi} \operatorname{Supp} \nu \ .$$

Assume that $F_0 \neq T$. Then there is an $x \in \mathcal{C}_0(T) \backslash \{0\}$ such that

$$\operatorname{Supp} x \subset T \backslash F_0 \ .$$

We get

$$\int x d\nu = 0$$

for all $\nu \in \mathcal{C}_0(T)^\pi$, i.e. $\mathcal{C}_0(T)_{\mathcal{C}_0(T)^\pi}$ is not Hausdorff.

Assume $F_0 = T$. Take $x \in \mathcal{C}_0(T) \backslash \{0\}$. There is a $\nu \in \mathcal{C}_0(T)^\pi_+$ with

$$\operatorname{Supp} x \cap \operatorname{Supp} \nu \neq \emptyset \ .$$

Then $\overline{x}\cdot\nu \in \mathcal{C}_0(T)^\pi$ and

$$\int x\, d(\overline{x}\cdot\nu) = \int |x|^2 d\nu \neq 0 .$$

Hence $\mathcal{C}_0(T)_{\mathcal{C}_0(T)^\pi}$ is Hausdorff.

We prove now the last assertion. We may assume that U is open and compact. Let $\nu \in \mathcal{C}_0(T)^\pi_+$ with

$$U \cap \operatorname{Supp} \nu \neq \emptyset .$$

Then $\mu := e_U \cdot \nu$ has the desired properties. ∎

Remark. $\mathcal{C}_0(T)^\pi$ comprises precisely the order continuous elements of $\mathcal{C}_0(T)'$.

Example 1.7.2.15 *Let T be a discrete topological space and βT its Stone–Čech compactification. Given $x \in \ell^1(T)$ and $y \in \mathcal{C}(\beta T)$, define*

$$\widetilde{x} : \mathcal{C}(\beta T) \longrightarrow \mathbb{K}, \quad y \longmapsto \sum_{t \in T} x(t)y(t)$$

and

$$\widetilde{y} : \ell^1(T) \longrightarrow \mathbb{K}, \quad x \longmapsto \sum_{t \in T} x(t)y(t) .$$

a) *$\widetilde{x} \in \mathcal{C}(\beta T)^\pi$ for every $x \in \ell^1(T)$ and the map*

$$\ell^1(T) \longrightarrow \mathcal{C}(\beta T)^\pi, \quad x \longmapsto \widetilde{x}$$

is an isometry of ordered Banach spaces.

b) *βT is a hyperstonian space.*

c) *$\widetilde{y} \in \ell^1(T)^\pi = \ell^1(T)'$ for every $y \in \mathcal{C}(\beta T)$ and the map*

$$\mathcal{C}(\beta T) \longrightarrow \ell^1(T)', \quad y \longmapsto \widetilde{y}$$

is an isometry of ordered Banach spaces.

Given $y \in \ell^\infty(T)$, let βy denote the continuous extension of y to βT. Then

$$\ell^\infty(T) \longrightarrow \mathcal{C}(\beta T), \quad y \longmapsto \beta y$$

is an isometry of ordered Banach spaces. a) and c) follow from Example 1.7.2.9 d) (and Example 1.7.2.8 d) and the last assertion of Proposition 1.7.2.5). By the above remark, $\mathcal{C}(\beta T)$ is order complete and so βT is a Stonian space (Proposition 1.7.2.13 d $\Rightarrow$ a). By a), βT is a hyperstonian space. ∎

Example 1.7.2.16 (0) *Let T be a hyperstonian locally compact space, $\mathfrak{U}$ the (order complete) lattice of clopen sets of T, and $\mathfrak{N}$ the (order complete) lattice of bands of $\mathcal{C}_0(T)^\pi$ (where we identify $\mathcal{C}_0(T)'$ canonically with $\mathcal{M}_b(T)$). Given $U \in \mathfrak{U}$, put*

$$\widetilde{U} := \{\mu \in \mathcal{C}_0(T)^\pi \mid \operatorname{Supp}\mu \subset U\} .$$

Then $\widetilde{U} \in \mathfrak{N}$ for every $U \in \mathfrak{U}$ and the map

$$\mathfrak{U} \longrightarrow \mathfrak{N}, \quad U \longmapsto \widetilde{U}$$

is a lattice isomorphism.

It is easy to see that $\widetilde{U} \in \mathfrak{N}$ for every $U \in \mathfrak{U}$. The above map is injective since T is hyperstonian. Let $\mathcal{N} \in \mathfrak{N}$. We define

$$V := \bigcup_{\mu \in \mathcal{N}} \operatorname{Supp}\mu\,, \quad U := \overline{V} .$$

V is an open set (Example 1.7.2.14 c)), so $U \in \mathfrak{U}$. Take $\mu \in \widetilde{U}$. We want to show that $\mu \in \mathcal{N}$. For this we may assume that $\operatorname{Supp}\mu$ is compact. Since $U \backslash V$ is nowhere dense, we may assume moreover that $\operatorname{Supp}\mu$ is contained in V (Example 1.7.2.14 a)). But then there is an $\nu \in \mathcal{N}$ such that

$$\operatorname{Supp}\mu \subset \operatorname{Supp}\nu .$$

From this relation it follows that μ is absolutely continuous with respect to ν, so $\mu \in \mathcal{N}$. Hence $\widetilde{U} \subset \mathcal{N}$. The converse inclusion is trivial. Thus $\widetilde{U} = \mathcal{N}$ and the map

$$\mathfrak{U} \longrightarrow \mathfrak{N}, \quad U \longrightarrow \widetilde{U}$$

is bijective. It is easy to see that it is even a lattice isomorphism. ■

Example 1.7.2.17 *If T is an infinite set, then the compact space $\{0,1\}^T$ is not σ-Stonian.*

For $t \in T$, define

$$\pi_t : \{0,1\}^T \longrightarrow \{0,1\}, \quad f \longmapsto f(t),$$

$$U_t := \overset{-1}{\pi_t}(0) .$$

Then U_t is a clopen set of $\{0,1\}^T$ for every $t \in T$.

Let $(t_n)_{n\in\mathbb{N}}$ be an injective sequence in T. For $n \in \mathbb{N}$, put

$$V_n := U_{t_n} \setminus \bigcup_{k=1}^{n-1} U_{t_k}.$$

Then V_n is a clopen set for every $n \in \mathbb{N}$. Hence

$$V' := \bigcup_{n\in\mathbb{N}} V_{2n}, \quad V'' := \bigcup_{n\in\mathbb{N}} V_{2n+1}$$

are exact sets. Set

$$A := \{t_n \mid n \in \mathbb{N}\}.$$

Then e_A belongs to the closures of both V' and V''. Since V' and V'' are disjoint, their closures are not open. ■

Name Index

Subject Index

NT means Notation and Terminology

Symbol Index

NT means Notation and Terminology